John Alexander Harvie-Brown

A Vertebrate Fauna of the Outer Hebrides

John Alexander Harvie-Brown

A Vertebrate Fauna of the Outer Hebrides

ISBN/EAN: 9783337320799

Printed in Europe, USA, Canada, Australia, Japan

Cover: Foto ©berggeist007 / pixelio.de

More available books at **www.hansebooks.com**

A

VERTEBRATE FAUNA

OF THE

OUTER HEBRIDES

BY

J. A. HARVIE-BROWN

AND T. E. BUCKLEY

Dedication

To R. F. H.

One classed the quadrupeds ; a third the fowls ;
Another found in minerals his joy.

And from my path I with my friend have turned,
A man of excellent mind and excellent heart,
And climbed the neighbouring hill, with arduous step,
Fetching from distant cairn, or from the earth
Digging with labour sore, the ponderous stone.

. . . . . .

POLLOK.

PREFACE

In almost every island of the group, and certainly in all the more important ones, we have been aided both by the correspondence and assistance of many kind friends, whom we here most heartily thank, but who are so numerous that to do more than designate each by name[1] would swell this far beyond the limits of a preface. The amount of help we have received from each and all will be apparent to those who condescend to follow us through the body of this work.

By the proprietors we have been afforded the fullest opportunities of observation and exploration, ably backed by the kindness of the shooting tenants, factors, gamekeepers, and servants, who were always willing to give us their aid and any information of which they were possessed.

[1] Lord Dunmore, Harris; Sir John Orde, Bart., North Uist (and his factor, Mr. J. Macdonald of Newton, now deceased); Lady Gordon Cathcart (through her factor, Mr. Ranald Macdonald, and the local factor, Mr. Birnie of Grogary); Lady Matheson (through Mr. Mackay, Chamberlain of the Lews); Mr. H. H. Jones, shooting tenant of Mhorsgail; Mr. A. Williamson, formerly tenant of Sobhal; the late Mr. Greenwood, Carn House; Mr. H. H. Hubback; Mr. B. B. Hagen; Mr. John Finlayson, schoolmaster, Mingulay; Mr. D. Mackenzie, gamekeeper, Stornoway; Mr. J. Henderson, late gamekeeper in South Uist; Mr. A. M'Lean, late gamekeeper in North Uist; Mr. Finlayson, head-gamekeeper in South Harris; Messrs. Macleod and Macaulay, foresters in North Harris; and Mr. Stoddart, the late Mr. John Macdonald's head-shepherd at Newton.

To the Commissioners of Northern Lights we are indebted for their readiness in granting us the use of the rooms set apart for themselves; and also to the light-house-keepers at the various stations for their aid, and the interest displayed in ornithology, as evidenced in the Migration Schedules returned by them.

We cannot omit a special mention of our late friend Mr. John Macdonald of Scolpig, Newton, North Uist, whose hospitality we have so often enjoyed; and whose keen interest in all matters relating to natural history and sport, and whose care and accuracy in recording facts, have ever been of the greatest assistance to us during the eighteen years of our attention to the Fauna of these isles. We mention with sincere regret that, almost as the last pages of our volume were passing through the press, his death took place suddenly, at Newton, on the 21st August 1888, at the comparatively early age of sixty-four years.

We wish to acknowledge our indebtedness to Dr. John MacRury of Benbeculay; and we regret not to have had an earlier opportunity of benefiting by his ability as an observer. At a late stage of our work, however, he sent us most interesting communications regarding the fauna of Benbeculay—an island which is less known to us, personally, than are any others of the group. Some of these notes arrived in time for admission to the body of the text, but others had to be entered in an appendix. Quite the most valuable and interesting of these communications has resulted in the insertion of extra pages in the text at page 36, and the coloured plate by Messrs. Peter Smit, and Mintern Bros.

We must not omit mention of the late Mr. Robert Gray, the well-known author of *The Birds of the West of Scotland*, a book which, we consider, certainly has done a great deal to stimulate the study of ornithology in Scotland. To Mr. R. Gray belonged the honour of having been the chief cause of renewed vigour and usefulness of the Royal Physical Society of Edinburgh, which, had it not been for his energy, would at one time have become moribund. We desire shortly and earnestly to acknowledge the many obligations under which he laid ourselves by his ready and free assistance at all times. At the same time, we would wish our readers to understand that Mr. Gray's personal experience of the Outer Hebrides was confined to two or three short visits made in his business capacity as bank inspector, and a fortnight's detention in North Uist, and that, therefore, his book must be looked upon, as regards the area with which we are now dealing, as almost entirely a compilation, unverified by personal experience.

To Mr. John Guille Millais, we owe our fullest acknowledgments and thanks for the lovely illustrated title-page which forms a companion plate to his title-page in the first volume of the series.

To Mr. Alexander Carmichael, for the trouble he has bestowed upon the Gaelic names of the animals, we are also greatly obliged.

To the Macleod of Macleod we are indebted for

permission to reproduce the estate map of St. Kilda—the only map of the island in existence; and, to Mr. William Mackay, the Chamberlain of the Lews, for so kindly placing at our disposal for purposes of our map, the estate map of the Lews.

Lastly, let us not neglect our kind friends who have assisted in completing the volume: Professor R. Forster Heddle of St. Andrews for his comprehensive account of the general geology of the Isles; Mr. W. Anderson Smith for his kindness in undertaking the fish fauna: and to Mr. William Douglas especially are our thanks due for the care and attention bestowed by him as sub-editor, and in revisal of proofs; and to Mrs. Beaton, who has enabled us to insert the likeness of the late Professor MacGillivray.

In an Appendix notice is taken of an annotated copy of Fleming's *British Animals*, containing notes relating to the Outer Hebrides and other parts of Scotland. There are also Dr. J. MacRury's notes from Benbeculay, already partly referred to, and of several additions to date to the avifauna of the whole group, viz.—The Arctic Bluethroated Warbler (*Cyanecula suecica*); The Common Redstart (*Ruticilla phœnicurus*); The Common Linnet (*Linota cannabina*); The Bullfinch (*Pyrrhula europæa*); The Kingfisher (*Alcedo ispida*); The Goshawk (*Astur palumbarius*); The Ruff (*Machetes pugnax*); The Barn Owl (*Strix flammea*); The Garganey Teal (*Anas querquedula*); The Spoonbill (*Platalea leucorodia*).

There are also a few notes in the Appendix upon

some of the other species which we considered are of sufficient interest to warrant their insertion.

It must be remembered that species admitted to our lists, both in this volume and in others, are only admitted upon what we consider conclusive statistics and reliable records. Did we not endeavour to adhere to a very hard and fast line in this respect, we might, we truly believe, swell the list to contain twenty or thirty more species. The knowledge of such a fauna is as distinctively progressive as we believe all local faunas to be; the unexpected may, almost at any time, occur; and therein lies the charm, we think, of progressive and chronological data; whilst, on the other hand, it is *vain* for any one to attempt, by indefinite delay, to become perfect. These remarks specially apply if we look to results of observations at lighthouses.

Of the 368 species comprised in the complete List of the Birds of Great Britain as given in our faunal chapters, 155 are now ascertained to be residents in or migrants to the area at present under consideration. This is inclusive of such as pass the lighthouses, and the discovery of which is due to their being caught or killed there. As we have just pointed out, the number of these latter must ever continue a progressive one, so long as observation at the lighthouses is continued and improved, and reported upon, and so long as specimens continue to be sent for identification; so that any present estimate of faunal values in this direction is extremely subject to change. In our Index we endeavour to show the presently ascertained faunal value by placing in square brackets all species *not* finitely placed upon the List of the Birds of the Outer Hebrides; and where species have been added only by the latest information available in the Appendix, the numbers of the pages in the Index have alone been similarly enclosed—thus [352].

J. A. HARVIE-BROWN.
T. E. BUCKLEY.

18th March 1889.

Lichia vadigo (*Risso*).

This species has been added to the British Fauna and to the fish-fauna of the West of Scotland, a single specimen having been taken off Waternish Point, Isle of Skye, by Captain MacDonald of Stein, on 17th September 1888. Of this species, only a few specimens were previously known, which had been taken in the Mediterranean and in Madeira. The earliest notice of this rare acquisition is given by Dr. Günther on 5th February 1889, at a meeting of the Zoological Society of London. This is one of the results of a continuous interest now being taken in the natural history of Skye, as at present wrought there by the encouragement and assistance of my friend, the Rev. H. A. Macpherson of Glendale.

CONTENTS

l 2

ILLUSTRATIONS

INTRODUCTORY.

For a number of years back Harvie-Brown has been in the custom of visiting many portions of the Long Island, and, since the year 1870, when he first visited these Western Isles of Scotland on a collecting trip, he may be said to have paid continuous attention to the Mammals and Birds of the group. Taken in connection with the observations of previous writers upon the area, it is hoped that this volume will present an almost perfect and chronologically exact account of the progress of our knowledge in the direction indicated in our title, and thus form a handbook from which to start afresh.

As there are always considerations of greater interest attaching to the study of an insular fauna than to that of a continental area, especially of Mammalian and Reptilian types, a considerable amount of our time was spent among these lonely sea-girt isles. Notwithstanding their proverbial sadness of aspect, their bleak and wind-swept barrenness, save where the "machar" or sandy meadows have been naturally reclaimed from the sea, there is a something which takes a strong hold on the imagination, and which woos the repeated visits of those who have once partaken of the charm of their solitudes; and this feeling has been, not once only, but repeatedly and vividly, expressed by authors who have travelled in and lived among these interesting isles.

Whether we have tossed about on the billowy seas to the west of Lewis in our yacht, or tacked out of the narrow sea-lochs like Loch Valamus, or lain becalmed in the shelter of hill-encircled natural harbours like Loch Hamanaway or Loch Eport, or dodged about in the offing for a night, in order to land at daybreak on some lonely isle, or been imprisoned on the savage shores of The Lews by consecutive gales of wind, we have always found renewed pleasure, year by year, in these old-world solitudes. Fantastic shapes and memories, even in the short span of one's own experience, flash like meteor-lights across the mind, recalling happy hours and days of health and pleasure, exciting scenes, triumphs, and disappointments—joys and sorrows too—and friendships formed and cemented. Often thus do the wondrous effects of light and shade, and the weird twilights of the long, long summer nights, recur to us in memory almost as vividly as they did in

PROFESSOR WILLIAM MACGILLIVRAY.

nature; while the mere naming of a well-known loch or isle or stream calls back incidents as freshly as if photographed before us.

We would fain recall in glowing words one wondrously weird rainbow and setting-sun effect, witnessed by ourselves and the yacht's crew, just after having slipped into the wide Loch Eport, between its high portals at the narrow entrance, only a faint "airie" filling our topsails; but we refrain, only recalling it here as a finger-post for memory.

The system of our work is chronological: and we desire to avoid personal criticism of others as much as possible, because we hold that a great deal of personal criticism is only personal opinion. We will say no more here, but only refer our readers to the prefatory note to the Fauna on page 12, which it is hoped will be found to cover all the ground necessary.

JOHN MACGILLIVRAY.

We are glad to have the opportunity of giving the accompanying woodcuts from the only known likenesses of Professor William MacGillivray, and his son Mr. John MacGillivray; and we are indebted for assistance in obtaining these to Mrs. Beaton, Paris, and to Mrs. MacGillivray of Eoligary, Barray.

Owing to their well-known connection with the Hebrides, and our quoting so frequently from their works, we have thought that their portraits would prove of interest to our readers.

GEOGRAPHICAL POSITION OF THE OUTER HEBRIDES.

THE Outer Hebrides, consisting of Lewis, or, "The Lews" and Harris—together called the Long Island,—and of the other principal islands of North Uist, Benbeculay, South Uist, and Barray, along with the innumerable smaller islands, islets, rocks, and skerries which lie along its shores; and including the more distant isles of North Ronay and Sulisgeir off the Butt of Lewis, the Flannan Isles or Seven Hunters to the West of Lewis, St. Kilda—the furthest of the group,—the Monach Isles, and also the Shiant Isles, which lie close to the south-east shore of the Park of Lewis, form what with justice may be looked upon as a separate Faunal area.[1]

Geographically, the Outer Hebrides are far removed from the mainland of Scotland, and only at one point are they approached by any other land, viz., Skye, where the latter island is only separated from North Uist by the Little Minch, some fifteen miles in width.

Between the Hebrides and the New World there is no land, only the great expanse of the Atlantic Ocean, unless we except the barren rock called Rockall, which lies 230 miles out in the Atlantic, and 184 miles nearly due west of St. Kilda.[2]

[1] Some of the more important of the smaller isles are Great Berneray of Lewis, Scarpay, Taransay, and Scalpay on the coasts of the Long Island; Pabbay, Bernoray, and Boreray in the Sound of Harris; Grimisay and Ronay, and Baleshare between North Uist and Benbeculay; Wiay in the Benbeculay Sound; and Eriskay in the Sound of Barray; and Vatersay, Pabbay, Hekla, and Barray Head or Berneray, which, with the main island of Barray, form the Barray group at the southern extremity of the Outer Hebrides. Of some of these we have more to say later on.

[2] See page lxxxvii for some more detailed account of Rockall.

The Butt of Lewis, or the northernmost part of The Lews, is distant 45½ miles from Cape Wrath, the extreme north-west promontory of the mainland of Britain, and lies almost due west of it in north latitude 58° 31′, and west longitude 6° 15′ 30″. The extreme south point of the Barray group or southern extension of the outer islands, or Barray Head,[1] is in north latitude 56° 46′ 30,″ and west longitude 7° 38′, or on the same parallel of latitude as Montrose on the east coast of Scotland, and Ardnamurchan Point, the extreme western point of the mainland. Barray Head is also 97½ miles from the nearest point of the north coast of Ireland. The whole range of the Outer Hebrides thus extends over 1° 44′ 30″ of latitude, or 130 miles in length. In breadth the Outer Hebrides vary in a marvellous manner—as will be shown later, when we describe the physical features,—from a mile or two, and even in many places a few yards between sea and sea, or between Minch and Ocean, to a distance of twenty-nine miles over land at the broadest part of the Long Island. This breadth is extremely difficult to define, owing to the ramifications of the sea-lochs, as will be at once understood by a glance at the map. We have drawn an imaginary line across Lewis, which seems to be the longest straight line that can be drawn anywhere in an easterly and westerly direction, which is not liable to intersection by salt water, viz., from Aird Bhreidhuis on the west coast to Kebock-head on the east coast. This line as defined is twenty-nine miles in length; and if another line be drawn, as shown on the map, in a northerly and southerly direction from near the Butt of Lewis to W. Loch Tarbert, such a line, with very approximate exactitude, divides the easterly and westerly flows of water, and is the longest walk in a straight compass-line obtainable in Lewis. The total area of the Outer Hebrides is 939$\frac{1}{16}$ square miles, as given by the last Ordnance Survey. This is *exclusive* of fresh-water areas. But as will be shown further on, this land is so honeycombed by fresh-water lochs, throughout a large extent of its area, as well as by sinuous and far-reaching arms of the sea, that exact measurement of land and water is a labour of no small

[1] Also known as South Berneray.

trouble and difficulty, and has been so even to members of the Ordnance Survey.

Of the fresh-water areas, perhaps the largest are Loch Langabhat in Lewis, Loch Langabhat in South Harris, Loch Scatavagh in North Uist, Loch Bee in South Uist (which, however, is tidal, but at high spring-tides only). Loch Obisary in North Uist is also extensive and tidal, but it is only tidal in spring-tides; and there are many other trout-frequented lochs, especially in North Uist, which at times and seasons admit the tides. Of some of these we will have more to say under "The Fish of the Hebrides and West of Scotland."

Of wooded areas there are almost none, if we except the plantations around Stornoway Castle, a small plantation in Rodel Glen, South Harris, and a few wind-topped trees within the garden walls of the larger farm-houses, and similar enclosures. Small as these are, their beneficial influence is evident, and within a very few years has brought about great changes in the distribution and extension of insect, bird, and other life, as we hope to show when we come to treat of the species further on.

A long reach of sandy, shelving, shore stretches south along the west coast of Lewis, from the Butt of Lewis to Barvas. To arrive at the acreage left dry at low tide is not easy, but it may be mentioned, in illustration, that the coast of Loch Maddy in North Uist, if all its sinuosities be minutely followed, gives a distance of something like 300 miles at the time of high-water.

Those lands, which are strictly denominated "meadows," extend to some 2500 acres throughout the Outer Hebrides, and are subject to submergence during high spring-tides, and to flooding by lakes, rivers, and torrents in many places (*vide* M'Donald's *Agriculture of the Outer Hebrides*, 1811, p. 306); of other pastures, excluding wastes, there are about 200,000 acres. "Gardens," says the same account, "are little more than an empty name." Of woodlands, it is stated that in 1540-49 Dean Munro found abundance, and so did Buchanan about 1600. At the present time there are almost absolutely none, but the evidences of timber-growth are perfectly apparent at more than one locality in Lewis and elsewhere, as observed by ourselves in 1886, and at other times. Thus, close to the road near Loch Erisort in Lewis, stumps and roots of trees are still dug out of the peat, and are used for primitive fencing purposes, or as cow-stoppers to the crofters' cattle.[1] In 1811 this barren aspect was even surpassed, as M'Donald tells us: "The whole Long Island, comprehending altogether about half a million of acres, contains not a single acre of wood which deserves the name of either copse or plantation" (*loc. cit.* p. 324).

With regard to replanting, though something has been accomplished at Stornoway and Rodel, it is extremely doubtful if success can be looked for except in particularly well-sheltered situations; and the ardent hope and encouragement held out by the writers of the "Agricultural Survey" above quoted, seem at present as far from being realised as ever. An attempt to plant the wind-swept sides of the hill on the south side of East Loch Tarbert, in Harris, about eight years ago (say 1879), has resulted in almost total failure. Indeed, it could hardly have been expected to succeed where the westerly winds rush through the narrow ravine over a low isthmus not quite quarter of a mile in breadth. A more unlikely place could scarcely have been selected for a first experiment; and the wind-crushed tops of the outskirting trees at the top of the Glen of Rodel might have taught the lesson. Only would dense and broad masses of trees be able to stem these wild, westerly, salt-laden gales. In the gardens, which are surrounded by high walls, the tops of the trees are cut as if with a scythe, or bent at right angles to the parent stems, whenever they venture to reach above the cope-stones.

The most persistent and heaviest winds in the Outer Hebrides are south-westerly, or from three to four points of south-west, a fact which it would be well to remember when planting larger areas of wood for shelter. Such, we are well aware, has been ably advocated before, but "out of sight is out of mind," and we make no apology for repeating the advice. Where statistical facts have been insisted upon, it is not easy to clothe them in new wording, nor where such are correct, does it seem to us altogether desirable to do so. Many of the above statistical facts, as well as others which follow, which are sufficiently accurate for the pur-

[1] *Auct.* Captain Thomas: 1862.

poses of a Faunal work, we have quoted freely from previous authorities.[1]

These authorities, in 1811, tell us that in that year waste-lands amounted to about 300,000 acres, or a little less than one-fifth of the whole extent—this, however, including the Inner Hebrides,—and of these, 100,000 acres, or one-third of the said waste-land, was moorland.[2] Mountains occupy about 120,000 acres, including the Inner Hebrides, and 80,000 of reclaimable land.

[1] A prevalent idea exists that not only were the trees at Stornoway Castle introduced by Sir James Matheson, but that even the mould in which they were planted was brought over from the mainland. We have several times had this related to us at Stornoway, and Mr. Hubback received the same statement from the under-keeper at Stornoway Castle in 1881.

[2] Three principal species of heather in the Outer Hebrides are *Erica vulgaris*, common heath; *Erica cinerea*, bell-heather; and *Erica tetralix*, rinze heather (*op. cit.* p. 363).

DESCRIPTION OF THE PHYSICAL FEATURES OF THE MAIN RANGE OF THE OUTER HEBRIDES.

THAT portion of the Long Island called Lewis, or "The Lews," is comparatively flat, and covered with innumerable lochs,[1] and is cut up in many directions by the far-reaching arms of the sea, especially throughout its more southern portion—*i.e.* from Loch Roag to Loch Reasort on the west side, and from Loch Luirbost to Loch Seaforth on the east. But, though comparatively flat, Lewis contains some elevations in the south-east, or that portion known as The Park, such as Ben Mhor (1750 feet) and Crionaig (1500 feet); while in the portion north of the neck of land between Loch Roag and Loch Luirbost there are minor heights, such as Monach (800), Ben Barvas (900), and Stacasthal (710); and in the west, in the Mhorsgail deer-forest, Suainabhal (1250), and Mhealasbhat (1750).

A great portion of the north-west coast of Lewis is open, shelving sand-beach, exposed to the full force of the Atlantic breakers, and offering no shelter until the snug anchorage of Carloway is reached. South of Carloway the coast is rocky and abrupt, though not of great altitude, and there are several excellent harbours for vessels of light draught. Remarkable amongst these are Loch Roag and the creeks around Great Bernerav Island and West Loch Roag, Uig Bay, and Loch Hamanaway (Thamanabhaidh ?), and lastly, on the west coast, Loch Reasort, which separates the shores of Lewis from Harris.

Nowhere we think, in broad Scotland, can be found a more

[1] Martin tells us that from this fact the island received its Gaelic name, viz. *Leog*, meaning *water lying on the surface of the land* (*Description of the Western Isles*, 2d edition, 1716, p. 1).

dismal and dreary, featureless waste of rolling bogs and rounded hillocks than in The Lews, unless possibly we except the central portions of Caithness, which we have already endeavoured to describe and illustrate in the previous volume of this series. On one occasion Harvie-Brown drove across The Lews from Harris to Stornoway; the day was fine and clear, and comparatively warm, but nevertheless all life seemed absent or asleep. After passing the long Loch Seaforth, and looking northwards and westwards, the eye traversed these wastes in vain for resting-places. Macculloch has described the solitudes of the Moor of Rannoch; but in The Lews even the hum of a bee is scarcely heard, and but comparatively rarely in a day's journey does the pipe of the plover break the stillness. Imagine the Moor of Rannoch divested of its many glistening lakes; its heather-clad "knobbies"—dear to the stalker—levelled or removed; the heather burned; and the moor rain-pitted, and growing tough, wiry, yellow grasses and red *sun-dews*, and increased to many, many times its dimensions, then rolled out, here flat, there rising into featureless, shapeless hillocks, and miles upon miles of this with endless repetition, and you can picture somewhat of the drear expanse of this large portion of The Lews. However numerous the lochs of The Lews are, but comparatively few are visible from the high-road between Loch Seaforth and Stornoway, and there is little to interest the traveller during a long, weary drive.

But, as we have already said, there are in other parts of Lewis redeeming features. Looking back upon the deep glens, the dark corries and high hills of Harris, and across the lovely Loch Seaforth, among the high hills of The Park of Lewis, there is much of grandeur and beauty, and Loch Erisort has some fine natural gems of landscape. "The Park" itself is a "confusedly-hurled" mass of mountains, seemingly, at a distance, continuous with the Harris hills, which are of still greater altitudes, but really separated by the long arms of Lochs Seaforth and Erisort from the rest of The Lews and from Harris, and forming in itself an exceedingly compact self-contained deer-forest, somewhat difficult of access either by sea or land, because terrific squalls sweep down the deep corries and expend their forces on the bosoms of the sea-lochs, often making it a matter of danger to cross the narrow sounds.

The great "Lochs" district[1] lies more to the westward of that range of low hills which commences near the south-west end of the great Loch Langabhat, and terminates in the tops of Roniebhal, near the north-east end of the same loch, which latter is about nine miles in length. This range is connected with lower ridges, and reaches an altitude of about 1000 feet.[2] Another great loch-besprinkled area lies eastward of the Park hills, both north and south of the imaginary line before referred to, and between Lochs Luirbost on the north and Loch Odairn on the south, and, to a less degree, between the latter and the entrance of Loch Seaforth; but it is perhaps invidious to particularise any given area of Lewis as excelling any other portion in this respect, and the extraordinary distribution of land and water is only surpassed by the more southern islands of the Uists and Benbeculay. Numerous as these lochs are, few, as we have already remarked, are visible from the high-road.

Approaching Stornoway the scenery changes for the better, and a pretty peep is obtained of the Eye Peninsula and Broad Bay, and the wooded heights and hollows around the castle—a refreshing relief after so many previous hours of monotony. Then, as we skirt the policies of the castle, and by the descending road approach the town, the busy harbour and its multitude of sails at the herring-harvest season, and the bustle of fisher-life, bursts somewhat suddenly on the gaze; the glistening waters of the Minch, occupying, with the Eye Peninsula, the middle distance, and the unmistakable mountains of distant Sutherland towering aloft in the clear sky—Suilbhein most prominent of them all.

A considerable extent of foreground in the immediate vicinity of Stornoway has been reclaimed from the universal peat mosses,[3] and converted into grass lands, and, as is mentioned by the late Mr. Greenwood: "It was curious to note in 1879 the extraordinary numbers of *Landrails*, calling continuously from every patch of grass; but they confined themselves entirely to the cultivated areas." The abundance of its favourite shelter-plant, the common

[1] Parish of Lochs. [2] *Vide* Map. Cearnabhal, 1000.

[3] The reclaimed and enclosed and cultivated portions are usually spoken of as "within dykes," and the unenclosed and uncultivated as "without dykes."

dock, is no doubt an additional inducement to the increase of the species in the Hebrides. Mr. Greenwood also speaks of terrestrial mollusca[1] as scarce, owing no doubt to the absence of lime in the geology of the Hebrides. We mention all these points as distinctly bearing upon the vertebrate fauna of the Hebrides, whether we take them merely as "side lights," or as having more direct bearings upon our subject. The absence of lime and of soil proper—the old gneiss being covered by a coating of peat and sphagnum, sometimes underlaid by a species of boulder-clay—is sufficient perhaps to account for this scarcity of shell-bearing molluscs; and, perhaps, from the same cold nature of the land, scarcely any water-plants are found in the lochs—a few water-lilies, some rushes, and occasional coarse grasses,—and these scarce,—are almost all that are found, except where shell-sand may to some extent have supplied the necessary phosphates. By liming and artificial manures, by admixture of guanos and phosphates and sea-ware, the better cultivated portions are now producing such plants as the common dock and nettle; and in the plantations around Stornoway Castle there is quite a luxuriant growth of underwood, scrub, weeds, and grasses. Captain Thomas (*Geological Age of the Pagan Monuments of the Outer Hebrides*) is careful to point out in this connection the *extremely* limited growth of *wood-peat* in the Hebrides, and also uses his illustration as showing the rapid growth of the superincumbent peat mosses, and the slow growth of ling and grasses.

[1] Mr. Greenwood has observed specimens of *Helix nemoralis* (one dead), *Pupa umbilicata*, and one living *Zonites* (sp. ?), and since then *Helix virgata* has been added.

THE SHIANT ISLANDS.[1]

THE Shiant Isles—the property of Lady Matheson of The Lews—compose a somewhat inaccessible little group, well worthy, however, of a visit either by the ornithologist, geologist, or mineralogist. They are situated about twelve miles north of Skye, and lie out in the North Minch, about five or six miles from the south-east end of Lewis. They are about twelve miles in a direct line from Scalpay Island at the entrance of East Loch Tarbert—the nearest land of Harris. There is no regular communication between the islands and the mainland, nor with the Long Island, and in winter sometimes two or three months may pass without their being visited. The only inhabitants are the shepherd and his family, who most hospitably entertained our friend Dr. Heddle and Harvie-Brown during a short stay of two days and one night on the island in 1879.

Casual visitors like ourselves often find considerable difficulty in reaching these islands, and in making arrangements to be taken off again. The fishermen and other inhabitants of Tarbert asked an exorbitant sum to land us there, and when we refused to be imposed upon, influenced all the others in the village, some of whom might otherwise have been satisfied with a more reasonable remuneration. Should any parties after reading this account of the Shiant Isles be inclined to pay them a visit, we would strongly advise them to leave Tarbert out of their programme, and rather endeavour to approach them from *any other* place on the coast, at least until the natives see that it is to their advantage to moderate their charges.

Occasionally there have been shipwrecks on the islands, but as they are not far from the Long Island, the crews have hitherto

[1] Partly reprinted from the *Transactions of the Norfolk and Norwich Naturalists' Society*, vol. iii. Read 25th November 1879.

easily reached The Lews, and found their way to Stornoway, which is the nearest post-town, and about twenty-four miles from The Park.

The Shiants lie in the track of Macbrayne's West Coast steamers, between Stornoway, and Portree in Skye; and not much out of the way of vessels passing down the coast of the Long Island from Stornoway. The proprietors also of the s.s. *Dunara Castle*, which plies between Glasgow and the Outer Hebrides, have of late years run their steamer in the summer season to St. Kilda. It might be worth the consideration of either Company to institute a similar trip to the Shiants, which only require to be better known to be appreciated. Their columnar basalt cliffs are vastly superior in height and grandeur to those of Staffa. The statement in Lord Teignmouth's account of the islands—referred to further on—that a landing "sometimes can scarcely be accomplished during ten days in the year," must, we think, be somewhat exaggerated, because, owing to the double bay on either side of the isthmus, there is usually shelter. The landing is infinitely superior to that at St. Kilda.

There has been no very full or connected account of this group of islands published that we are aware of. Martin has a few remarks upon the islands which he calls, "Siant, or, as the natives call it, Island More," but there is little to be gathered from his very superficial description.[1]

Macculloch at a much later date (1824) gives a short account,[2] but goes off into a learned disquisition upon the food of the Highlanders, useful and interesting in itself, but as may be imagined not of much count in this connection. Macculloch names the islands, Gariveilan, Eilan na Kily, Eilan Wirrey, and Eilan More.

Lord Teignmouth perhaps gives the fullest account of the group in his *Sketches*, published in 1836.[3] The "tours," as stated in the preface, were performed in the years 1827 and 1829.

[1] *A Description of the Western Isles of Scotland*, first edition, 1703; second edition, 1746, both of which editions we have consulted.

[2] *The Highlands and Western Islands of Scotland*, in 4 vols. (London, 1824), pp. 323-325.

[3] *Sketches of the Coasts and Islands of Scotland, and of the Isle of Man* (London), 1836, pp. 166-172.

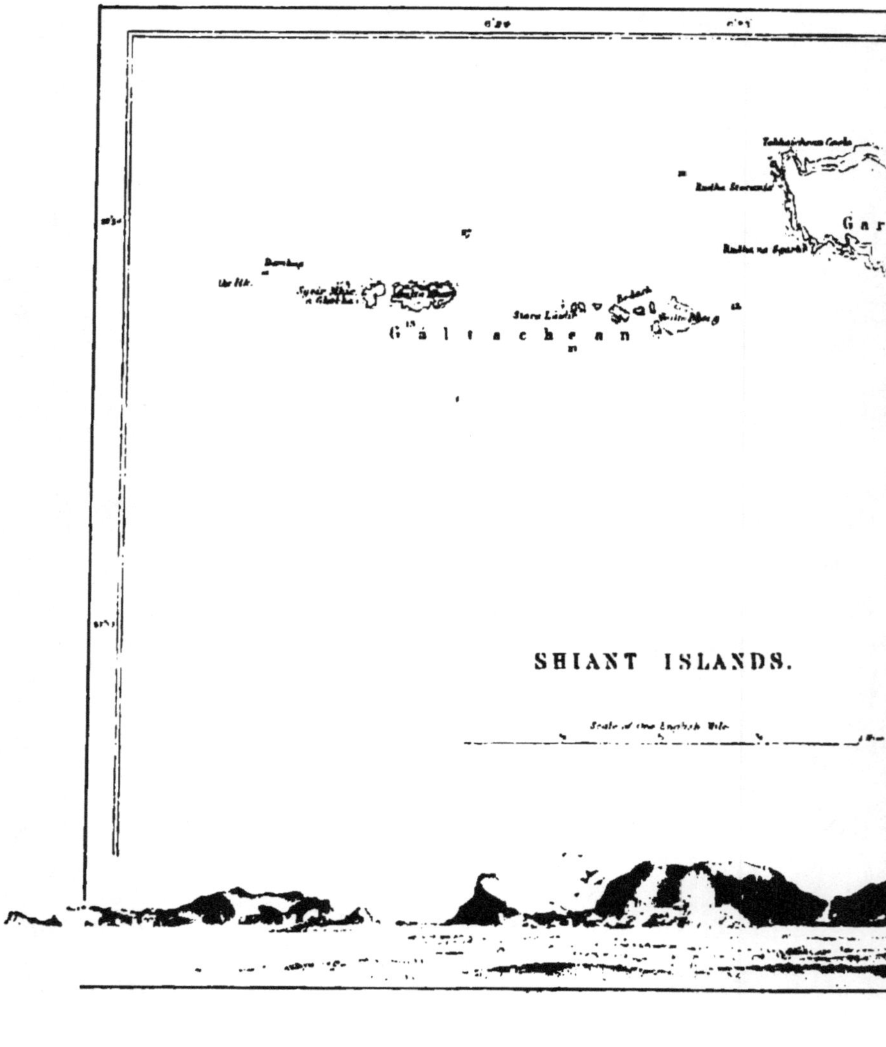
Gar
Gàltachean
SHIANT ISLANDS.

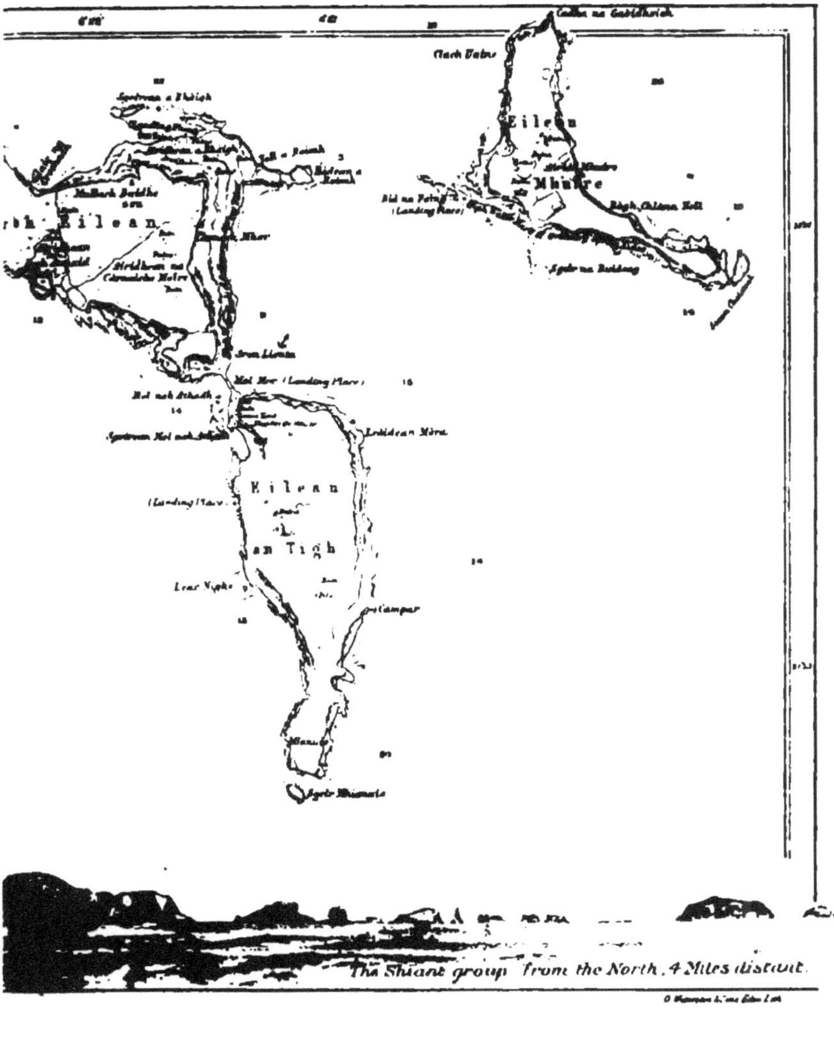

The Shiant group from the North, 4 Miles distant.

Wilson in his *Voyage*[1] has a very short notice of the Shiant Isles. He underrates the columnar cliffs. He says: "They exhibit (especially Garbheilan) a fine cliff-like columnar coast on one side, showing strong indications of that basaltic structure of which Staffa and the Giant's Causeway are the crowning glories." In one direction certainly,—that of simplicity and grandeur,—the Shiants far surpass Staffa.

One very general interpretation of the word *Shiant* is *green lump* or *hill*,[2] which certainly admirably describes the general appearance of the main island from the south. But the name of the main island is Garbh-eilean, *the rugged* or *rough island*, which is even more descriptive of its aspects if viewed from the north, east, or west, or of its shores and precipices, débris-slopes, and stony strand. The group of the Shiants consists of two considerable islands and several outlying rocks. The main island, as before stated, is Garbh-eilean. Connected with Garbh-eilean by a narrow ridge of shingle—over which in unusually high seas and tides the water flies in foam from one bay to the other—is Eilean-a-chille (or Eilean an Tigh), upon which, facing the south-west, is the shepherd's house, and also the remains of the cell or church. The *next largest* island is Eilean-Mhùire (called also Island More), which lies about a quarter of a mile to the north-east, and is of triangular shape with a fringing of mural precipice, and a grassy top. The rocks which lie off the south shore of Garbh-eilean are called Galta More and Ox-rock. Garbh-eilean is about four miles in circumference, if the cliff and coast line be followed and Eilean-a-chille be included as part; and there is, besides, a long-stretching jagged line of reefs and rocks which extends for nearly a mile from the western extremity.

Viewed from the southward, and as we saw it upon this occasion, the main island rises with a steep slope from the sea, and is

[1] *A Voyage round the Coasts of Scotland and the Isles*, 1842, vol. i. p. 386.

[2] Mr. Macpherson adds: "We have also *Sian* (plu. *siantan*), a scream, a shriek, a roar; also *Sian* (plu. *siantan* or *siantaidh*), any storm, as rain, snow, or wind. This latter would also suit Ben Hiant in Ardnamurchan, which, however, we have been told also means "green hill."

covered in places almost from base to summit with fine green grass but at almost all points this slope ends abruptly at the base in a low but inaccessible precipice—a fringing wall of rock. In many places, rugged cliffs and high knees of rock project over the coast line; and as we approached—on the occasion of our first visit—close to the narrow isthmus or shingle connecting Garbh-eilean and Eilean-a-chille, we felt somewhat at a loss to know how we could possibly ascend to the summit after landing.

Very different is the view of the group from the north, as we have seen it on some occasions. On a somewhat hazy day, the sheer front of the basaltic cliffs of the north shore, when viewed from some little distance, imparts to the whole group a most singular wall-like appearance, the upper edge sharply defined against the sky. It appears like the side-wall of a gigantic square-built house with the roof and gables removed; like, in fact, the sole remaining wall of some vast ruin.[1]

There are three landing-places on Garbh-eilean,[2] viz., on either side of the isthmus of shingle and round water-rolled boulders, which connect the two portions of the main island, and one on the extreme north point. The shepherd's house stands in a sunny nook facing the south and west, and close to the isthmus. The wind being favourable on our arrival, we landed in the south-west bay on the rocks a little to the south of the shepherd's house, but on leaving next day, the wind having veered round, we got on board our chartered fishing-smack, where she lay anchored in the north-east bay.

On the 19th June 1879, Professor Heddle and Harvie-Brown having failed to come to terms with the clique of Tarbert fisher-

[1] Since writing the above we read for the first time Macculloch's account of the islands—before referred to (*antea*, p. xiv),—and we find he describes the same view as follows:—"But these islands are nowhere more striking than when viewed at a sufficient distance from the northward; the whole of this lofty range of pillars being distinctly seen rising like a long wall out of the sea, varied by the ruder forms of the others which tower above or project beyond them, and contrasted by the wild rocks which skirt the whole group" (*op. cit.* vol. iii. p. 325).

[2] These are marked on the map of these islands accompanying this chapter.

men and boatmen, hired, along with Mrs. Thomas,[1] a small boat which took us to the island of Scalpay[2] before mentioned, where we intended to try our further luck; here, most fortunately for us, the Fishery yacht *Vigilant*, which looks after the fisheries of the Long Island, arrived, and, thanks to Mrs. Thomas's introduction, Captain MacDonald most kindly offered to land us on the Shiants, which offer we gratefully accepted. We also succeeded in making arrangements—for a more moderate sum than that demanded by the Tarbert people—for a fishing smack to come and take us off the following afternoon.

With a light and favourable breeze we soon reached across and lay-to opposite the shepherd's house. A boat was manned, and shortly afterwards we and our small kit were landed, and the white sails of the yacht disappeared round the south end of Eilean-a-chille, bound for Stornoway. We received a hospitable welcome from the family, which consists of mother, two daughters, a deaf-and-dumb son, and a little girl. The father of the flock was away on the mainland attending the "clipping" on one of Mr. Sellar's farms. The son was a fine-looking, stalwart, fair-haired giant, who acted as our guide to the cliffs, and during our ramble over the island. Dr. Heddle was in search of minerals, and evidence of the direction of ice-action; Harvie-Brown in search of the picturesque, and bird-life. In both objects our guide proved of great assistance, notwithstanding his misfortune.

[1] Mrs. Thomas is widow of the late Captain Thomas, R.N., who surveyed for the Admiralty a large portion of the Long Island and the surrounding seas. Mrs. Thomas has for many years done much to assist the natives of Harris and the Long Island, in obtaining a market for their worsted and other work, such as Harris tweeds, etc., and she has taken a fast hold upon the affections and respect of the native population. It is, however, to be regretted, that with their advanced knowledge of money's worth, they should give way to such marked exhibitions of cupidity as was witnessed on this occasion.

[2] Scalpay Island is a fishing station, and contains a resident population of about 500 souls, for the moral and intellectual training of the younger portion of whom quite a magnificent Board school-house has lately been erected. Here Dr. Heddle discovered a very remarkable ice-boulder, which had evidently been borne from a seam in a hill-side on the mainland of Harris, about three miles distant to the north-west, which latter was clearly visible from the spot where this huge boulder now lies, in the centre of the fishing village.

After taking some refreshment, we ascended Garbh-eilean to the north-west of the cottage, crossing first over the shingle isthmus, where Heddle, with our guide's assistance, found water-rolled pebbles foreign to the geology of the island. We reached the top by a rude flight of steps, partly cut, partly worn in the lower part of the decayed basaltic rock, and by a steep slope of some 300 to 350 feet. We walked through an almost continuous colony of Puffins, which bird inhabits all the slopes and tufty patches, ledges of the cliffs, and crevices of the débris round the island, occupying every available nook and cranny from the tops of the cliffs almost to the sea-level.

The cliffs along the north side of Garbh-eilean are 499 feet high, and the summit of the island 23 feet more.[1] The cliffs are basaltic,[2] and more than twenty-seven times higher than those of the Cnoc of Staffa, and very grand indeed in their massive simplicity and swelling contours. In two places the pillars reach in unbroken grandeur from the summit to the sea, but in most parts are interrupted by a long slope of fallen débris, and broken columns at the base. At one place where the cliffs form a semicircular concave sweep, with a dip in their height—or rather to the west of this point—the Sea-eagle has long remained undisturbed in its eyrie, which is said to be—and has every appearance of being—perfectly inaccessible.[3] Captain Elwes visited the place in 1868, and con-

[1] By Dr. Heddle's aneroid: thus showing a difference of only two feet from MacCulloch's estimate (*op. cit.* p. 324).

[2] Lord Teignmouth says (*op. cit.* p. 168): "The merit of the discovery of the basaltic character of the Shiant Isles has been already attributed to Dr. Clarke. Chalmers describes these islands without referring to it, and Pennant says that the most northern basalt which he was aware of was that of the Brishmeal Hill in Skye. It is remarkable that the basaltic stratification proceeds almost in one meridian from the Giant's Causeway in Ireland, through Mull, Staffa, and some smaller islands; Skye from its southern to its northern coast, and the Shiant Isles to the distant Ferro."

[3] Of this eyrie Martin wrote more than 170 years ago (*op. cit.* 2d edition, 1746, p. 26); and Lord Teignmouth also notices the fact of eagles breeding on the islands in 1827 and 1829 (*op. cit.* p. 169). Martin also describes the ascent from the isthmus as "narrow, somewhat resembling a stair, but a great deal more high and steep, notwithstanding which the cows pass and repass safely, though one would think it unsafe for a man to climb" (*loc. cit.*).

sidered that it was "as perfectly inaccessible as any nest can be, owing to the way in which the rock overhangs."[1]

There are cliffs at Staffa of basaltic formation which reach an altitude of 150 feet, but unbroken pillars are nowhere in Staffa higher than about 18 feet.

On this same side of the bay formed by the curve of the cliff, the columns are not so regular, but, near the base, in one or two places, they show with unusual symmetry. At the foot of the columns a little further on, and on the east side of the bay, is a long slope of débris and turf, honeycombed with crevices and holes. From its highest point at the base of the cliff to the sea, on the slope, is probably 250 feet. Enormous are the legions of Puffins breeding here, filling the air, and covering the sea with their hosts. Compared with all other rock stations of the Puffin which we have seen, the 3½ miles[2] of Puffins of Garbh-eilean of the Shiant Isles far and away carry off the palm in numbers. The face of the basaltic precipices is also covered with Puffins, which nestle amongst the long loose tufts of grass and bunches of bright green sorrel which cap each broken column. They tunnel deep into these tufts, and between them and the rock, and lay their eggs at the far extremities. On the summit of the cliffs the grass-slopes are tunnelled in every direction for many yards back from the cliff-edge. On the east side of Garbh-eilean is another slope equally populous with Puffins. Coming, as we then did, with all the memory of the great St. Kilda fresh upon us, nevertheless we can safely say that, for Puffins alone, the Shiant Isles will prove hard to surpass.

Next morning, the 20th of June, saw us both early afoot, and having viewed the cliffs from the tops the previous day, we took boat and rowed round to the north side. The gale blew strong across the north-east bay, and we could not venture with safety round the point, nor through the sound between Garbh-eilean and

[1] Vide *Ibis*, 1869, p. 25; Captain Elwes on the "Bird Stations of the Outer Hebrides."

[2] There are but few Puffins comparatively on Eilean-a-chille. A few nests close to the shepherd's house; but they are equally abundant on Eilean-Mhàire as on Garbh-eilean.

Eilean-Mhûire. A long promontory juts out in this direction, and near its base or junction with the main island it is pierced by a remarkably fine sea-worn arched tunnel, fully forty yards in length, and about four times the breadth of a good-sized row-boat. Guided by our good friend, the shepherd's son, we decided upon shooting through this tunnel into the calmer water on the north side. Wind and tide being in our favour, we darted through with such velocity that we had scarcely time to glance upward at the arch forty feet above our heads, but only to attend to the guidance of the boat. The arch acted as a perfect funnel, and the wind, which rushed impetuously through, caught up the crests of the waves, which swelled grandly through it, and bore the water away in clouds of spin-drift towards the sea to the northward. Once through, we lay in calmer water, gently rising and falling on the swell. Around us flew a few Kittiwakes, and on the ledges a small colony of Guillemots, and a few Razorbills had their eggs, but we searched every ledge in vain for a single specimen of the Bridled Guillemot.[1]

It was not, however, until the summer of 1887 that an opportunity was afforded us of landing upon and examining Eilean-Mhûire. On the 25th June in that year, our friend—and again our companion in our further Hebridean investigations, Professor Heddle of St. Andrews—and Harvie-Brown landed on Eilean-Mhûire. Close to the landing-place on the south side is an apparent raised beach or kitchen-midden of limpet shells. We were unable to decide as to its antiquity, or as to whether it had been brought there by human agency or otherwise. There are several other places where lesser accumulations, always of limpet shells, had been

[1] See also MacCulloch's description of this nature-worn arch (*op. cit.* p. 320). He says: "The velocity with which we entered this dark and narrow passage, the shadowy uncertainty of forms half lost in its obscurity, the roar of the sea as it boiled and broke along like a mountain torrent, and the momentary uneasiness which every such hazardous attempt never fails to produce, rendered the whole scene poetically terrific. As we emerged from the darkness of this cavern, we shot far away beyond the cliffs, whirled in the foaming eddies of the contending streams of tide." As with all of MacCulloch's rhapsodies, we must allow for exaggerated description. Such have almost invariably a contrary effect upon the reader, from that intended by the writer.

made. Mr. Norrie, to whose skill we are indebted for a fine series of photographs of the Hebrides, soon after landing, took, amongst others, a view of a fine aiguille of rock with Eilean-Garbh beyond looming darkly through the haze. Nearly the whole of the top of Eilean-Mhŭire is one great expanse of old *lazy-bed* potato or cropped ground, and here and there a few remains of dwelling-houses are still traceable. A somewhat ruinous old sheep-fence separates the level or undulating grass land from the cliff edge. There is little to be seen of the cliff face from the landward side, except the broad fringe or circle of three more miles of Puffins, which tunnel deep into the rich mould above the rock, and populate the whole cliff and faces of the slopes.

The sward is deep in luxuriant grasses, as is the case with many sea-girt pastures, and bespangled with flowers and plants, amongst which may be noted purple orchis, purple vetch, buttercup, and crowsfoot, sorrel in great luxuriance, tormentilla, nettles in patches where the old crofts were, white clover and daisies, and here and there wild thyme. On the faces of the cliffs and slopes, bladder campion, sea pink, wild celery, and large coarse camomile gowans, and here and there a patch of lovely scarlet lychnis, which, however, is much more abundant under the basalt cliffs of Garbh-eilean, not far from the Eagle cliff.

Approaching the eastern horn of Eilean-Mhŭire, we passed over a col or narrow ridge, with precipice on either side. The breadth of the path worn by tread of sheep and shepherd is about 15 feet, and it may be from 30 to 40 yards long. The whole south face of this is populated by Kittiwakes and other rock-birds; indeed, this, and a large extent of the cliffs of Eilean-Mhŭire, especially on the south side, afford nesting-ground for quite one of the largest colonies of Kittiwakes in the west of Scotland. A few days later than the above visit, we sailed close along the north side of both Eilean-Mhŭire and Garbh-eilean, and had a splendid view of the rock faces and bird-colonies.

The whole circumference of the precipices of Eilean-Mhŭire is densely populated, but, as already noted before by us, and now still more accurately observed, the northernmost cliff face holds considerably fewer birds of any kind than the rest. The whole stretch from the col westward is one vast colony, principally of Kittiwakes and Puffins, but there are also great numbers of Guillemots and Razorbills on the ledges or in crevices. Passing along the north side of Garbh-eilean we had also a fine view of the columnar structure, though deep in shadow, and of the rocks and skerries of fantastic forms which stretch away to the westward from Garbh-eilean.

THE FLANNAN ISLES.[1]

REGARDING the Flannan Isles, or, as they are also called, "The Seven Hunters," or "Seaforth's Hunters," several authors have written, but none so fully as these islands seem to merit. The difficulty of landing upon their almost precipitous sides, which can only be accomplished in the calmest and finest weather, and then only with ease upon the two largest, has no doubt interfered greatly with any attempt to survey them thoroughly.

Dean Munro devotes rather more space to them than he usually does, and his account is not without interest. He writes as follows:—

"SEVEN HALEY ISLES.—First, furth 50 myle in the Occident seas from the coste of the parochin of Vye in Lewis, towarts the west northwest, lyes the seven isles of Flanayn, claid with girth, and Haley isles. Very natural gressing within the saids isles: infinit wyld scheipe therein, quhilk na man knawes to quhom the said scheipe apperteines within them that liues this day of the countrymen; bot M'Cloyd of the Lewis, at certaine tymes in the zeir, sendis men in, and huntis and slayis maney of thir sheipe; the flesche of thir sheipe cannot be eaten be honest men for fatnesse, for ther is na flesche on them, but all quhyte like talloune, and it is verey wyld gusted lykways. The saids isles are nouder manurit, nor inhabit, but full of grein high hills full of wyld sheipe in the seven isles forsaid, quhilk may not be outrune; they perteine unto M'Cloyd of the Lewis." So much for Dean Munro's quaint account.

Next comes the still more curious account of Martin.[2] He

[1] Partly from the *Proceedings of the Natural History Society of Glasgow*, January 31st, 1882.

[2] Martin's *Description*, etc. (p. 16).

says, "To the North-west of *Gallan-head* and within 6 leagues of it, lyes the Flannan Islands, which the Seamen call *North-hunters.*" He adds, "they are but small Islands, and six in number, and maintain about 70 Sheep yearly." He relates also how the natives of Lewis, "having a right to these Islands," visit them every season, "and there make a great purchase of Fowls, Eggs, Down, Feathers, and Quills." The natives never attempt a landing in a west wind, and a novice "not vers'd in the Customs of the place, must be instructed perfectly in all the Punctilio's observed here, before Landing," which punctilios Martin, in his own quaint language, proceeds to describe. "This superstitious Account," as Martin justly terms it, he had received *viva voce* from two fishermen who had visited the Flannans the previous year.

In connection with the sacred character of these isles, Buchanan writes:[1] "There are Seven Islands at Fifty Miles distance above *Lewis*, which some call Flavanae, others the *Sacred*, or sanctuary islands." They would appear to be sacred or St. Flann's isles, just in the same way that the Shiants were sacred, being the Virgin Mary's isles. They were, as Dean Munro tells us, "claid with girth and Haley isles," or they may possibly—as suggested by Mr. James Macpherson to me, *in lit.*—have been sanctuary isles in a legal sense as well. Islands far out at sea, and difficult of access, often seem to have been held as holy isles or places of veneration.

Wilson, in his *Voyage round the Coasts of Scotland*, designates the feeling which induced the early Christians to settle upon these outermost isles, "the pertinacity of devotion." St. Flann was "a patron saint said to have flourished in the ninth century. Some regard them (the remains of buildings) as Druidical, and therefore of more ancient date. These small islands are the *Insulæ Sacræ* of Buchanan." *Vide* also Martin (p. 19), where he relates the fact of these and other remote islands being considered "places of inherent Sanctity."

MacCulloch gives Flann as meaning *red* or *blood*; "possibly from the reddish colour of the cliffs of gneiss. It was," he adds, "also the name of some Irish chieftains." He gives a fair descrip-

[1] *History of Scotland*, 1751.

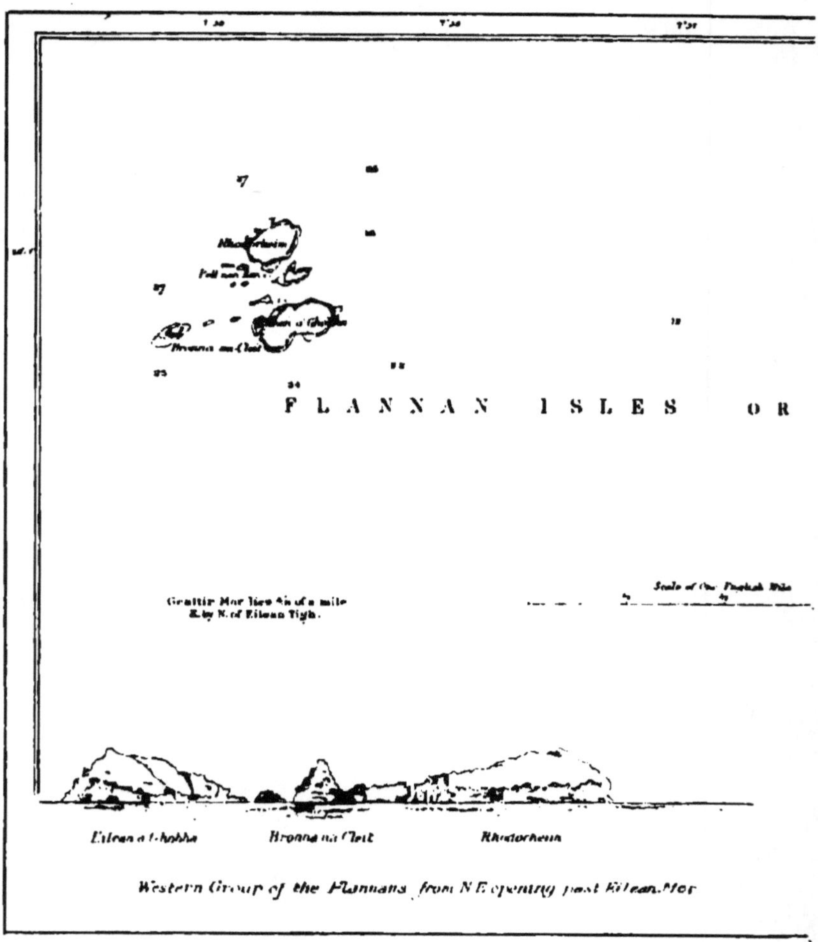
FLANNAN ISLES OR
Bronna na Cleit
Eilean a Ghobha
Bronna na Cleit
Western Group of the Flannans from N.E. opening past Eilean Mor

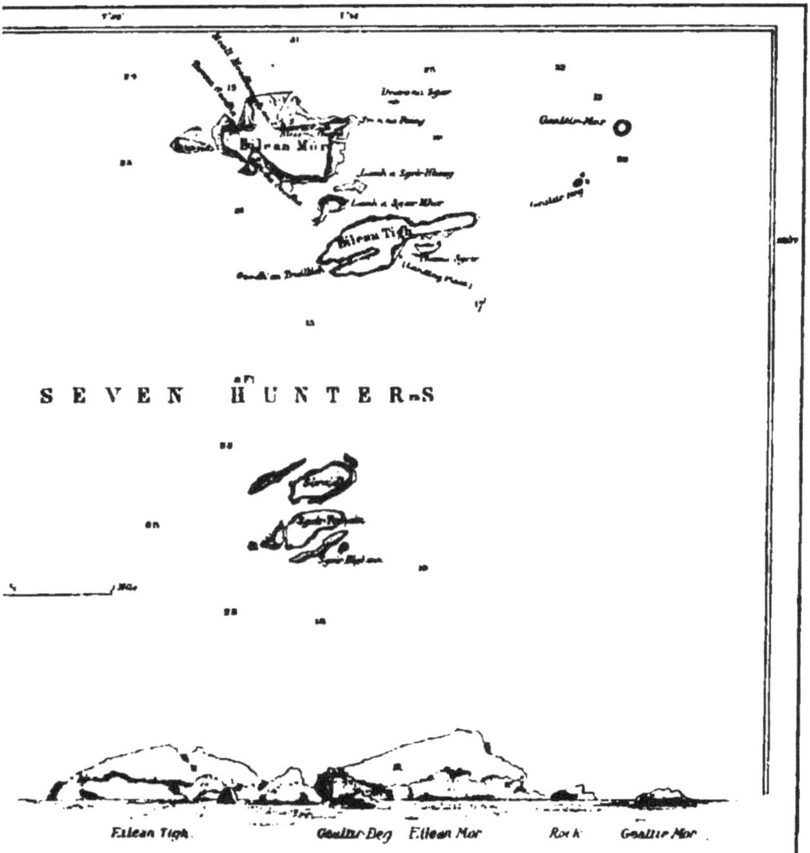

Eastern Group of the Flannans from the S.E.

tion of these isles, which "are seven in number, and lie seventeen miles to the north-west of the Gallan Head in Lewis. . . . The annual rent of the whole is £10." He thus describes their coast-line: "These islands are bounded all round by cliffs cut sharply down to the sea, and almost all bearing the marks of recent fracture and separation; an appearance arising from the little-wearing which they undergo from atmospheric action, and from the obstinacy with which they seem to resist the growth of lichens. . . . Their average height appears to be about 100 feet."

Talking of their geology, he says: "The Flannan Isles are all composed of gneiss traversed by numerous granite veins of different sizes, and ramified in all directions. . . . Here everything appears as if it had been cut and polished by a lapidary." He remarks upon the utter absence of lichens, so common in most other islands of the Hebrides, and the consequent ease with which the "disposition of the rocks" can be traced, and the abruptness and sharpness of the rock scenery, indicating, as already noted, recent fracture and separation.

The late Mr. H. Greenwood, who rented Carn House and shootings in Lewis, wrote us that, in June 1879, his gamekeeper visited the Flannans for eggs, but, unfortunately, the weather was so bad that he had to leave almost as soon as he landed. He procured several Eider-ducks' and some Peregrines' eggs, besides a young Peregrine, and some Razorbills and Guillemots. He told Mr. Greenwood that the Herring-gull bred there.

Mr. H. Heywood Jones, lessee of Mhorsgail shooting in The Lews, visited the Flannans in 1880. He landed on Eilean-Tigh—the same that we landed upon,—and his experiences were somewhat similar as regards the unsatisfactory state of the weather, for it began to blow hard about an hour and a half after landing, so that he was obliged to leave hurriedly, and it blew a heavy gale for some days afterwards. We shall refer later to Mr. Heywood Jones's notes, which he kindly sent us, along with lists of birds of The Lews.

Dr. Heddle and Harvie-Brown made two attempts to land upon the Flannans, and it was only with considerable difficulty that they managed to succeed the second time, and then only upon the easiest of access, Eilean-Tigh. They had previously, whilst interrogating fishermen and others who had landed on these islands, found that there was considerable difference of opinion as to the situations of the landing-places. Some said there was only one landing-place between the two principal islands; others, that there were several suitable according to the direction of the wind. Some reported that the outer islands of the group were never landed upon; others—and one most persistently—that they could all be landed upon, and that a north wind was the worst for landing, and that a west or a south-west the best.

"Coming out of Loch Tarbert, and, after clearing the island of Scarpay—whence fishermen go out to the neighbourhood of Flannan for deep-sea fishing—we had a pretty stiff breeze and a good tumble of a sea on. In order to avoid the long reefs which lie off Loch Resort, and over which a high surf was breaking, we had to tack well out before we could run into our night's anchorage at Loch Hamanaway in Lewis. The next day we spent in fishing, as it was too stormy to attempt the Flannans.

"On the 9th June we made a fair start. A gentle summer 'airie' took us out of Loch Hamanaway about 7 A.M., but we were becalmed soon after till 11. With another almost imperceptible land-breeze, we came across the track of a large fishing-boat. They had a few Ling, Skate, one Tusk, and a large Halibut—about 80 lbs.,—and we laid in a stock. Soon after this, we got again into the 'doldrums.' The men had been cleaning the fish on deck, and Kittiwakes had gathered round us, attracted by the offal. While lying in the sun, Harvie-Brown jumped up, with an exclamation, 'There 's a Skua, and a Pomatorhine too,' rousing his companions—U. and Dr. Heddle—from their novels or their slumbers. He rushed down-stairs for a gun, and put the fact on record beyond cavil. 'A bird hit 's history—a bird missed 's mystery.' How curious to find the great invasion of Skua Gulls still lingering on our shores and seas far into our now almost arctic summers. The same day we saw others, and next day two more in the small harbour of Carloway, while, on the following one, numbers were seen by us out at sea. Fulmars, too, were not uncommon, and a few Gannets hovered round. We heard a report that a branch colony of Fulmars had reached the Flannans, and our pilot—MacDonald—told the same story; but this statement would require accurate and careful investigation, which we were not, at the time, able to give to it."

The chart of the West Coast of Scotland and the Hebrides (Imry, 1881) gives the following islands of the group. We arrange them from east to west as shown on the chart, but some confusion seems to exist as to their names and relative positions.

East.

Gealtir-Mòr.	Rhodorheim.	Soraidh.
Eilean-Mòr.	Gealtir-Beg.	Eilean-an-Gobha.

West.

Broun-na-cleit.

In all, *seven* islands; but Gealtir-Mòr and Gealtir-Beg are insignificant rocklets, with scarce enough breeding-ground for a pair of Oyster-catchers. This Survey Map makes no mention of the second largest island which lies close to Eilean-Mòr, called Eilean-Tigh, and perfectly known and recognised by the natives of Lewis. The "Directions"[1] say, "*Flannen Islands*, or *Seven Hunters*, are a group of islands 3 miles in extent, the highest and largest of which, Eilean Mòr, lies W. ¾ N., 44 miles from the Butt of Lewis; N.N.E. ½ E., 46 miles from Monach lighthouse; E. by N., ½ N., 38 miles from Borreray, St. Kilda, and 16½ miles from Eilean-Molach, the nearest land of Lewis. The group consists of seven islands and thrice as many rocks, divided into three clusters, taking a triangular shape. Eilean-Mòr, 282 feet high, and a quarter of a mile in extent, produces rich grass at an early season, so that the sheep left here for pasture are fattened before any in The Lews. . . . The best landing-place is the south-west side of Eilean-Mòr; but it should only be attempted in moderate weather." This last statement is at variance, as regards the situation of the landing-place, with the reports of some of the natives, and does not seem to be in accordance with the conclusions we drew after cross-questioning a number of them. As will be seen, we landed

[1] Otter's *Admiralty Directions*. Part I. Hebrides. Potter: London, 1874.

without very serious difficulty on the east side of Eilean-Tigh, and the weather could hardly have been designated moderate. Imry's "Directions," however, which profess to be copied or compiled "from recent Admiralty Surveys," as stated on the title-page (1875), mentions Eilean-Tigh as "possibly of 50 acres." They also recommend the south-west side of Eilean-Mòr as a landing-place.

Approaching the Flannan Islands from the E.N.E. their first appearance strikingly reminded us of Haskeir, but a little later this view quickly changed, and it was seen that the group was of more circular or triangular shape, unlike the long single strait ridge which the Haskeir island and rocks present.

Gealtir-Mòr, translated *the big white* (*bright* or *clear*) *land*, and Gealtir-Beg, *the lesser white land*, are only small spray-washed rocks, not a cable's length from the larger masses of Eilean-Mòr and Eilean-Tigh, which latter is nearer to them than the former.[1]

Eilean-Mòr—*the big island*—is the highest and largest of the group, and has a fine precipice facing E.S.E., and a deep göe from top to bottom, with what appears at a distance a feasible but difficult landing-place on the left or south side of the göe, and another equally difficult, still more to the south, and beside a small triangular-shaped cave. These, however, are *not* landing-places, though they appear possible from the sea, and the recognised landing-place is on the north side.

Eilean-an-Tigh—*the island of the house*—is the second largest; and is next to the two Gealtirs in position from the Lewis. Upon it are the remains of an old house, standing at the inner extremity of a large göe, which cuts into the full height of the cliffs—about 120 feet?—from the west side, and of which we could see the further dark cliff-top above the top of the nearer and somewhat lower cliff facing it.

The island is a roundish "lump" with a fine green sloping top facing E.S.E. We are now describing the aspect from the sea, but will say more of it later after landing. One more point, however,

[1] The best map we have seen of the group is a very careful one in Muir's *Characteristics of Old Church Architecture of Scotland*, p. 178, which, along with others, is reproduced in this volume, with some slight additions on the stone.

should be noted now, and that is, the long peninsula which juts out towards the E.S.E., and which, however, is better seen to be a peninsula when approached from a more southerly direction.

As we kept on along the north side of Eilean-Mòr, passing two or three göes where apparently there were fewer birds than on the E.S.E. side, the outer portion of the group began to open out; but before speaking of them, Soraidh—*farewell island*—comes next to be mentioned. It is composed of three main and several smaller semi-detached "lumps" at either end, residences of sea-fowl. These are separated by apparently deep water channels, and the landing appears to be difficult on any of them. They form rounded "lumps," rather than stacks, with precipitous sides. Soraidh is the furthest south of the group, and hence, perhaps, has received the name from mariners leaving the islands.

Besides the above, between Eilean-Tigh and Eilean-Mòr, there is a stack—a square, comparatively low, mass some 60 or 80 feet in height—in mid-channel,[1] and beyond appears an island of somewhat similar shape, which forms a stack close to the west side of Eilean-Mòr, but has no name on the chart, though from its size it appears to have equal claims with the outlying skerries of Gealtir.

The next which come into view are the following:—Rhodorheim, probably an adaptation of *Rudha dòruinn*, meaning *anguish point*, which lies most to the northward of the western group. Bronn-na-cleit—*the protuberance* (or *belly*) *of the reef*—which, agreeably with its name, rises in peculiar shape and abruptness from the centre of what, from this view, appears a long reef, is really three detached islands, and many skerries, of which Bronn-na-cleit is the westernmost. Eilean-an-Gobha—*the smith's island*—is the southernmost of the western group. These three open out past the end of Eilean Mòr, and, as we have said, appear at first like one large island. Penetrating the lower end of Rhodorheim is a small tunnel in the shape of an erect parallelogram, bridged over by apparently a square block of rock, and rising from the sea-passage beneath is a small pinnacle of rock in the centre.

[1] Làmh-a-Sgeir Mhor on the Map, *q.v.*

There are thus, therefore, six islands, two named rocks, and numerous skerries, which really constitute the group.

In Maps.	*Correct Gaelic.*	*Meaning.*
Flannan Isles. Seven Hunters. Seaforth's Hunters.	Eilean-Flannan *or* Eilean-an-Flannan.	St. Flannan's Isle. The Flannan Islands.

Easternmost of the group lie the skerries of

Gealtir-Mòr and Gealtir-Beg.	Gealtir-Mòr Gealtir-Bheag.	The Big White (bright or dear) Land. The small do. do.

Next, and more to the north, lies

Eilean-Mòr.	Ant Eilean-Mòr.	The Big Island.

Next it, on the left, is

Eilean-Tigh.	Eilean-au-Tigh.	Island of the House.

At some distance off, to the southward, are several rocks constituting the Isle of

Soraidh	Soraidh	Farewell Island,

which is the most southern of the group. Then about a mile further west lies the western group. The northern island is

Rhodorheim.	Rudha-dòruinn.	Anguish Point.

Between this and Eilean-Gobha appears

Bronn-na-cleit.	Bronn-na-cleit.	The Protuberance (*or* belly) of the reef.

But this, however, is really the westernmost of the group. To the south then comes

Eilean-Gobha.	Eilean-an-Gobha.	The Smith's Island.

And between Bronn-na-cleit and the other two islands are innumerable reefs and skerries lying in a narrow sound, perhaps 150 yards in width.

Writing to us concerning the names of these several islands, Mr. James Macpherson says: "There is no reason to doubt Martin's assertion that the island (or islands) derives its name from St. Flann. The martyrologies record no less than half-a-dozen Saints of the name from which to choose. *Flann* is a word that I have never heard used for *red*, although the dictionaries give it in a very restricted form. *Gealtire* (a coward) appears to me to be a very absurd name for an island. The nearest approach to the word is *Geal*—literally, *white*. This is often used in a much more extended sense, indicating intense fondness and affection, and, applied to an inanimate object, might also mean veneration—*vide* Martin's notice on the inherent sanctity of the islands. Gealtire may also mean fair and fertile, as opposed to black, bare, and barren rock." [1]

"Soraidh," continues Mr. Macpherson, "is a very appropriate name, though the termination *ay* or *ey* has a suspicious sound. I read in your writing *Rodhorcim*, and I give the only Gaelic form which occurs to me, which is almost similar in sound. There can, I think, be little doubt as to the correctness of the name of Bronn-na-cleit. . . . 'Hebridean names are very puzzling, there are so many elements of corruption.'" [2]

Our view was, of necessity, a hurried one, as we had shortly to put about, the wind springing up fast from the southward and rapidly freshening, and the south sky looking dark and "dirty." We ran for it, and made Loch Carloway early in the morning. On June 10th we fished, as it continued stormy. In the harbour, as already noted, two Pomatorhine Skuas were busy. Some boys,—questioned,—said they had never seen such birds here before, and wondered what they were, although some men affirmed that they were always seen about these seas in summer; but did these fishermen not confound them with Richardson's Skua? Were

[1] In the case in question, this last cannot be the meaning, as both Gealtir rocks are mere skerries *showing white if there is much surf on them*; and thus, it appears to us, they probably deserved the name of *the white land*.

[2] Mr. Macpherson of course alludes to the Scandinavian spelling—Soray or Sorey. We prefer the Gaelic name here, because the other islands of the group appear to be all named from the Gaelic.

these barren birds, or young birds not yet breeding? All we saw, however, that summer, appeared to have reached the adult stage, and the peculiar tail-feathers were fully developed. Going out again to the islands, we saw more, and one followed the vessel for miles, flying on ahead, and then resting on the water, until we again came abreast of it.

We interviewed Mr. Macaulay, postmaster and fisherman at Carloway, who has often visited and landed on the Flannans, and we arranged that he should accompany us as pilot and guide next day if the weather turned out to be suitable. But during the afternoon it blew harder than ever, and by night he said it would be impossible to land, and that it would take two or three days to lull the new swell, which being from N.E. by N. would blow and roll right into the south-east landing-places, which were known to, and approved of by, him. Two cross seas had got up the day before, from north and south, and now there was a newer swell blending with the old north one, and coming from N.E. by E. Mr. Macaulay said it would be of no use, but we arranged to pay another visit, and go as close round the islands as possible. This we did, with the unexpressed hope that "after all we might manage to land." On the 11th June, therefore, about 8 A.M., we again laid on a course for the Flannans, with light airs of wind, and it was 3 P.M. before we reached off the nearest isles, and then stood away past Eilean-Mòr by the northward, with a heavy swell from north-east, and a fresh breeze. Captain MacGillivray would not stand closer to the lee shore than half a mile. We noticed, as we passed, a colony of rock-birds near the eastern extremity of the peninsula of Eilean-Tigh, and several considerable colonies on Eilean-Mòr, and we saw the landing-place on the latter, and how utterly impracticable it was to-day.

The average height of the whole group is about 100 feet, but Eilean-Mòr reaches an altitude of 280 feet, and has an area of grass of some 80 acres; whilst Eilean-Tigh is computed to have about 50. Only three sheep were visible on the former.

We then tacked through the sound between the two groups, in a channel about 1½ miles wide, and ran close under the cliffs of the windward shore of Eilean-Gobha, so that we could hear the loud mutterings of the rock-birds, and could easily identify the species with our binoculars. On Rhodorheim—the most northerly—two wild-geese were distinctly made out, standing in profile against the sky. The gneiss rock was curiously seamed, especially on Eilean-Gobha and Bronn-na-cleit, by numerous felspar and quartz veins; and on the summit of Eilean-Gobha, above a round-edged göe, a huge boulder about fifteen tons in weight, or possibly much more, lay a few feet from the edge of the cliff. Numerous rocks and skerries lie off the isles of Rhodorheim and Eilean-Gobha, and between these and Bronn-na-cleit, and are visited by seals and innumerable Scarts.

After sailing southwards, clear of Soraidh, we tacked again to try to effect a landing on Eilean-Tigh. On the grassy slope of Eilean-Mòr, we saw the remains of the chapel of St. Flann, and on Eilean-Tigh those of the dwelling-house formerly mentioned, at the head of the western göe. On Soraidh, on the south side, were several colonies of rock-birds, and a few Kittiwakes. We got into the gig, taking a few provisions, in case of landing and not getting off easily, and approached the reputed landing-place on the south-east side, which is sheltered by the projecting spur, or peninsula, and by a stack of some altitude. We then ran in on the curl of the surf, between the stack and Eilean-Tigh, into an embayed piece of smooth water; but here we found no landing practicable, so pulled out again by rapid, strong oar-strokes on the slack of the waves, and lay off the only apparently feasible landing-place to watch the surf. After a time, our men backed the boat in, and, watching the wave, gave U. a chance, and he landed easily. Next rise Harvie-Brown landed, carrying the bag and geological hammer; but Dr. Heddle remained in the boat, as the waves now again resumed their turbulence. We climbed up and spent about an hour, or less, on Eilean-Tigh, and traversed the whole top surface.

We took a few eggs of Razorbills, Puffins, Eider-ducks, Oyster-catchers, etc., and could have taken many more had time permitted, but any change in the tide might, within a few minutes, alter the conditions of our embarking. We got off comfortably; but Dr. Heddle, in landing to chip off a bit of a quartz vein, and, having *leather-soled boots without nails*, instead of india-rubber-soled shoes, slipped and came upon his knee, hurting it rather severely.[1]

Hereafter the breeze sprung up briskly, and we had a splendid run for the Sound of Harris, pleased, but not fully satisfied, with our experiences of the Flannan Isles.

[1] Query: As the "Punctilios" of Martin were not adhered to on our first landing, was this not just retribution upon our heads? Heddle had slightly damaged the knee-cap, and was more or less lame for six weeks after.

NORTH RONAY AND NORTH BARRAY, OR SULISGEIR.

Of the former accounts which have appeared of these islands, we cannot do better than reproduce here the chronological record as supplied by our friend Mr. John Swinburne, the year before Harvie-Brown visited them for the first time. Mr. Swinburne writes as follows, and by permission we quote very fully from his account.[1]

"On Monday, June 18, I left Stornoway, in the yacht for Ness Harbour, near the Butt of Lewis, where I wished to get a pilot, and at 2 A.M. on Tuesday morning I was awakened and told that we were off Ness.

"Going on deck, I found we were off the Butt of Lewis, in a dead calm, with a ground-swell rolling in from the north-west, such as I have seldom seen. The boat was lowered with great difficulty, and I proceeded ashore.

"Ness is a very strange place, being a large bay open to the south-east, the top of which is formed by a stretch of beautiful white sand, on which a heavy surf breaks continuously. The sides are formed by broken cliffs, and on the north side of the bay is the so-called harbour, which is merely a slip where the fishermen haul up their boats.

"I had considerable difficulty at first in getting any one to volunteer to act as pilot, as all the men were just going out to haul their lines, but at last a man of the name of Norman MacLeod

[1] *Proc. Royal Phys. Soc. Edin.*, 1883-4, vol. viii. p. 51 (pp. 52-63 quoted almost in full).

came forward. I give the name in case any one intending to visit Sula Sgeir or Rona should want a good pilot who knows the ground thoroughly. I should strongly advise any future visitor to get a pilot from Ness, as the Stornoway pilots are a terrible set of land-sharks.

"Leaving Ness about 5 A.M. on the 19th, we steered north-north-east for Sula Sgeir.

"On the way out, I got all the information I could out of the pilot. He said that some years ago he had lived for some time on Rona, tending the few sheep that are on the island, and that a great many birds bred there, and among them a bird which answered to the description of a petrel of some sort.

"About 5 P.M. we sighted Sula Sgeir from aloft, a little on the weather bow, and shortly after Rona appeared further to windward. About 7 P.M. we were close to Sula Sgeir, so I ordered the boat to be lowered, at the same time telling MacLeod that we did not want him ashore. He smiled rather grimly without answering, but just as we shoved off he suddenly fell into the boat with a mighty crash, very nearly capsizing her. Evidently he had not the slightest intention of letting me loose among the gannets by myself. He seemed to look upon them as his own private property.

"The landing-place we made for is on the east or south-east side of the island. I believe there is another and a better one on the north-west side, but there was a very heavy swell rolling in from that quarter.

"The landing had to be effected on a very steep rock by watching for an opportunity, and then making a wild leap at the shore before the boat fell for the next sea.

"As soon as we were safe ashore, MacLeod made straight for the south-western end of the island where the gannets were, and he stuck to me like a leech the whole of the time I was anywhere near this enormous gannet nursery.

"The nests were placed chiefly on the level top of the island, and not on the cliffs, and covered the whole of the western portion, being placed so close together that it was difficult to avoid stepping on the eggs. The greater number of the birds got up from their

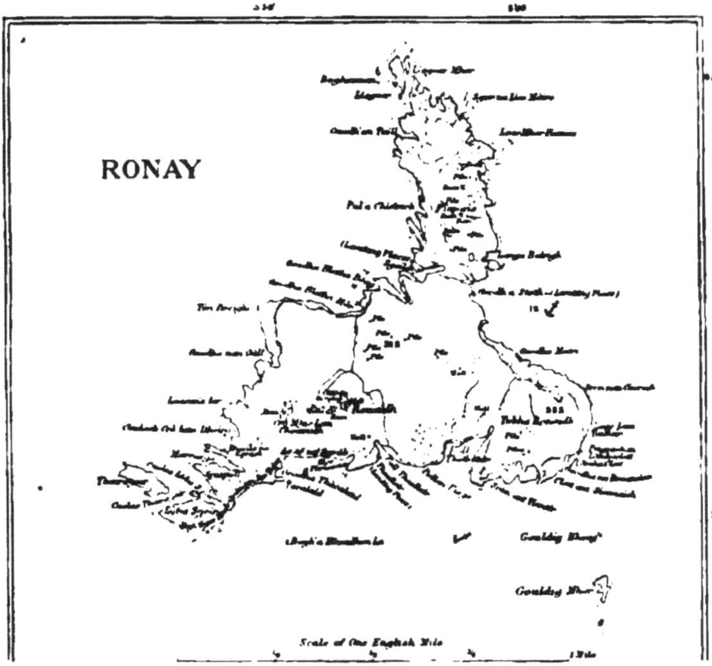

Ronay from the S.E.

nests when I got within about three feet of them, but a good many remained sitting, and bit at my trousers as I passed.

"MacLeod had a very strong objection to my touching the gannets' eggs, because, as I afterwards found out from him, a boat comes out from Ness every autumn about the beginning of September, to take back a cargo of the young birds, which are salted down for winter food.

"I was surprised to observe at this late date that only about 10 per cent. of the gannets' eggs were hatched, and I even took fresh ones, although most of the guillemots were hatching, and at the Bass most of the gannets are sitting by the end of April. On leaving the gannetry, I went down a nasty cliff on the north side to get some guillemots' eggs, in spite of the warning yells of MacLeod, who thought it a very mad exploit. I was fortunate enough to take several eggs of the bridled guillemot (*Lomvia troile*, var. *ringvia*) from under the sitting birds, thus thoroughly authenticating them.

"Seeing that MacLeod had retired, either not wishing to behold my sad end, or intent on the slaughter of the luckless puffin (at which he was a great adept), I reascended the cliff, and made a second attack on the gannets, after which I turned my attention to the island itself.

"The earliest account of the island of Sula Sgeir of which we have any record is that of Dean Munro, who visited it in 1594.

"He says,[1] 'be sexteen myle of sea to this ile (*i.e.* Rona), towards the west lyes ane ile callit Suliskerry, ane myle lang without grasse or hedder, with highe black craigs and black fouge thereupon part of them.' He goes on to say how the Lewis men come out for the feathers and young birds, and enters upon a long description of the manners and customs of a bird which he calls a 'colk,' and which is evidently the eider-duck (*Somateria mollissima*), which I found breeding there in limited numbers. A subsequent visitor to Rona, Dr. MacCulloch,[2] who visited it in 1819, but owing to a heavy sea was unable to land,

[1] "Description of the Western Isles of Scotland, called Hybrides," printed in *Miscellanea Scotica*, vol. ii. p. 153.

[2] *Description of the Western Islands of Scotland* (London, 1819), p. 204.

gives a very short account of Sula Sgeir, only mentioning its being a great gannet resort. The last account we have is that of Mr. T. S. Muir, who seems to have visited both islands about 1872. I believe only about twenty copies of his notes were privately printed.[1]

"He describes Sula Sgeir as 'a narrow stripe of rock, little more than one-third of a mile in length. Rising in most parts to a considerable height, and everywhere ruggedly indented by gloomy chasms, pools, and creeks, it presents a very naked and repulsive appearance.'

"He then proceeds to state that the southern end, however, is in part grassy. This, I think, is an error, as I only saw a few patches of the common sea-pink (*Armeria maritima*). He also refers to the existence of the ruins of several stone huts or houses near the south-east end, and says that near the same spot are those of a small cell called Tigh Beannaichte (Blessed House), 14 feet long, the width varying from 8 feet in the centre to 6 feet 4 inches at the ends. The roof is curved, and covered with stone slabs. The door, which is placed on the south-west side, is 3 feet 5 inches in height. There is also a small window in the east end, under which is an altar-stone 2 feet 8 inches in length. MacLeod informed me that the huts were built by the Lewis men, who live in them for the seven or eight days they take in killing and packing the young gannets, or any longer time they may have to stay on the island, owing to the sudden springing up of a gale.

"The cell mentioned by Mr. Muir was probably erected for the use of those who came from Rona or Lewis for a similar purpose in Roman Catholic times. On one occasion, now some years ago, a crew from Ness in the latter island, had their boat wrecked in landing at Sula Sgeir in the month of June, and lived on the island for several weeks, sustaining themselves on the flesh of birds. Captain Oliver, who commanded the Revenue cruiser *Prince of Wales*, visited Sula Sgeir in the month of August to look for the lost boat. He found the wreck of it, also an oar on

[1] *Inchcolm, Aberdour, North Rona, Sula Sgeir*: A Sketch, addressed to J. Y., Minster Yard, Lincoln, pp. 34.

end with an old pair of canvas trousers on it, and over the remains of a fire a pot containing birds' flesh; but there being no trace of the men, it was thought they must have been picked up by a passing vessel. Nothing more was heard of them until the month of October following, when a Russian vessel on her homeward voyage met a Stornoway craft in the Orkneys, and informed the crew of the latter that they had taken the men off Sula Sgeir and landed them in Rona. Captain Oliver at once went to Rona, and found the crew consuming the last barrel of potatoes which the poor shepherd had. He took away the former, and left the latter sufficient provision for the winter.

"Dean Munro's ideas of distance seem to be a little vague, as the chart makes Sula Sgeir about 10 miles west by south of Rona.

"Sula Sgeir is about half a mile long by about 30 yards wide at the narrowest part, which is just opposite the landing-place we used. It lies about north-east and south-west. The western end forms a steep, rocky bluff, tolerably level on the top. The island slopes downward in the centre, and rises again into a rounded mass of rock at the eastern end. The whole of the western end is surrounded by steep cliffs, and has its upper surface covered with huge blocks and slabs of stone, among the crevices of which a few sea-pinks grow. It is on this part, and not on the east end, that the huts, about six in number, are situated. They are indeed curious-looking erections, being built of huge blocks of stone piled up together, and generally having no other opening than the door. Most of them were about 8 feet by 5 feet, by 4 feet high. At the time of my visit they were tenanted chiefly by cormorants, which built in them. . . ."

"Speaking of St. Rona's Chapel, Dean Munro mentions the following curious superstition of the natives: 'unto quhilk chapell, as the ancients of the country alledges, they leave an spaid and ane shuil quhen any man dies, and upon the morrow findes the place of the grave markit with an spaid as they alledge. In this ile they used to take many quhailles (whales) and uthers grate fisches.'

"'Rona' means the 'Island of the Seals,' therefore the saint must have taken his name from the island, and not the island its name from the saint, as might possibly be thought.

"'There are also, 17 leagues from the Lews, and to the north of it, the islands called Suliskerr, which is the westmost, and Ronay, fyve myls to the eas[t] of it; Ronay (onlie) inhabited, and ordinarlie be five small tennents. There ordinar is to have all things common: they have a considerable grouth of victuall (onlie bear). The best of their sustinance is fowl, which they take in girns, and sometimes in a stormie night they creep to them where they sleep thickest, and throwing some handfulls of sand over there heads as if it wer hailes, they take them be the necks. Of the grease of those fowlls (especiallie the Solind Goose) they make an excellent oyle, called the gibanirtick, which is exceeding good for healing of anie sore ore wound ore cancer, either one man or beast. This I myself found true by experience, by applying of it to the legg of a young gentleman which had been inflamed and cankered for the space of two years, and his father being a trader south and north, sought all phisicians and doctors with whom he hade occasion to meet, but all was in vain. Yet in three weeks tyme, being in my hous, was perfetlie whole be applying the forsaid oyle. The way they make it is—they put the grease and fatt into the great gutt of the fowll, and so it is hung within a hous untill it run in oyle. In this Ronay are two litle cheapels where sanct Ronan lived all his tym as an heremit.'[1]

"About the end of 1600, Sir G. Mackenzie of Tarbert gave a not much longer account to Sir Robert Sibbald,[2] of the island, in which he states that for many generations the island had been inhabited by about five families or about 30 individuals, and that these numbers never increased, because if any one man had more children than another he gave some to his neighbours, and any surplus above 30 souls was sent to Lewis by a boat which went for the rent paid to the Earl of Seaforth in the form of meal packed in sheep-skins, and sea-birds' feathers. The inhabitants seem to have met twice or thrice daily in the chapel, and to have

[1] Sibbald MSS., Advocates' Lib., Edin., xxxiv. 2-8.

[2] *Miscellanea Scotica*, ii.

been Roman Catholics. They had evidently become more pious since Dean Munro's visit.

"Martin,[1] writing about the same date, gave a curious account of the island and its inhabitants as related to him by Mr. Daniel Morrison, minister of Barvas in Lewis, who then appears to have possessed it as part of his glebe, and who had visited it in person. The minister mentions the chapel as being in use by the natives, and that they kept it very neat and clean. The houses he described as being thatched with straw. The next account is that of Dr. MacCulloch the geologist, who seems to have visited the island in 1819. At the time of his visit there was only one cottar family left.

"The Doctor, referring to Rona as it was about the year 1670, says:[2] 'Some years have now passed since this island was inhabited by several families who contrived to subsist by uniting fishing to the produce of the soil. In attempting to land on a stormy day all the men were lost by the upsetting of their boat, since which time it has been in the possession of a principal tenant in Lewis. It is now inhabited by one family only, consisting of six individuals, of which the female patriarch has been forty years on the island. The occupant of the farm is a cottar, cultivating it and tending fifty sheep for his employer, to whom he is bound for eight years, an unnecessary precaution, since the nine chains of the Styx could afford no greater security than the sea that surrounds him, as he is not permitted to keep a boat. During a residence, now of seven years, he had, with the exception of a visit from the boat of the *Fortunée*,[3] seen no face but that of his employer and his own family.'"

"In a note he also says, 'On the appearance of our boat, the women and children were seen running away to the cliffs, to hide themselves, loaded with the very little moveable property they possessed, while the man and his son were employed in driving away the sheep. We might have imagined ourselves landing in an island of the Pacific Ocean. A few words of Gaelic soon

[1] *Description of the Western Islands of Scotland* (London, 1703), p. 19.
[2] *Op. cit.* vol. i. p. 206.
[3] Then employed in cruising after the President in 1812.

recalled the latter, but it was some time before the females came from their retreat—very unlike, in look, to the inhabitants of a civilised world.'

"Speaking of the houses, Dr. MacCulloch says, 'Such is the violence of the wind in this region, that not even the solid mass of an Highland hut can withstand it. The house is therefore excavated in the earth, the wall required for the support of the roof scarcely rising two feet above the surface. The roof itself is but little raised above the level, and is covered with a great weight of turf, above which is the thatch—the whole being surrounded by turf stacks to ward off the gales. The entrance to this subterranean retreat is through a long, dark, narrow, and tortuous passage, like the gallery of a mine, commencing by an aperture not three feet high, and very difficult to find. With a little trouble this might be effectually concealed; nor, were the fire suppressed, could the existence of a house be suspected, the whole having the appearance of a collection of turf stacks and dunghills. Although our conference lasted some time, none of the party discovered that it was held on the top of the house. . . . The interior strongly resembles a Kamtschatkan hut, receiving no other light than that from the smoke-hole, being covered with ashes, festooned with strings of dried fish, filled with smoke, and having scarcely an article of furniture. Such is life in North Rona, and though the women and children were half-naked, the mother old, and the wife deaf, they appeared to be contented, well fed, and little concerned about what the rest of the world was doing.

"'There were then about six or seven acres under cultivation, the surplus crop of which was paid by the cottar to the tacksman in Lewis, and which amounted to 8 bolls of barley. He (the cottar) was also bound to find 8 stones of feathers, the produce of the sea-fowl, which, with the produce of the sheep, was to the tacksman the value of North Rona. To the latter the land was let for £2 a year.'

"Such is Dr. MacCulloch's account of North Rona.

"Captain Burnaby, R.E., who had charge of the Ordnance Survey in Lewis about thirty-six years ago, gives the following description of the physical features of the island: 'This island is situated in the Atlantic in latitude 59° 7′ 15″ 48, and longitude 5° 48′ 50″ 45 west, and forms part of the Lewes property, and lies about 38 miles north-east of the Butt of Lewes, with which and Cape Wrath it forms a triangle, which is very nearly equilateral. From its highest point, which is nearly 360 feet above the level of the sea, Cape Wrath, a considerable portion of the neighbouring shore, and some of the Lewes and Harris hills, can on a clear day be distinctly seen without the aid of glasses. In figure it bears a striking resemblance to a long-necked glass decanter, with the neck towards the north. Its greatest length is nearly one mile, its greatest breadth the same. At its north end there is a portion about half a mile in length, which varies in breadth from ten to twenty chains. About half of this portion is composed of stratified rock without a particle of vegetation. This is the lowest part of the island, its eastern shore sloping gently to the sea, and its western one, though rugged and broken, not more than 90 feet in altitude. The southern portion is broader and more elevated, the largest part of it being three-quarters of a mile broad, and the two hills on the east and west not less than 350 feet high, that on the east being the higher of the two by about 40 feet; the seaward bases of both these hills form steep, precipitous cliffs, which in many places are inaccessible.

"'The rocks around Rona are few and small, the only ones which are more than two chains from the shore being Gouldig Beag and Gouldig Mor; the latter is about half a mile south of the south-east point of the island, and the other is between that and the shore. There is another small rock seen only at low-water near the south-west point, which is dangerous to navigators who may attempt to cast anchor in its neighbourhood. The soil of Rona is good, and the pasture, though not luxuriant, is beautifully green; indeed, the whole island, with the exception of about 50 acres, may be considered arable land, interspersed with a few small rocks and numerous small piles of stones. . . . There are neither rats nor mice on it. It has no peat moss, and not much seaweed. There is a sufficiency of spring water on its southern shore.

Seals are very numerous here, but not easily killed, and cod-fish abound around its coast. The tides rise from 5 to 10 feet, and the prevailing wind is from the south-west. The best landing-places are Poul Houtham on the south, Skildiga on the west, and Geodh Sthu on the east—the first and last being much superior to the other, both for safety and accommodation. The most favourable winds are—for Poul Houtham a northerly or easterly wind, for Geodh Sthu a southerly or westerly one, and for Skildiga a southerly or easterly one. So well sheltered is Geodh Sthu that three vessels have been known to cast anchor at its mouth about six years ago. They remained during one night, but it is said that such had not previously occurred, nor has it been since repeated. Articles of any weight may be safely landed at Rona, providing the weather is moderate, but the small boat, which must be used on such a duty, should invariably be drawn up on the shore after use.'[1]

"The last published account of Rona is that by Mr. T. S. Muir, previously referred to, bearing date 1872.[2] He seems to have visited Rona on two occasions, on the last of which (in 1860) he took careful measurements of the chapel referred to by Dean Munro and Mr. Morrison, and to these measurements he has added a plan and sketch of this ancient building, which he describes as a roughly-built cell, 11 feet 6 inches by 7 feet 6 inches, and 9 feet 3 inches high, with the side walls gradually sloping inwards until they are only 2 feet apart at the roof. There is a low square doorway in the west end, through which it is necessary to creep on one's hands and knees; there are also two windows—one over the door, the other near the east end of the south wall. The altar-stone lies near the east end, and is 3 feet long. The remains of another chapel are attached to the building, forming a sort of nave to it, the internal measurement of which is 14 feet 8 inches by 8 feet 3 inches. The masonry is very rude, and seems of great age, certainly of some centuries. A burying-ground, surrounded by a low stone wall, is attached to the building, and contains several plain stone crosses, the tallest of which

[1] Original, dated 1850, in Ordnance Survey Office, Southampton.
[2] *Op. cit.* p. 16.

is about 30 inches high. I shall now proceed with my own short notes.

"Mr. Muir's description of the chapel is so accurate that I need hardly say anything more about it. The appearance also of the ruins of the houses fully corroborated Dr. MacCulloch's description. I was struck at once by the great thickness of the walls, as well as by the fact that the terreplein of the interior was sunk below the level of the surrounding ground. The last family which lived upon Rona was that of a shepherd named Donald M'Leod, otherwise the 'King of Rona,' who returned to Lewis in 1844; since which time it has been uninhabited, except for a few days at the annual sheep-shearing at the end of July.

"The island has been rented for upwards of 230 years by a family named Murray from Lewis, who only gave up their tenancy in the present year. They usually kept from 100 to 160 sheep of the blackfaced breed upon it, but occasional losses from their falling over rocks or being stolen by the crews of passing vessels were not unfrequent. On one occasion, however, the owners of a vessel whose crew had, through scarcity of provisions, been tempted to 'lift' a few sheep, honourably sent a sum of money in payment. It has now been let to Mr. Finlay Mackenzie, Habost, Ness, in the island of Lewis. The late tenant informed Mr. Muir that the rent paid by the resident subtenants of Rona at an early period considerably before his family became tenants was partly in the form of seal oil. I saw on the island about forty or fifty sheep, which seemed to be very wild.

"As soon as we landed, I made straight for the place where the pilot said the petrels bred. This turned out to be the spot where all the ruins are situated, namely, pretty low down on the grassy slope near the western end of the island. We were all soon at work hauling out large stones, and scraping with our hands, guided by the strong musky odour which pervaded the inhabited burrows which run through and through the thick walls of the old buildings, the latter of which, mixed with earth and turf as they were, afforded unequalled facilities to the birds for the purpose."

Harvie-Brown visited North Ronay on two occasions, the first time in 1885, in company with Mr. Hugh G. Barclay, of Norwich,[1] and again, on a considerably more extended survey in his own yacht the *Shiantelle*, along with Professor Heddle, in June 1887. The second of these visits having been much more thorough, we give the account of it here, adding such observations as occur to us, resulting from the experience of the former visit.

We think it better here to quote Harvie-Brown's own words:—

"About 3 P.M. of June 18th, 1887, we landed for the second time on North Ronay with perfect ease, even easier than in 1885. Meanwhile the yacht stood off and on in the East Bay, opposite the landing-place known as Geodha-Stoth. On this occasion, being anxious to complete my previous survey of 1885, which was a very hurried and unsatisfactory one, I turned my back upon the Fork-tailed Petrels' end of the island, and struck away across the rich carpet of sea-pink and short sweet grass of the lower northern peninsula. The sea-pink, which grows in continuous profusion over the whole surface, filled the air with delicious fragrance, faint but sweet. The rich but short pasturage is strewn with scattered boulders, and in places these have been piled together, no doubt by many previous generations of shepherds and crofters, and formed into many rough sheepfolds and shelters. Oyster-catchers were abundant and aggressive; perhaps nowhere have I seen them so numerous and tame. Gulls were constant in their cries and attendance upon us, the species being Greater and Lesser Black-backed Gulls and Herring-gulls. Eider-ducks were constantly crossing my path, lumbering along in heavy flight, or squattering off their nests, or swimming in sheltered creeks along with their newly-hatched young. Puffins were bobbing about, or ducking head first into the crevices of the cairns, every loose heap of boulders holding some proportion of the general colony. Hundreds of these birds, disturbed by Dr. Heddle from favourite resting-places alongshore, streamed continuously past. Shags innumerable lined the lower talus of débris close to the rocky coast, or sat fanning their wings on the rocks themselves. These ranges of loose stones are arranged in two distinct tiers, one along the higher or westward cliff-tops of the peninsula, the other just above

[1] *Proc. Royal Phys. Soc.*, Feb. 1886.

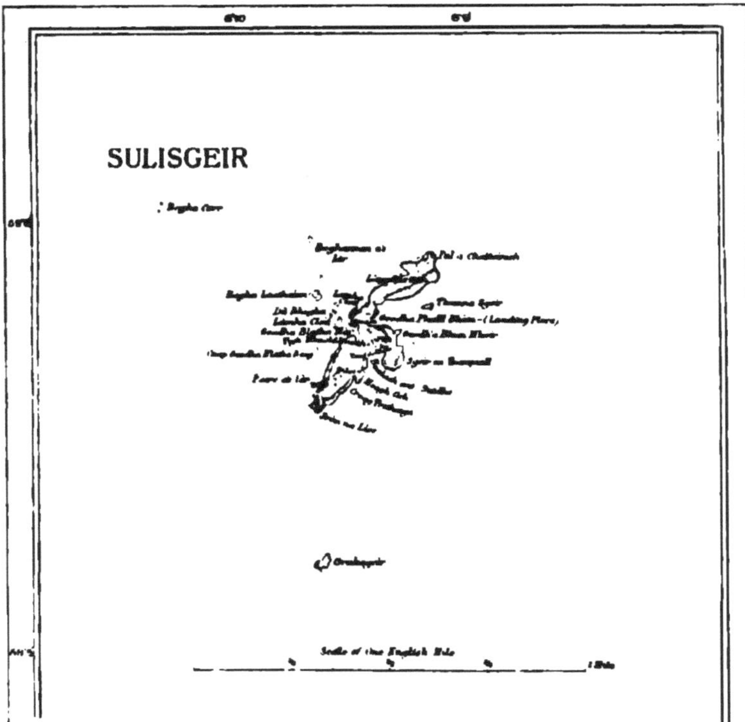

Sulisgeir from the South West

the solid rocky shore of the eastern slope. The Shags were the inhabitants of the gloomy caves of the more lofty southern portion of Ronay, which penetrate the northern faces of the East and West Horns. Rock Pipits were not so abundant as observed elsewhere, though there appeared to be no lack of suitable nesting-ground. Crossing to the west side of the peninsula, I walked over a great stretch of unbroken and continuous bare gneiss, which held in its hollows wind-caught spray-pools, covered with green sea-weeds carried up from the shore. On one of these pools of considerable size, which occupied a hollow close within a ledge of rock, not more than five feet from the cliff-edge, an Eider-duck and her young were swimming about, and Black Guillemots seemed to use it as an occasional playground. On this side, huge caverns, göes, gloups, and rock-arches, stacks, and detached masses of rock, abound, and at once attract attention; and the booming of unbroken Atlantic waves, and giant rollers lashing deep into their recesses, and filling often to the roof some of the great arches, proved a very fascinating scene to me. Here the actual element of water was seldom visible, save glistening with a wondrous green lustre through the white foam of the retreating surges. A Dunlin or two sprang from the sides of the rocky, spray-holding hollows, and a flock of Turnstones circled round the island out of sight. The highest portion of the cliff here is not more than 60 feet, but the extreme wildness of the scene rather gained than lost by the lesser altitude, and the strength of ocean's waves seemed almost to crush their insignificance. On a flat-topped summit of the cliff was a large resting colony of puffindom, which evidently had its nesting holes amongst the great tier of disintegrating gneiss, which ran parallel with the cliff-top, but at a slightly higher elevation, about 50 yards further inland, and which ridge forms the summit of the seapink-covered slope, of which we have already made mention.

"After some biscuits and potted meat, and a pull at Silver's water-bottle, Heddle and I climbed the eastern top of Ronay, 355 feet, but mist obscured the view. We then crossed over the green valley and slope facing the south, and so along the second hillside to the ruins of the old village, the church mentioned by Muir, the underground houses, the crofting lands of the former inhabitants, and the breeding-place of the Fork-tailed Petrels. In the ruined masonry of the old church we heard the churring of the Fork-tailed Petrels, probably 'churching their wooing,' but a desecrating blow from the doctor's hammer upon some garnet-holding lump of rock silenced their revelry, and they stopped, much as a clock would do when 'run down,' and I was left lamenting.

"Coming round the high cliffs to the west—or on the Western Horn,—the Peregrine soon asserted himself, his fine wild challenge being the first notice of his presence. The cliffs of this portion are very fine indeed, and have a peculiar grandeur, partly owing to a gigantic gable-end of MacCulloch's red granite seam, which, stretching across the whole southern portion of Ronay in a vast dyke some 20 feet thick, and interrupting the darker coloured masses of hornblende, stands out of the cliff with a noble curve, giving rich colour to the precipice, and frowning defiant o'er the deep. This is best seen from a point to westward, but a very fine general view of these cliffs facing the north-west can also be had from a spur of the north peninsula, close to a vast cave which runs almost through the rock or neck of this peninsula, and is joined about half-way across by a deep creux, or 'swallow,' which descends to the sea-cave at an angle of about 65°. The following day—being the 19th June—we landed again, principally to allow of photographs being taken, and Mr. Norrie was successful in obtaining several fine views. Among the old ruins, C.,—our steward and cook on board—and I dug away for an hour. The Fork-tailed Petrels' eggs were much harder to reach than on the occasion of our former visit in 1885, being in much more secure retreats, and deeper in the masonry. We took six eggs, and I kept three birds.

"On the face of the N.W. precipice or Horn of Ronay, where there is a considerable broken surface suitable for such birds nesting, I saw six or eight Fulmar Petrels skimming, as is their wont, close to and fro past the cliff-face and top. I saw one alight twice at the same place, about 50 yards west of the granite cliff before mentioned. On returning to this place in the afternoon, but from a further off vantage spot near a large Kittiwake colony in a cave, I could not see the Fulmars, though at once, on going to the west of the granite cliff, I saw them again, and several flew very close past where I was sitting, just as I have seen them do also on St. Kilda; but they rarely flew *over the land*, almost always over the sea. It seemed quite evident that this part of the cliff is the only bit frequented by these birds. I did see *one* Fulmar fly with tremendous speed from west to east across the neck of the peninsula, as if he did not feel himself at home over the land; I never saw one fly at such speed over the sea. After making some of the above notes I crept on hands and knees to the edge, and craned over to get a better view of the face; and my delight was great when I saw one Fulmar sitting, apparently on its nest. Whether there were any more or not, I cannot say, as buttresses of hornblende and granite intervened and obstructed my view. Wishing to see if she were breeding, I threw down several small stones, and, not without some trouble, managed at last to dislodge her. My disappointment was as great as my previous delight, when I saw an empty nest. But the grassy nose on which it was placed showed a perfectly formed cup or wide depression about the size of a soup-plate, scraped or dusted out of a flat piece of seapink turf, and apparently ready for the reception of an egg. I looked around for a long, long time, trying to catch sight of a herring-gull, lest possibly this might have been its nest; but no, not one was visible. Whether it really is a nest of the Fulmar, or merely a resting-place, must, however, still remain undecided.

"I examined the east cliff-face also, and found it populated with the same species as the west, though less densely, and with the addition of Shags in the caves and on the undercliff. The scenery of the West Horn and side of Rona is very fine, and though inferior in altitude to Handa in Sutherland, there are many parts quite as picturesque.

"On the occasion of a former visit in 1885, we made direct for the Fork-tailed Petrels' breeding-place among the ruins of the old village, and worked nearly an hour and a half at excavating the Petrels' eggs, but were not quite so successful as Swinburne, who obtained rather more specimens than we did. Amongst them were three Stormy Petrels, which we caught in the holes, but they did not appear to have eggs. We kept these as specimens, along with seven examples of Fork-tailed Petrels, each of the latter caught on its own egg. Others we let away, and I tossed several up in the air in order to observe the flight. It was very curious in the broad daylight—strange, graceful, zigzag, uncertain, wavering, part bat-like or butterfly-like, part swallow-like or pratincoline, part snipe-like, if I may be allowed to attach so many adjectives. They flew first in circles round their breeding haunt, and in a few seconds made away down the slopes towards the sea, the light-coloured smoky wing-coverts showing to great advantage in contrast with the other dark plumage, and the white tail-spot very handsome and distinct. I should have liked to toss up the Stormy Petrels too, in order to compare the flight of the two species, but as I only obtained one for my share, and did not feel sure that the species had been previously recorded from North Rona, I preferred to keep the specimen. The single egg of each bird lay at the extreme end of the tunnelling, deep amongst the stones, and the seapink-covered turf walls of the long-since deserted village. Our men, as well as ourselves, worked heartily with the spades, but we took with our own hands all the eggs except four. In all cases of dissection the birds proved to be females, both those taken in 1885, and three more I took in 1887. Indeed, the papas do not usually 'lie in' like Donelly's Antediluvians in *Atlantis*. Only on one occasion did we find a pair of birds in the nesting-hole. The ruins are on the southern slope of the island, as shown in the plate, and are of considerable extent, surrounded by a large number of old cultivated patches of land, showing the 'lazy-bed' method of cropping, but now again covered over with good grass and seapink mounds. There appears to be about nine inches of good mould, and what, in a good season, would be a luxuriant growth of herbage. At a little distance from the houses the turf had been cut off the surface for fuel or for building purposes—a custom also adopted by the natives of St. Kilda. Here the surface-wounds soon heal up, and become covered with turf again; but in St. Kilda the contrary is the case, and often the denuded patches do not again bear any turf-covering. From the whole southern slope, loose stones have been gathered off, and these now stand in cairns here and there, affording shelter and nesting-ground for a few pairs of Wheatears.

"North Ronay carries some few scores of sheep, which appear to be in excellent and thriving condition. The habit of sheep-stealing is still carried on by passing ships or fishermen. In the summer of 1885 both sheep, and the oil-barrels and the plenishings of the house belonging to the men who died there the previous April, consisting of tea, sugar, butter, soap, a grinding-stone, etc., were stolen by some Grimsby fishermen, who were, however, shortly apprehended in their own homes in Grimsby, and taken prisoners to Stornoway, where they were tried before the Sheriff, and sentenced to imprisonment—the crew to two months, and the master to eight months. All these articles were upon the island when we were there in June, and the theft must have been committed about the end of the same month, and within a very short time of the date of our visit. Another theft of four sheep was committed whilst Malcolm MacDonald and Murdoch Mackay were alive, as reported by them to their friends in August 1884. They could not read the name of the boat, though there was one on the stern."

SULISGEIR OR NORTH BARRAY.

We continue to quote Harvie-Brown's own words :—

"Not till the remarkably fine summer of 1887 did I obtain a chance of landing on this desolate rock 'for ever fixed in the solitary sea.' After our visit to Ronay, Heddle and I landed on Sulisgeir the same evening, with perfect ease, and scrambled up the tilted strata of the old-world gneiss. All the slopes on either side of the highest ridge and plateau are populated by many Razorbills and Puffins, the former being more than usually abundant, and even the Guillemots were laying where any child could scramble with ease and safety. To the very edge of the precipice the island is covered with wonderful tree-like tussocks of seapink, often developing single stems like tree-ferns, and their roots and gathered earth binding together the great, loose, weathered slabs and boulders, which strew the whole upper surface, and rattle beneath our footsteps as we pass along. Nearer the cliff, and between the looser fragments and the edge, the rocks are white with, and deep in, Gannets' excrement, and the big clumsy nests, some few with fresh, clean, many more with rotten or hard-set eggs, or young in various stages, squelch beneath the tread, almost ankle-deep. It is easy to clamber, and even to run here, at least in dry weather, because the foothold is good. The stench was overpowering, but what it must be in wet weather would beggar description. The nests were masses of putrid seaweed.

"Sulisgeir is indeed a desolate isle. To the left of the landing-place, a big spur runs out, populated all over as above described. A number of rudely constructed dry-stone huts, tolerably comfortable and wind-proof—about a dozen in all—have been erected for refuge-houses by the Ness men, who come over to take the eggs and birds. Their visit this season must have been recently paid, as close around the huts lay innumerable Gannets' heads; and a few peats were left in a dry crevice near at hand. An iron ring has been sunk into the rock to assist in securing the boats, and to this we attached our painter.

"To attempt to give an idea of the Gannet population would, we fear, be hopeless, unless we can believe the accounts related of the annual slaughter, some putting the estimate as high as 3000 birds killed during the few days the Ness men spend on the island. These accounts cannot always be relied upon, as a certain amount of secrecy seems judicious, on account of rents and privileges being subject to alterations or amendments. We have the figures with some degree of accuracy for such better-known resorts as the Bass Rock and Ailsa Craig, but Sulisgeir must far surpass both of these colonies put together, in the amount of its annual yield of birds.

"Quite a number of Fulmars were circling around and even resting on the rock, but I searched vainly for eggs or young. I shot one bird afterwards whilst rowing round the island. The whole extent of the rock is not more than half a mile in length by perhaps 150 yards in width, and nowhere exceeds an elevation of 229 feet.

"The göe to the left of the landing, between the spur and the landing-place, runs right through, by a long arched cavern, to the west side, and contains one of the finest colonies of rock-birds known to me. At the discharge of my gun, dense masses barely escaped immolation as they dashed out in a vast cloud, only just missing the heads of the astonished onlookers; and the noise of their wings echoed back into the great gloomy cavern. Many similar places are to be found on our wild west coasts, but none that I have ever visited compared with this in size and grandeur, and the denseness of its population.

"The Leac rock or flat-headed spur of the southern gable-end of Sulisgeir is covered with Gannets, as is the whole east face and the somewhat rounded off or terraced tops of the solid cliff, and certain spots also of the western face.

"The top of the island is, as I have said, a confused chaos of stones bound together by seapink roots and fibres, sticky with bird-excrement, and most assertive of its smells, if stirred up in its fœtid hollows and dark green spray-pools, which latter are usually covered with green slime and feathers, and surrounded by dead young birds, rotten or highly incubated eggs, and old saturated nests from 6 to 8 or more inches in depth. After our run on shore we again went on board, and kept 'dodging' about during the night instead of anchoring. Next day, the 20th June, we landed Mr. Norrie and C. to take photographs of this extraordinary place, whilst Heddle and I circumnavigated the rock, and its outlying skerries and rocks on the N.W. in our ship's boat. It fell pretty calm, and Mr. Norrie was also able to take a view or two from the deck of the yacht afterwards, before we set sail for Stornoway.

"I forgot to mention that, inhospitable and barren as this isle appears, I saw close to the landing a wretched-looking half- or wholly-starved sheep, and Heddle the piled bones of another. These probably have been left on the island by the birdmen from Ness, to supply their wants when remaining on the island.

But it would have been a mercy to the wretched bag of bones if we had carried it away and restored it to more succulent pastures. Here it literally could have had nothing but sea-pink tufts and seaweed to eat, 'herring-guts to handle.' At every second Gannet's nest lay a herring or two fished up from the deep."

Cave on Rona
Near the Fulmar-haunted Cliffs of the West Horn

HARRIS.

VERY different from the flat monotony of The Lews is the rugged and grand mountain scenery of Harris, particularly of the northern portion. Especially grand are the mountain outlines and bold grouping of the Harris hills, when viewed from the lower-lying heights of S.W. Lewis, near Loch Hamanaway, on a fine warm summer evening. The marvellous colouring at times displayed by the sun-lit ridges, catching the long rays of the Atlantic setting sun, and the deep black shadows of the less favoured and buttress-hidden hollows, produce a picture, once seen, not easily forgotten. Such an one Harvie-Brown witnessed from the above-named vantage, on the glorious evening of the 9th June 1881. That morning we had run in *The Crusader* yawl from W. Loch Tarbert, and after passing in safety long ridges of breakers off the entrance of Loch Reasort, we sought shelter from a stiffish and strengthening breeze within the welcome harbourage of Loch Hamanaway. At this point the marked difference between the lands of The Lews on which we stood,—being rolling or flat and mossy, interrupted by shallow lakes and rocky boulder-strewn moorland,—and the sudden transformation scene of rapidly rising tiers of steep land, and great terraces, ridges, and mountains which rise from the south shore of Loch Reasort, and from the deer-forest of Harris, is perhaps nowhere more apparent. But another and not less remarkable scene, in some respects perhaps, can be had by reversing the view, *i.e.* by taking up a position upon the top of the cliff-like shores of Loch Reasort on the south side, and looking far away over the lower-lying lands of Lewis to the northwards, whence also a very fine seaward view is obtained looking down

Loch Resort and over the island of Scarpay. Such a view was enjoyed by us on a previous occasion, on the 20th April 1870.[1]

Many other magnificent views of the grand valleys and hills of Harris we have enjoyed, having traversed on foot a large portion of its northern half, when collecting there by permission of Lord Dunmore, before this part was sold to its present possessor. The interior of North Harris is weird and wild, typical of a rugged Highland deer-forest. South of Tarbert (which means a place across which a boat can be dragged, and where East and West Loch Tarbert are only separated by a very short neck of land, two to four hundred yards in width), the scenic effects, especially of West Loch Tarbert, are very fine, the great range of the Luscantire hills stretching along the southern shore. We can recall an occasion when, as we were scrambling along the steep heather-slope, near the summits of these hills, in April 1870, a gale of wind sprang up from the westward, and several times we had to throw ourselves on our faces, and clutch hold of the heather, the fierce whirling eddies descending the rifts and gullies with tremendous force, and nearly lifting us off our feet. The west coast of Harris, after the Luscantire range is passed, and where the road follows the coast-line, is low, with a great waste of sand at low tide, over which—so compact and firm is it when still damp—it is preferable to drive, and is much shorter than to follow the road; but the force with which a westerly gale can make itself obnoxious to the traveller was experienced here by us also. Indeed, at times, when the gusts were heaviest, the highland pony which drew our dog-cart could hardly stagger against it.

The east coast of Harris is similar, in great measure, to that of Lewis, being abrupt, rocky, barren, and of no great height, though

[1] A most excellent account, well worth reading, yet too long for insertion here—of a panoramic view from the summit of Cleisham, the highest hill of the Hebridean group—is given by Mr. William MacGillivray in his admirable work on *British Birds*, vol. i. p. 306; and this is described under the more unusual aspects of whirling mist and snow-wreaths, and "giant clouds advancing in disordered groups" over the Western Ocean. Following this is his paper on the *Winter Birds in the Outer Hebrides*, which originally appeared elsewhere. MacGillivray tells us he counted 120 lochs in Harris, the largest three miles long (probably Loch Langabhat).

Roniebhal—a fine mountain behind Rodel—rises to a height of 1506 feet; but there are few patches of cultivation, all the arable land being naturally on the west side. Rich grazing for cattle and sheep occurs upon the green islands of the Sound of Harris, such as Ensay and Killegray, and on the large island of Berneray, which latter holds a considerable population. On Berneray is a wonderful bank or mountain (!) of dazzlingly white sand, which, in size and extent, can only be compared with those of the Culbin Sands on the Moray Firth; and a sand-surrounded, shallow, weed-grown loch, communicating with the sea by a tiny burn and sand-stream, is populated by innumerable and delicious silvery eels, which, after a north gale, descend the burn, and are thrown out on the sandy bay in myriads,—a rich harvest for the appreciative natives, who luxuriate in them on such occasions. It was here, also, that a rare bird-visitor to the Hebrides, a Little Gull—*Larus minutus*, Lin.—found a congenial resting-place for two or three days. Harmetray, another of the islands of the Sound, lately held a few Fallow Deer, which were introduced by Lord Dunmore. At different times we have visited many of the islands of the Sound of Harris, and have landed on Pabbay and Shillay, shot seals on some, taken Greylag Geese and other eggs on others, or hunted the Otter among the many lochs of North Uist. Very great improvements have been effected in Harris with respect to the admission of Salmon and Sea-trout to lochs in the interior, often at considerable elevations, in one case at nearly 500 feet above the sea, close to the base of the Hill of Cleisham (2622 feet), which is the highest hill in the Outer Hebrides. The course of the burn communicating with the loch has been straightened at some places, cut round obstructions in others, and regular ascending steps cut or blasted out of the solid rock. This ascent for Salmon rises rapidly from sea-level to a height of 600 feet, in a distance of less than two miles, and the success of the enterprising experiment is already assured. To the energy of Mr. Hornsby, the late landlord of the Tarbert Hotel, will anglers be indebted for extension of the sport they love so dearly. Much more still remains to be done in this way, and in many places could be done, all through the Outer Islands, at comparatively trifling expense. The lochs of Lacusdale, near

East Loch Tarbert, are also annually visited by shoals of Sea-trout; and a new path has lately been made over the intervening three miles between them and the hotel. It passes through a dreary wilderness of glaciated rock and stone, amongst which a few diminutive patches of crofters' crop- and lazy-bed potato-land intersperse to an altitude of, perhaps, 200 feet above sea-level. Across the Lacusdale river the pathway climbs precipitous hill-faces and knolls, and in less than a quarter of a mile rises from nearly sea-level to upwards of 300 feet. Nor ought we to omit from mention altogether the extremely intricate navigation of the Sound of Harris,[1] which forms one of the principal impediments to the opening out of the great cod and deep-sea fisheries to the west of The Lews. Either must fishing-vessels reach a market *via* the Sound of Harris, and that often in most critical weather and states of tides, or they must weather the almost equally dangerous Butt of Lewis. The distance off, and the danger in fishing these great and rich banks, render them, at the present time, comparatively of little national value. We have ourselves known, when out at sea on the cod-bank, a Halibut, having a money value in London of £4, 10s., to be cut up for bait for Coal-fish and Ling. Yet about one hundred crews are engaged in fishing upon these banks every summer. (See Mr. Anderson Smith—chapter on the Fish, *infra.*)

As already incidentally mentioned, many of the lesser islands in the Sound of Harris are greatly frequented by seals and birds; but while those of Pabbay and others toward the western extremity of the Sound are fairly well inhabited, Coppay appears more destitute than most, though frequented at a certain season by a considerable herd of wild Greylag Geese.

Whilst wood-peat is scarce in Lewis, and almost equally so in Harris, still, subsidence of the land is clearly proved by Captain Thomas, who notes the ancient quotation of Martin. The latter says, whilst speaking of Pabbay, one of the islands of the Sound of Harris: "The west end of this Island, which looks to St. Kilda, is

[1] Witnessed first by Harvie-Brown and Professor Heddle, whilst on their first visit to St. Kilda, from the deck of the s.s. *Dunara Castle*, about 3 A.M. on a summer morning in 1879, and made careful note of for future use. For notes on the intricacies of Harris Sound, see Captain Otter's Charts and Instructions.

called the Wooden Harbour, because the sands at low-water discover several trees that have formerly grown there." Captain Thomas speaks of this as an argument in favour of actual, and not very remote subsidence.

During the past few years Lord Dunmore has introduced a great number of foreign water-fowl to a small loch close to Rodel. He has also enclosed the loch, with part of the wood of Rodel, and the hill-face above it, with a wire fence and netting; and additional ground was planted in 1886. In this year we visited the loch for the purpose of seeing the birds and making a list of them, with the idea of keeping an eye on future occurrences in the Hebrides of rare waterfowl. This list, which was rendered more complete by the assistance of Mr. Finlayson, gamekeeper, we reproduce here :—

In 1885—

2 Egyptian Geese, *Chenalopex ægyptiaca* (Gm.).
4 Brent Geese, *Bernicla brenta* (Pall.).
4 Canadian Geese, *Bernicla canadensis* (L.).
4 Golden Eyes, *Clangula glaucion* (L.).
4 Pochards, *Fuligula ferina* (L.), two of which have since disappeared.[1]
Several Widgeons, *Mareca penelope* (L.), (all gone).
4 Teal, *Querquedula crecca* (L.).

And in 1886 were added as follows :—

2 pairs Carolina Ducks, *Anas carolinensis sponsa*, Gm.
2 pairs Mandarin Ducks, *Æx galericulata* (Lin.).
1 pair Blue-winged Teal, *Anas discors*, L.
2 pairs Bahama Duck, *Dafila bahamensis* (Latham).
2 pairs Yellow-billed Ducks, *A. xanthoryncha*, Forster.
1 pair Spotted-billed Ducks, *A. poecilorynchа*, Pennant.
5 pairs Teal, *Querquedula crecca* (L.), (all gone).
2 pairs Black Swan, *Cygnus atratus* (Latham).
2 pairs Storks, *Ciconia alba*, Bechst. (dead).
1 pair Pelicans, *Pelicanus sp?* (dead).
1 pair Chilian Pintails, *Dafila spinicauda* (Vieillot).

[1] A Pochard shot in the north of Mull is in the possession of Dr. P. M'Bride, Edinburgh. The Pochard is somewhat rare in Mull.

2 pairs Shielducks, *Tadorna cornuta* (Gm.).
2 pairs Mallards, *Anas boschas*, L.
1 pair Bar-headed Geese, *Anser indicus*, Gmelin.
1 pair Ruddy Sheldrake, *Tadorna casarca* (L.).
1 pair Falkland Island Geese, *Bernicla magellanica* (Gmelin).[1]
1 pair Falkland Island Geese, *Bernicla rubidiceps*, Sclater.[1]

Besides these there are a flock of about twenty Greylag Geese, and many curious crosses between wild and domestic ducks.

Outlying Isles of Harris.

Of Gaskeir and the Gloraigs su Taransay we have little of interest to relate, or of the islands of Scarpay off Loch Reasort, or Scalpay at the entrance of East Loch Tarbert, all of which, however, we have from time to time visited, except the Gloraigs. In 1887 we discovered that these rocky islets were inhabited by a colony of the Great Cormorant, but could not make out whether this was a *nesting* or only a *resting* colony; but we think the former, as the birds were crouching, not standing.

On the 4th July 1887, we landed on Gaskeir, finding an easy place in the north göe with a light north wind; the göe is protected by a long point of land, and faces really westward. We spent an hour upon it, and got off with equal ease at a göe on the south shore. A fine view of the Long Island as far as Aird Bhredhuis on the Lewis coast to the north, and of the mountains of Lewis and Harris, was obtained, as well as of the western isles of the Sound of Harris and southwards.

Of the islands of the Sound of Harris we have already spoken casually. We once wasted some valuable time in closely inspecting Coppay and others of the islands at the western end of the Sound, without much ornithological results, but we added to our larder, from one of these islands, a glorious supply of most excellent mushrooms. Passing south on the course to Uist (west side), the big sandhill of Berneray, on this and other occasions, shone out with dazzling whiteness in a dry clear air, with very hot sun, but with little refraction.

[1] A pair were shot at Bunchrew, near Inverness, and are now in Macleay's shop.—T. E. B.

NORTH UIST.

North Uist presents to the eye a most extraordinary mixture of land and water, in this respect to be compared only with Benbeculay and South Uist. On the mainland of Scotland there is only one district we know of which can compare with, or in some measure approach it, viz., Assynt, the Stoir Peninsula, and portions of Edderachyllis, in Sutherland. Looking down upon the great central hollow or plain of North Uist from the sides of any of its higher elevations which surround this portion, it appears to be a perfect network of lochs, islands, and arms of the sea, and it is impossible in such a view to decide where land or water is continuous. Arms of the sea stretch inland from the east coast till they nearly grasp hands with those of the Atlantic. Remarkable amongst these are Loch Maddy and Loch Eport, the former of which, at one point, is only twenty yards distant from the Sound of Harris, and, though this is not very clearly rendered in our map, not a much greater distance from the Atlantic, at a point close to a dune or fort on the road to the farm of Newton. The fresh-water sheet of Loch Scatavagh cannot be very far behind Loch Maddy in the extent of its ramifications. Some days spent amongst these curious creeks and lochs, accompanied by a shepherd and two ably trained collie dogs, in pursuit of the wary Otter, offer experiences not easily forgotten. We have traversed it in many directions, but only a person living constantly on the island could pretend to any knowledge of all these intricacies and curious crooks and corners.

Ben Eabhal and Ben Lee are the only hills of any magnitude, but the lower range of the Grogary Hills bounds the above described country, and encircles the western portion, shutting out in that direction the view of the Atlantic. Beyond these low rolling hills a rich pastoral country, with fine grazing farms and rich "machar" or sand-pastures, growing succulent clovers and grasses under the genial influence of the Gulf Stream, yield wealth of Highland cattle. This rich land stretches as an encircling rim all round the west coast from near Baleshare to Newton on the Sound of Harris opposite Berneray. The rich farm lands of Balranald, Scolpig, and Newton can testify to the mild climate of this western paradise.

Looking out eastward over the Little Minch, the bold rugged outlines of Skye are seen with great distinctness fifteen miles distant; and the magnificent range of the Cuchullin Hills tower over all, still, on 8th May 1870, capped with snow. Looking northward, but avoiding the partially intervening hills of Grogary, the forest-hills of Harris, Roniebhal behind Rodel, and the round-topped but lofty Cleisham still further off, all the ranges to the west of the mainland, and their clear, rugged, and grand outlines, are clearly visible from any elevated land near Loch Maddy. Haskeir rocks, and even the far-distant St. Kilda, are visible from the side of the Grogary Hills above Newton.

The east coast of North Uist is rocky, and of no great elevation, but honeycombed with caves and fissures, and indented by the deep-reaching arms of the Minch, the favourite haunt of Blue Rock-doves and Black Guillemots; but around Loch Maddy are many green knolls and pasture and cropped lands skirting the central moor. Two peculiar rocks lie off the southern horn of Loch Maddy, called "Madday-mhor" and "Madday-gruamach." On their west side, *i.e.* facing North Uist, they are precipitous, and their height from 100 to 150 feet; but on the east side, or the side facing the Minch, they gradually slope down to the sea. They are about 300 yards apart, and are only separated from the main island by a narrow belt of deep sea. The precipitous faces rise from the water-line on either side to a rounded cone, at the apex of which the cliffs culminate in their greatest height. From the shore opposite these curious rocks, and looking out over their tops, the Shiant Isles are seen, rising in the distance like a giant wall, and, owing to refraction, seeming to hang in air.

As has already been said, the north part of North Uist, and westward and southward from Newton, the coast is sandy and level, a stretch of turf-covered downs extending for miles, covered with bent grass near the shore, or among the sandhills, and with

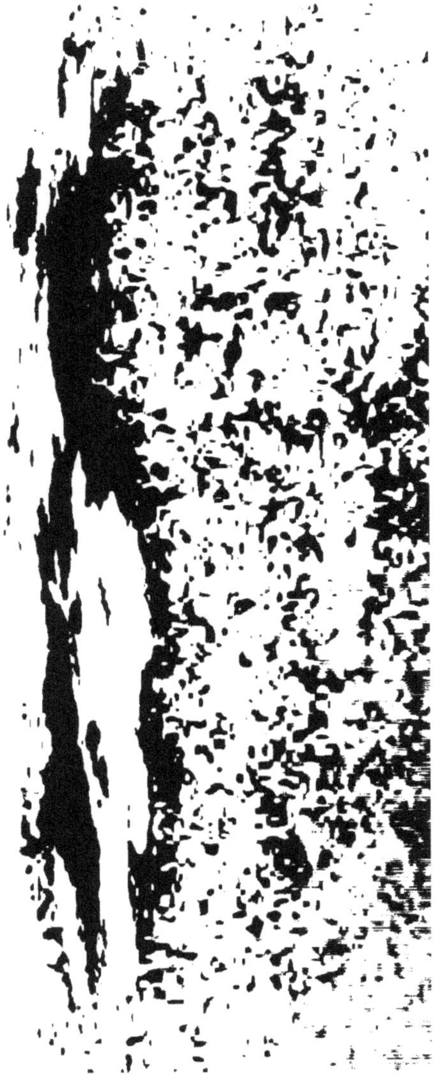

On ... that

... forty miles distant ... Hills tower over ... looking north-... many intervening hills of Gregory, ... behind Rodel, and the noble ... farther off, all the ranges to the ... their clear, ... and grand outline ... elevated land near Loch Maddy, ... ton ... are visible from ... hills above Newton.

... North Uist is rocky and of no great elevation ... caves and fissures, and indented by the deep ... the favourite haunt of Blue Rock ... but around Loch Maddy are ... and ... lands skirting the central ... lie off the ... horn of Loch ... and "Mac-lay-gruamach." On ... North Uist, they are precipitous, and ... east side, or the ... gradually slope down to the sea. They are ... only separated from the main island by a narrow ... sea. The precipitous faces rise from the ... line ... to a rounded cone, at the apex of which ... cliffs ... their greatest height. From the shore ... and looking out over their tops, the ... are seen, rising in the distance like a giant wall, and ... to hang in air.

... already been said, the north part of North Uist, and ... from Newton, the coast is sandy and ... of turf-covered ... extending for miles, covered ... at the shore, or among the sandhills, and with

INTERIOR OF NORTH UIST. LOOKING NORTH

short, clover-mingled grasses more inland—a rich pastoral district. In high winds from seaward the finely comminuted shell-sand rises in great clouds, and is borne far inland, covering in some places the newly sown fields amongst the machar. Where eddies of these winds reach round the intervening sandhills, the turf is torn, and the bare sand exposed. It is in such spots that the Ring-plover finds concealment for its eggs, whilst in the moister meadows inland, Lapwings, Dunlins in small numbers, Larks and Corn-buntings, find shelter and breeding-ground, and in the stagnant, weed-grown ditches, and shallow pools or lakes of water, are Water-hens and Coots. Seaward are some sandy islands, one of which at least, Vallay, is connected at low tide with North Uist. On the intervening stretches of smooth, damp sand, innumerable razor-fish and cockles are gathered for food by the natives, and there are miles and miles of good shell-fish ground.

Probably the contour map—Ordnance Survey—will supply a more definite and lasting impression of the scenery of North Uist than more elaborate illustration, but an accompanying plate is introduced to show its characteristics.

THE ISLANDS AND ROCKS OF HASKEIR, OFF NORTH UIST.[1]

"No special papers have appeared upon Haskeir, that we are aware of, but numerous short notices of it are scattered about in the works of authors who have written on the Hebrides generally.

"Thus in 1540, Dean Munro mentions it as a great resort of 'selchis,' but dismisses it in four or five lines of manuscript.[2]

"Martin,[3] in 1703, writing of these isles, says: 'About three leagues and an half to the West, lie the small Islands called Hawsker-Rocks, and Hawsker-Eggath, and Hawsker-Nimannich, *i.e. Monks'-rock*, which hath an Altar in it, the first called so from the Ocean as being near to it, for *Haw* or *Thau* in the Ancient Language signifies the Ocean: the more Southerly Rocks

[1] In part, from the *Proceedings of the Natural History Society of Glasgow*, December 27th, 1881, and slightly amplified.

[2] We are obliged to Mr. James Macpherson for the loan of a most beautiful MS. copy of Dean Munro's account—*Description of the Western Isles of Scotland, called Hybrides*, by Mr. Donald Munro, High Dean of the Isles, etc.; Edinburgh, 1774. This work is printed in *Miscellanea Scotica*, vol. ii., 1818.

[3] *A Description of the Western Islands of Scotland*, etc., 1703, p. 66. Eagach, *notched, indented*. [*Vide* Macleod and Dewar's Gaelic Dictionary.] The meaning *toothed* was also given to us, an adjective admirably descriptive of these isles when approached from the main island of Uist. The New Stat. Account says: "On Husker (anciently named Iollen na Moinich, or Island of the Monks) are found several crosses, rudely cut in stone." This Iollen na Moinich is the "Hawsker-Nimannich" of Martin, as quoted above, and the Helscher-Vetularum of George Buchanan (English translation of 1751), and are the islands now known as the Monach Islands, which lie 10½ miles S.S.W. ½ W. of the Haskeir group.

Haskeir is variously spelled by authors—Hawskeir (Martin), Hyskere (Knox's Tour), Heisker (M'Donald's Agricultural Survey), Havelschyor (Buchanan), Haveskera (Monipennie), Helskyr (Irvin), and Haskeir (modern writers).

are six or seven big ones, nicked or indented, for *Eggath* signifies so much. The largest island, which is Northward, is near half a Mile in Circumference, and it is covered with long Grass. Only small Vessels can pass between this and the Southern Rocks, being nearest to *St. Kilda* of all the West Islands; both of 'em abound with Fowls, as much as any Isles of their extent in *St. Kilda*. The Coulterneb, Guillemot, and Scarts are most numerous here, the Seals likewise abound very much in and about these rocks.'

"In 1751 Haskeir is spoken of by George Buchanan as 'Havelschyer, to which, at certain seasons of the year, many *Sea-calves* (*i.e.* seals) do resort, and are there taken.'[1] They are spoken of also in somewhat similar terms in Monipennie's *Abridgement of the Scots Chronicles*, under the name *Haveskera*.

"This Irvin repeats in 1819, and speaks of them thus: '*Havelskyra:* Havelskyre; an island. *Helskir Vetularum*, vel *Hlkyr Naimonich: Helskyr Egach;* and *Helskir na Meul*: these are three little islands that lie three miles to the west of Uist, belonging to the lairds thereof."[2]

"They are also referred to by Macaulay, who had landed in 1764 there when on his way to St. Kilda. He speaks of them as 'accessible in a single place only,' but this is hardly correct, as we shall show later on. He refers also to the Great Seals, and describes the method by which many Cormorants were taken in the western caves which terminate the deep göes, and he speaks of the great abundance of the wild-fowls' eggs.[3]

"Capt. H. J. Elwes visited Haskeir in a boat from North Uist on June 30th, 1868, and he gives us a short notice of it in his admirable paper in the *Ibis* for 1869.[4] He found breeding there a large colony of *Sterna arctica*, of which, in the beginning of the same month in 1882, we saw no trace whatever. In Newton Bay,

[1] *History of Scotland.* English translation of 1751.

[2] *Historiæ Scoticæ Nomenclatura Latino-Vernacula*, 1819.

[3] *The History of St. Kilda*, containing a description of this remarkable Island, etc. etc. By the Rev. Mr. Kenneth Macaulay, Minister of Ardnamurchan, Missionary to the Island from the Society for Propagating Christian Knowledge. London, 1764.

[4] "Bird Stations of the Outer Hebrides." By Henry John Elwes, Lieut. and Capt. Scots Fusilier Guards, F.Z.S. *Ibis*, 1869, p. 20.

however, one of our men found an egg the day previous to our visit. Capt. Elwes succeeded in landing on Haskeir Aag, and on this second island from the south found fresh eggs of Cormorants, while on the rock also were a good many Herring Gulls and Great Black-backed Gulls, and one pair of Oyster-catchers.

"The Island of Haskeir, with its outlying skerries of Haskeir Aag, as shown upon the Admiralty Chart,[1] is described in the accompanying Sailing Directions[2] in the following terms :—

"'Haskeir Islands, two in number, are distant from each other one mile in an E. by N. $\frac{1}{4}$ N. and W. by S. $\frac{1}{4}$ S. direction. The easternmost and highest, which lies N.W. $\frac{1}{3}$ N., $6\frac{1}{3}$ miles from Griminish Point, North Uist, and N.N.E. $\frac{1}{2}$ E., $10\frac{1}{3}$ miles from Monach Lighthouse, is one mile in circumference, and rises at the West end to 120 feet; the East end is nearly as high, and between the two the land is very low and nearly divided by a remarkable cave or basin, 140 feet long and 34 feet broad, so that from a distance of 5 or 6 miles the island shows two flattish lumps. Towards the West end are 3 or 4 acres of rich soil and coarse grass, but in winter the waves cast their spray over the whole surface; no springs could be found, but there are several pools with brackish water, where the seals resort in autumn with their young. Rocks dry half a cable off the West and South-West points, but the East side is '*bold-to*': the best landing is on the North or South side of the East lump according to the wind, but it can only be effected with safety during fine weather.'

"'*Haskeir Aag*, the western of the two islands, may be said to be composed of five bare rocks, with deep water-channels between; they are without a blade of grass or any fresh water, and can only be landed on in fine weather. The highest is 83 feet above the sea.'

"Besides the above, sundry sunken rocks are indicated on the Chart and described in the Directions, and it is stated that there is no anchorage in the vicinity, except on a rocky patch, with five to

[1] Published and sold by James Imry and Son from the latest surveys (1881). Two landing-places are shown on the Chart.

[2] *Sailing Directions for the West Coast of Scotland.* Part I. Hebrides or Western Isles. London, 1874.

seven fathoms of water, 'which lies S. by E. ½ E., two cables from the highest part of Haskeir [where] an anchor might be let go in fine weather.' This description, though necessarily concise, appears to be very correct, and well conveys the general aspect and situation.

"Our object in this, as in other similar papers on our bird-stations, is to describe their physical features somewhat more minutely than has hitherto been done; and to treat of the feathered inhabitants as fully as our materials and opportunities of observation permit in the faunal lists.

"Our first attempt to land upon the group was made upon the 30th May 1881. Previous to this date, we had enjoyed what was, perhaps, the only week of real summer weather Scotland had seen that year. Even when tempered by the light sea-airs which gently wafted our good yacht *Crusader* northward from Tobermory—but failed us in the narrow Sound of Harris—the heat was great, and was felt all the more, perhaps, that we were forced to be inactive, and to spend our time in reading and reclining on deck.

"With a fair but very light air of wind, we left our anchorage at Obb in Harris, about 8 A.M. on the 30th May, to thread the somewhat intricate Sound of Harris; the bearings, landmarks, and beacons appearing familiar to us from a previous acquaintance.

"After clearing the westernmost beacon, and avoiding certain sunken rocks, we stood away for the Sound of Shillay, between the two islands of that name and the isle of Pabbay. The two Shillays are the westernmost of the islands of the Sound of Harris. The wind freshened, and being favourable, we bowled along merrily at about six knots an hour. But with this freshening, the dread came that the wind would raise an angry sea on the Haskeir rocks and prevent our landing; and so indeed we found, a few hours later, when, having reached close up to them, we stood off and on for a time to enable us to judge of our chances. A score or so of the Grey Seal (*Halichærus gryphus*) tumbled off the easternmost points of the main island into the white churning surf, which, in a few short hours, had transformed the silent summer sea into a vast heaving caldron. Seeing that no more could be done that day, reluctantly we gave it up, and ran for anchorage to Newton Bay, North Uist.

Some days before, we had met Mr. MacDonald, of Newton, at Loch Maddy, and he imparted to us certain *ins and outs* of the Seal Rocks at Haskeir, one of our objects in reaching them being, if practicable, the 'annexation' of a good specimen of the Grey Seal. Though not successful in this quest, as we afterwards rowed past the places he described, whilst unable to land, we recognised the spots which he had indicated to us with such close precision.

"On 1st June we left Newton Bay about 7 A.M., and soon hove-to near the western rocks of Haskeir Aag. Captain MacGillivray, finding a considerable surf still running, and being apprehensive of more wind, would not allow our pilot—Mr. MacDonald—to land along with us, *his* duty being on board the yacht; but our friend—U. and Harvie-Brown, accompanied by two sailors—both "Dan" by name—got into the gig and rowed over close to the S.E. side of Haskeir Aag, leaving the yacht tacking off and on to await our return. Alas! the surf was quite too heavy to admit of any landing here, so we had to give up thoughts of a Grey Seal from Haskeir. We saw some *great grey* monsters bobbing about in the white surf where no boat dared venture; one even came within 30 yards of the boat, when one of the 'Dans' saluted him with a double discharge, all at once, of an 8-bore gun! but did no damage except to himself. Cormorants were perched and breeding in large numbers, and on No. 3 about 50 pairs of Shags had covered the rounded and sloping top with excrement, whilst in the caves and on the ledges all along the group many more of the latter had their nests.

"After reluctantly leaving behind the rocks of Haskeir Aag, we dropped down upon the main island of Haskeir. On the S.E. side the surf was not so heavy, and more shelter was afforded from the Atlantic swell. Rowing past just round the N.E. extremity, where but few seals were seen, we turned back and easily effected a landing, just below the high 'lumps,' at the East end, by picking an easy place on the rocks, though we could have landed with almost equal ease at many points along the shore. In a few minutes we stood on the rounded tops amidst forests of most luxuriant seapink and bladder-campion; but of grass, as related in the 'Sailing Directions,' we saw nothing.

"To Haskeir, which lies about 13 miles from our anchorage of the two previous nights, Griminish Point—6½ miles off—is the nearest land, and on our voyage back, after leaving Haskeir on our first attempt, we distinguished, at a distance of some 10½ miles towards the S.S.W., the pillar of the Monach Islands lighthouse. Although Griminish Point is nearest, most of the natives visit Haskeir from Obb in Harris, and Berneray in the Sound of Harris, and occasionally from Hogary, near Balranald in North Uist. After Sir John Orde prohibited the men of North Uist from killing the seals, boats for a time came from Lewis, and, we were told, far more were annually killed then than before. About six or seven years ago, a boat's crew when out at Haskeir were obliged to run before an easterly gale for forty-eight hours, and provisions not holding out, they were four days without food, before getting back. For landing on Haskeir the S.E. side is usually the best, as the Atlantic swell is least felt there, but in easterly winds the N.W. side is preferred, and the landing is generally effected at the head of a narrow göe, in which the seals used to be intercepted and killed with clubs as they sought to make good their escape from their breeding haunt near the summit—a stagnant pool of rain-water or sea-spray, of only a few yards in circumference, situated above the west end of the island.

"Once landed on Haskeir on the narrow lower level between the E. and W. 'lumps,' one can traverse the whole island on foot, with scarcely any climbing at all, except of course on the sea-cliffs, and on a few detached rocks at the eastern extremity which are at times frequented by the seals at low water, some of which animals, it will be remembered, we saw plunge into the surf two days before.

"The higher cliffs of Haskeir are on the N.W. side facing the Atlantic. Two ranges of high cliffs—about 80 to 100 feet—face the west, and run nearly across the island; but the summits of these are accessible towards their eastern extremities. At their base are deep gullies or göes, in one case terminating in a large Shag's cave, and in the other running right through the island, but bridged over by a natural arch of rock by which one can easily cross and ascend to the summit of the cliff. From seaward, two openings through the rock are visible, one being the tunnel above mentioned, the other a small hole caused by the falling in of a mass of loose rock.[1] About the centre of the island, which is the lowest part—30 to 40 feet elevation—the rock on the N.W. side slopes steeply, evenly, and smoothly seawards, and up this long incline the great waves rush headlong nearly to the summit, making the rock slimy with green seaweed, and most treacherous footing, as we nearly found to our cost. The S.E. side of this central portion is steeper but more broken, and on many parts a landing might be effected during westerly winds. The tops of the higher portions—E. and W.—are clothed with dense hummocks of seapink, sea-campion, and other rock-plants, forming admirable ground for the innumerable Puffins which burrow in every conceivable direction beneath. The seapink is particularly luxuriant, covering often patches of half an acre or even an acre in extent. Amongst these hummocks also, as well as on the barer and more rocky portions, Eider-ducks breed in numbers. From under a bunch of a dark-green rock-plant—the name of which we do not know—a Rock-pipit fluttered off its nest, which contained four eggs, and an Oyster-catcher ran off hers, which held two. This nest was quite 80 to 100 feet above the sea, and is the highest we remember to have seen. One pair of Wheatears was observed a little above this. At the west end and above the seals' pool, before noticed, a colony of Herring-gulls and Lesser Black-backed Gulls had their widely scattered nests, and Eider-ducks, even at this considerable altitude, were breeding commonly. In the various göes of smaller dimensions than the two before mentioned, Shags, Guillemots, and Razorbills were breeding, often in perfectly accessible situations, and Rock-doves were also seen. Some of the Razorbills had laid their eggs in very simple places, under big boulders[2] or stones on the level or sloping rocks; and by lying down on one's

[1] This is not shown in our drawing in the *Proc. Nat. Hist. Society of Glasgow*, *loc. cit.*, being shut out from view when the sketch was taken.

[2] "Perhaps worthy of the notice of the Boulder Committee. Such, however, were scarce. We only remember two of any size, and these lay on the lower part of the island, near, or on the summit of the sloping rocks, and facing the west. At the time, we neglected to take more careful stock of their situation and appearance, and we do not now feel certain that they were ice-carried boulders."

stomach, close to the edge of the cliff, and reaching down, others could easily be taken without a rope. On the lower rocks along the sea-margin, great numbers of Shags sat in groups, and in the surf below, Eider-drakes and -ducks and various rock-birds swam about. One pair of Great Black-backed Gulls had their nest somewhere near the west end, above the seals' pool, though we could not find it.

"Our friend U. scrambled over the eastern part of the rock which rises above the lower central portion, above a range of cliffs facing west, the climb being easy, and having evidently been often done before. He found the same birds breeding here as on the other portions, and took a few more eggs. By this time, Captain MacGillivray began to feel a little anxious, and showed signs of impatience by blowing the ship's horn. We heard it the third or fourth time and complied, but were deaf to the first two appeals, as we were determined to do the island as thoroughly as possible before leaving. Besides, we were keeping an eye upon the western ocean, and there appeared no increase of wind or swell to prevent our getting off again. Our own experience in such matters we confided in.

"We landed on Haskeir about 12 o'clock, and boarded the yacht again at 3.30 P.M., and in this time believe we saw all that could be done during a land-survey. We also saw the S.E. side well from the boat, and a portion of the S.W. side from the deck of the yacht on the 30th May, but we were not able narrowly to inspect the S.W. shore from the boat, owing to the surf."

BENBECULAY.

THE following notes have been given to us by Dr. J. Mackury:—

"The main difference in physical features between the island of Benbeculay and the adjacent islands of North and South Uist is in its almost universal flatness—there being only two small hills of any size on the east side, viz. 'Rueval,' and 'Wiay' or 'Fuiay Island,' whereas the whole of the east side of South Uist, and the east and middle of North Uist is hilly—some of the hills being of considerable altitude. There are also an unusually large number of fresh-water lochs in Benbeculay, more, I believe, than there are in North and South Uist and Barray put together, and when standing on the top of 'Rueval' (409 feet) on a fine day, when one can have a good view of the whole island, the area of water appears equal to, if not larger than, that of the land. The seashore, like that of the two Uists, is rocky and precipitous on the east side, flat and sandy on the west. Eels, trout, and sticklebacks, but no other kind of fish, are found in nearly all the lochs. The only mammals are *Otters*, which are getting rather scarce, *Mice*, *Rats*,—the *brown* abundant, and the *black* rather rare. Some ten or twelve years ago I killed a large black rat—a male—on the west side of the island. Round the island are considerable numbers of the *black* and *grey* Seal, the latter chiefly, if not wholly, on the *west* side. Whales and porpoises are also frequently seen."

SOUTH UIST.

IN the scenery on the eastern seaboard of South Uist we find much of the mountainous rigour which we met with in Harris—the high mountains of Hekla and Ben More (2035) and their connecting ridges giving a greater appearance of backbone to the long narrow island. The central portions are mossy moorland, and the west side machar and pastoral, fringed with sandhills—an admirable protection against the encroachment of the Atlantic waves.

The grand east coast of South Uist, and the gloomy gorges of Ben Hekla and Ben More, are doubly dark and deep in shadow as the steamer passes close alongshore, after the mists and shades of evening have fallen. The wild fantastic tops are usually obscured by driving mist and cloud that cling and roll before a southerly wind, but the strange umbrella-shaped cap of Ben More reveals itself black as ink through the occasional openings in the mist. Such weird appearances often reminded us of similar scenes we had read of in accounts of the vast Jan Meyen peaks, and of what we ourselves had seen of the Gannet-haunted stacks of St. Kilda—Stack-an-Armine, and Stack Lii. It is thus, by sailing close under these majestic hills, just out enough from shore to see their summits, that by far the grandest idea of their vastness, and their glories of mist and shadow, can be obtained. We have long looked upon this near view of the South Uist ranges as one of the finest effects to be found in the whole Long Island.[1]

Since the body of the text and footnote was first penned, once more have we had the opportunity of noting a phase of the same phenomenon of "Ben Hekla and its smoke-clouds," viz., on May 18th, 1888. We had run north from Otter Sound that morning with the intention of landing in Loch Eynort, to scale both Hekla and Ben More, but when we came abreast of the entrance their canopies of mist were far down their sides; and more and more smoke-clouds were rolling over the lower elevations to the south, "advancing in disordered masses," in turn to be caught and held by the twin summits. At one time the top of Ben More became clear, but all the rest of the mountain was obscured by denser and denser accumulations of mist. These slowly pushed their way upwards on Ben More until at last the same strange appearance—like an umbrella suspended in the air, jet black against drifting scud beyond—alone remained visible. For a time it remained so,

[1] "Once more, and as we have several times witnessed them before, the grand corries and peaks of South Uist came into view—Hekla and Benmore, as usual, carrying their scarfs of fleecy drifting cloud. Hekla has its name of Icelandic or Scandinavian origin, and as we have several times seen it, it is indeed well named. The drifts of cloud merely capping the extreme summit, and re-forming as the last ones leave, give startling effects, suggestive of a burning cone or volcano. It is indeed curious how often this effect is produced, and how seldom the extreme summit is free of mist, catching as it does in almost any wind the moisture from the sea, and being the highest land in the south of the O. H."—(Quoted from J. A. H.-B.'s Journal, 1887, under July 14th.)

no doubt owing to peculiar eddies keeping it clear, caused by the very strange projecting cap of its summit; but even this had to yield at last, till the twin summits became densely enveloped in one vast bank of cloud, whilst still more of the smoke-clouds incessantly rolled up from southward. We knew it was vain now to hope for its clearing; indeed, some time before it finally became blotted out, we had given the order "South about for Mingulay, and lose no more time here," and added the remark—"If rain comes, the change of wind may follow, and enable us to land there." The wind had been strong south-east by south all of these two days.

Since the above was written, our attention has been attracted by a passage in Mr. W. Jolly's interesting account of *Flora Macdonald in Uist* (S. Cowan & Co., publishers, Perth, 1886, p. 12). He says: "We had climbed the volcanic-looking crest of Hekla, the name suggestive of old Viking days," etc. On corresponding on its appearance with Mr. Jolly, and comparing notes, he further says: "Its top, as viewed from the machars of the west side, especially from near Askernish, shows a crater-like outline, and a general volcanic look as viewed from the plains." Referring to a hill in Mingulay of the same name, Mr. Jolly asks me, "Has the Mingulay Hekla any such cloud-flag?" (*i.e.* such as I describe above and in the footnote.) "If so," continues Mr. Jolly, "you are probably right as to the origin of the name." Of course no actual volcanic top occurs, as all the rock is of the old Hebridean gneiss. It seems very probable that these cloud-effects suggested the name, in combination with the otherwise volcanic-looking crest. It appears to be most noticeable with a moisture-laden wind from the south.

From the western slopes of these mountains run many brooks, and the famous Howmore river or Big burn. Interminable mosses, studded with the characteristic lochs and tarns covered with water-lilies, stretch westward to the machar and to the sand-hills. The long Loch Bee, connected with the sea at both ends, yet mostly fresh water, famous alike for its trout and as a resort for wild swans and other water-fowl, reaches towards the north-west—a noble sheet of water, some four or five miles in length.

MONACH ISLES.

CONCERNING these sandy low-lying outspurs of the Western Isles, we have not much to relate beyond the curious fact that they seem to act as a decided turning-point of many migratory birds in autumn, which, previous to arrival here, are seen to pursue a north to south course, but on arrival, and after resting, fly away upon an easterly or north-easterly course. From evidence given in the Migration Schedules,—from Mr. Joseph Agnew in 1887,—we are pretty fully convinced that these outlying spurs of the Hebrides indicate to the migrants the amount by which, further to the north, they had previously *overshot* the land.[1] The westernmost of the Monach group,—Shillay,—on which stands the lighthouse, intercepts their north to south route, and *re-directs* them into the more densely followed inner channel of migration which (as principally shown in the Migration Committee's *first* (1879) Report, but fully borne out by their later ones) crosses Tiree and the Ross of Mull, in a general south-easterly direction. In fact they "hark back" and rejoin the main stream which they had before overshot.

We visited the outer island of the Monach group in our yacht in the summer of 1887. Breakfast-time on the 6th July found the ship lying snugly in Shillay harbour, guarded—*as she was before threatened*—by an encircling rim of reefs, breaking the force of the great Atlantic swell. All the Monach Isles are low and grassy, with rocky or sandy edges, and having a perfect maze of outlying reefs at all stages of the tides. The lighthouse is a plain circular red brick column of somewhat graceful proportions. The view from the balcony is very fine and extensive, as far south as

[1] *Vide* Migration Reports.

Mingulay; and to the eastward over the intervening land in the far distance appear in clear weather the tops of the mountains of Rum and Skye. Nearer are the Uists and Benbeculay, Harris and part of Lewis, and to the westward St. Kilda and Haskeir out at sea. In vain during our short scramble over the island did we search for evidence of the presence of the Common Tern, though Harvie-Brown shot one or two blackish-billed immature birds of the allied Arctic Tern.

BARRAY.

BARRAY is somewhat similar to Uist, but mostly pastoral, and there are no very high hills. An admirable anchorage for the fishing fleet is taken advantage of at Castle Bay at the south end of the island, and thence, on one occasion, we accompanied some 360 sail of fishing-boats on their way to intercept the herring shoals to the south of Barray Head. In 1886, upwards of 700 sail of herring-boats and their attendant "puffers," steamboats, and various trading craft found ample harbourage in Castle Bay, the "harvest of the sea" being reaped many miles to the westward, out in the Atlantic Ocean. Many improvements have taken place at Barray since our earliest visit in 1870, *i.e.* by the date of July 1887, when we last visited it.

MINGULAY AND BARRAY HEAD.

WHEN we come to visit the lesser islands of Barray Head or Berneray, and Mingulay, however, it is then that we are reminded of the stupendous cliffs of St. Kilda. The cliff of Aoinaig, in Mingulay, which frowns down on the wide Atlantic, is a sheer black-faced precipice of 753 feet in height, with two or three green ledges. On two occasions on which Harvie-Brown visited Mingulay—viz., in 1871 and again in 1887,—he came to the conclusion that the Stack of Liánamull is the most densely packed Guillemot station he had ever seen.

This fact is no doubt owing to the unsurpassed suitability, regularity, depth, and number of the breeding ledges, along many of which two men could crawl abreast on hands and knees, with a roof of solid rock above, and a floor of equally solid and horizontal rock beneath. Deeply cut into the cliff-face are these great horizontal and parallel fissures: there is no tilting outward of the strata, no 'fault' in their regularity, while the top of the Stack is feet-deep in rich sorrel and seapink-covered mould, the accumulation of many years of birds' excreta, and which is tunnelled and honeycombed in every direction by Puffins.

At one time Manx Shearwaters inhabited the summit of Liánamull, but latterly they have been completely ousted by puffindom. Formerly the Stack of Liánamull was reached by a suspended rope-bridge from the mainland across the great chasm, but for many years back it has been reached at a landing-place, and by a severe climb, only in exceptionally fine weather, from the seaward side, near the southern extremity.

Mingulay, or Mingalay, as it is also spelled, has aptly been termed by Mr. Jolly "The nearer St. Kilda." Mr. Jolly gives an admirable detailed account of the island, but we think he surely overstates its proportions and the height of its cliffs and hills.

As we have said, we paid two visits to Mingulay, in 1871 and 1887. Mingulay at present grazes about 200 sheep, but could carry more. A score of these are grazed upon the summit of the Stack of Arnamull, and about five head on Liánamull. It also carries about a score of good-looking Highland ponies, and some cattle. The ponies have only about a fortnight's work to do in each season carrying down the peat, cut high upon the hills, the tallest of which reaches 752 feet, and on the summit of which rest the scattered remains of Admiral Otter's beacon. These have been thoughtlessly removed or turned into a rude sheep-shelter by the natives, forgetting the necessities of fellow-beings far out at sea, it may be, when Barray Head light is obscured by mist, though Mingulay hills be clear, and thus their only guide.

Lobster-fishing is well prosecuted by the people of both Mingulay and Barray.

St. Kilda is visible in clear weather, usually *before rain.* On the occasion of our visit in 1887 (16th July), it was indicated by a cloud of its own parentage. The view of the Barray group of islands was very clear and distinct. Also Tiree and Coll were seen away to the S.E., and Ben Hiant in Ardnamurchan was recognised, as well as Ben More in Mull, and the Cuchullin Hills of Skye, Rum having his night-cap on. Skerryvore Lighthouse was distinguished by one of our party from Barray Head the day before, but in the haze we could only make it out with our binoculars.

On the island of Mingulay there is still—judging from a superficial examination, yet we flatter ourselves a tolerably accurate one—plenty of good peat; and new faces or "hags" are being opened out; but the older peat-hags are nearly worked off, and have thus temporarily (or permanently?) influenced the local distribution of the Stormy Petrel. Stormy Petrels used to breed in the dry cracks of the old peat-faces, and were often found by the natives when they were cutting their peats; but they are now driven away, and Mr. Finlayson has not seen or heard of an egg being found for a great number of years, though he does still occasionally see a bird in the twilight.

The Stack of Arnamull is much more easily accessible than that of Liánamull. These great Stacks are separated from the main island by deep chasms, caused by the disintegration or washing away of a great whinstone or trap-dyke,[1] which cuts perpendicularly from summit to sea-level, and which is visible, not only through the greater part of N. and S. extent of Mingulay, but is also clearly seen to have already caused a deeply receding fissure in the north cliffs of Berneray. It also cuts deeply down between the "dune" at the S.W. end of Mingulay, though not to sea-level. On the neck of trap-talus so formed, the ancient inhabitants had erected a considerable fortification or defensive wall, and, retiring to the "dune," protected themselves against the occasional raids and attacks of enemies.

Standing at the head of the great göe which runs in from the westward under the vast Aoinaig or Baulacraig precipice—which by actual survey is 725 feet of sheer, and even overhanging, hard rock—a fine view is obtainable of the wide valley which holds the clustered and primitive dwelling-houses and the crofts of the inhabitants, the silver strand of the bay, and the far-off mainland hills of Scotland. Westward, from our feet, downwards and outwards, the vast sheer precipice of the Aoinaig stretches—terraced by several steeply sloping ledges of green grass, down and along which the sheep scramble, and from many places are unable again to ascend, and where adventurous natives, more often formerly than now, went in search of birds' eggs. A great cave pierces the rock beneath our feet, at the extreme end of this great göe, said to be of unknown extent.

Over the N.W. promontory we found immense numbers, rings upon rings, stripes and patches of snow-white and delicious mushrooms.[2] Yet this delicious, succulent, health-giving fungus is contemptuously left alone by the natives, and called "Balagan buachair," *i.e.* "*spots made by dung.*" Amongst many other wild

[1] See Heddle, article *infra*, p. 227.

[2] We lunched on raw mushrooms and "Crawford biscuit," with a drop of "mountain dew" to wash it down. We filled our handkerchiefs, and a large mineralogical bag of the Doctor's, full to overflowing, and left clothes-basketfuls—nay, cartloads—unplucked.—*Ex* Harvie-Brown's *Journal*, 1887.

flowers and plants which we found on the islands, growing most luxuriantly on the slopes above the village, facing east, were Centaury (*Erythræa centaurium*), Wild Carrot[1] (*Daucus carota*), Meadow Rue (*Thalictrum minus*), a Moonwort, Betony, Primrose, Dwarf Willow, Dandelion (dwarfed), Tormentilla.

We have been at pains to describe Mingulay at some length, as we found it of fresher interest, and much more primitive than St. Kilda, especially as regards the cottars' and crofters' houses. The picturesqueness of St. Kilda bay and village will not for a moment compare with that of Mingulay. We spent a happy and enjoyable day in Mingulay, in company with our friend Mr. Finlayson, to whose good companionship and genial conversation we owe a good deal of the information gleaned otherwise than by eye. Other items of information will be found under the paragraphs referring to the rock-birds of Mingulay, for which we are also obliged to Mr. Finlayson.

[1] Sheeproot, or Gaelic "Briogan."

BERNERAY, OR BARRAY HEAD.[1]

"BARRA or BARRAY, according to one reading, means—*the island of the point or extremity*, from *bar*, a head or point, and *ay* or *i*, an island (*v.* Robertson's *Gaelic Topography*, p. 211, and *New Stat. Acct.*, vol. xiv. p. 198), and this appears to be the one usually accepted. But other authors, as early as the days of Martin, assigned the name as in honour of a tutelary saint, St. Barr, which has been contradicted, and again reasserted since that time, by different writers, some stating that no such saint existed or appeared in the Roman calendar, but others referring to 'St. Barr, Bishop of Cork, in the 6th century, whose commemoration-day, as Martin correctly states, is the 25th September.' Those who uphold the latter derivation in conjunction with the Norse termination *ay* (*oe* of the Northmen), would appear to have at least an equal share of the argument. Although not admitted in the Roman Calendar, many names of saints, Irish, Scottish, and English, to whom that honour has not been accorded, are to be found in native calendars, and the more prominent of these are now generally embodied in the calendars published in this country, which are often prefixed to books of Catholic devotion.

"If the first derivation be accepted, it admirably describes the geographical position of the island; if the second be taken, we find much to support it in what is stated,—that the 25th September was kept as the commemoration-day of the tutelar saint by the natives, and that the practice is only now dying out, while it is still further supported by the fact that churches on the islands are also called after the saint, *e.g.* Killbar. In 'The Story of Greltir the Strong,' translated from the Icelandic by Magnusson (1869), 'Barra' is several times mentioned, and, in the index, the name

[1] Reprinted from *Proceedings of Natural History Society, Glasgow*, 27th April 1880.

is bracketed '(Barrey) one of the Hebrides,' an evident recognition of the Norse origin of the name.

"Barray is also called Bernera or Berneray. This name Mr. James Macpherson[1] considers 'has a suspiciously foreign appearance, but Gaelic writers appear to have assimilated it, and write it, "Bearnairidh." *Bearn* means a gap or notch, and *airidh* (as they spell it) hill-pasturage, or a level green among hills,' which describes the features of the interior of Barray Island, north of Castle Bay, though scarcely of the island upon which the lighthouse is erected.

"Scattered notices of Barray Head occur in the works of our earlier authors. Martin only shortly refers to it; MacGillivray gives a good description of it in his *British Birds* (vol. v. p. 351);[2] Captain Elwes describes it faithfully (*Ibis*, 1869); and Mr. Theo. Walker goes somewhat minutely into a description of it (*Zoologist*, May 1870, p. 2117). Incidental accounts occur throughout the literature connected with the Hebrides, and Highlands and Islands of Scotland. Muir's *Barra Head* may be instanced as one giving illustrations of the scenery. The following is from Harvie-Brown's Journals:—

"When Captain Feilden and I visited Barray Head in 1870, we were not greatly impressed by the numbers of sea-fowl. Our first and last impression was that Barray Head, as a breeding station, has been lauded far beyond its merits, and we have, since that time, as well as prior to it, satisfied ourselves that there are many rock-bird stations holding a larger population of birds than Barray Head. The birds are much scattered upon small ledges all over the face of the cliffs, leaving great spaces quite untenanted and untenantable, owing to the irregular nature of the rock-strata. Very different is the cliff of Mingulay, with its regular and parallel ledges closely packed with birds. But of Mingulay we have already taken occasion to speak more fully."

[1] "I am obliged to Mr. James Macpherson, of Edinburgh, for assisting me in the above notes on the derivation of the name, as well as for much kind help in similar directions elsewhere."

[2] "This is evidently rewritten and somewhat compressed from his earlier 'Account,' in 1830 (*Edin. Journ. of Nat. and Geog. Science*, vol. ii. p. 331)."

Perhaps the finest view of the rock-birds' haunts on Barray is from the old ruinous Keep behind the lighthouse, where a deep gully in the rock runs inland about 100 yards, almost to the base of the lighthouse itself. From the southern side of this gully the whole face of the opposite cliff is seen. There, in 1869, the Peregrine Falcon had its eyrie, and Mr. Theo. Walker, with the assistance of Mr. Maclachlan,[1] procured the young. Captain Elwes, in his able paper on "The Bird Stations of the Outer Hebrides" (*Ibis*, 1869, p. 26), gives a somewhat full account of Barray Head, referring also to the previous accounts by MacGillivray (*British Birds*, vol. v. p. 351).[2] Captain Elwes, amongst other matter, describes the method adopted by cragsmen on Barray for killing the birds as they fly past, by an upward stroke of a long pole "resting, end downwards, across the thigh." A similar method is adopted at Ailsa Craig, as noted by Mr. R. Gray (*Birds of West of Scotland*, p. 436).

Scarcely so high, but equally grand with those of Mingulay, are the cliffs of Barray Head, with the lighthouse crowning the summit, having a total height of 690 feet; and from the balcony round the lanterns one can look almost perpendicularly down into the seething sea below. We have elsewhere described these cliffs and bird-life, partly the result of our own visit in 1870, and partly derived from the daily, weekly, and monthly observations of the previous lighthouse-keeper, Mr. Maclachlan. Some notes on the bird-life will appear through the text, and it is not perhaps necessary here to describe again the cliff scenery, which has been so often done before by Elwes, Walker, and others. We visited Barray Head again in 1887.

[1] Mr. George Maclachlan spent over four years as lighthouse-keeper on Barray Head.

[2] Rewritten, as before mentioned, from his own earlier notes, in 1830.

ST. KILDA.

Of the much-bewritten St. Kilda, we have little of personal experience to relate, as only on one occasion did we land upon it, and spend there some short hours in climbing to the top of Counachar and looking down over the great cliff, which is only some 90 feet less in altitude than the hill-summit. On that occasion, by aneroid, Professor Heddle made out the summit to be 1462 feet, and the cliff from top to sea-level just 93 feet lower or 1369 feet. Captain Otter made it only 1262 feet in total altitude. We had a fair view of the Fulmars on their nests, and afterwards, as we sailed close round the cliffs and coast, found the green slopes and terraces of the higher portions densely tenanted by these interesting birds, whilst the lower cliff-points and slopes were left to the Puffin. Our view, however, of the cliffs was obscured by fog, which hung, alas! all day over the upper half of the height, and in this canopy we were enveloped until we descended to the lower level of some 600 or 650 feet. During our walk we found peat up to the very summits of the hills in abundance, and to a considerable depth. Our view of Borreray from the deck of the *Dunara Castle* was very grand, the high-flying thin mist partially obscuring the tiers of cliff and lower basements of the pinnacles, yet leaving the higher tops only partially obscured, and the hive-like swarms of birds circling above the mist-clouds in countless legions.

In all the many accounts published of this far-out group of islands, few authors have almost anything to relate descriptive of the coast-line and its caves and göes and sea-worn recesses. To our friend Mr. Henry Evans, present lessee of Jura deer-forest, who has visited St. Kilda very many times, and circled round the very bases of its giant cliffs, penetrating into its deep caves and exploring these wild grottos even to a far greater extent, with little doubt, than have its own inhabitants, we are indebted for a slight sketch of his experiences, as well as for some interesting statistics.

Mr. Evans writes us that the largest number of Gannets he ever remembers to have been taken in one year is about 2000 birds. He adds, there are 17 houses, and each family for its share has some 120 birds. These are obtained from the adjoining islands of Stack-an-Armine and Stack Lii. Of sheep—small, brown, and active as goats, among the precipices—there are in all about 1200, and the three principal islands contribute to their sustenance. Of milch cows, there are about 30 on St. Kilda alone, and a stock of poultry. There are about 600 sheep on St. Kilda proper, the rest being divided between Soay and Borreray. The arable land, oats, and potatoes occupy only a small area near the town and landing-place.

"There are several sea-caves in the cliffs of St. Kilda, and perhaps two or three in Borreray. Some of these contain perpetually rolling boulders, whilst others are clear of them. The caves with boulders have usually the characteristic rounded edges to their rock fractures, the others are sharp-edged and unabraided. The sound, even in the calmest days, of the everlasting thunder in these caves is magnificent, and no doubt the perpetual motion of the boulders is wearing and boring the caves deeper.

"There are also several curious little blow-holes in St. Kilda and Borreray, the steam issuing from which might be fancifully compared to the breath of the mythical St. Kilda giants. None of the caves are very extensive, but some are very beautiful, partly owing to the clearness and colour of the water.

"There are only two easy landing-places in St. Kilda. The village has one, and the other is in the west bay, where a large corrie slopes gradually down to the water. Here is the ledge of rock named after the Garefowl. As the slopes of this corrie are pastoral and the milch cows feed there in summer, cheese is made there.

"The cliff scenery, if you row round the islands, is grand; many detached rocks have fallen into the sea, and the stack-rocks are magnificent. Stack Lii, Borreray, must be 400 feet high, and affords nesting-room for thousands of Gannets, and

Stack-an-Armine is scarcely inferior: a vast number of Gannets nest on the cliffs of Borreray itself. No Gannets breed on any other part of St. Kilda, and they do not fish much there. Borreray has far the grandest scenery of the three chief islands. I have been round the islands and stack-rocks of St. Kilda and Borreray at my leisure, and into every cave in a rowing boat. There is no easy landing-place on any of the islands besides those I have mentioned. Even in the finest weather there is a surge, but by choosing one's time when the sea is settled one can go anywhere in a row-boat. In the narrow passage between St. Kilda and Soay is a handsome stack-rock, called the Stack of Difficulty, and also some extremely picturesque rocks which form a beautiful natural archway, through which one may pass in a boat amid the croaks and cries of the Guillemots. The Rock of Levenish, about a mile from the entrance of the bay of the landing-place, does not compare in beauty with the chief stack-rocks of the islands, and few, if any, birds nest on it. There are a good many Ravens in St. Kilda. The factor says they have increased of late years."

As Mr. Evans has visited St. Kilda no less than nine times in his yacht *The Erne*, the above notes convey a freshness of description usually, as we have said, wanting in other accounts. With further details we do not occupy more space here, but recommend our readers to scan the well-worn pages of innumerable other writers, from Martin downwards.

St KILDA FROM THE SOUTH EAST

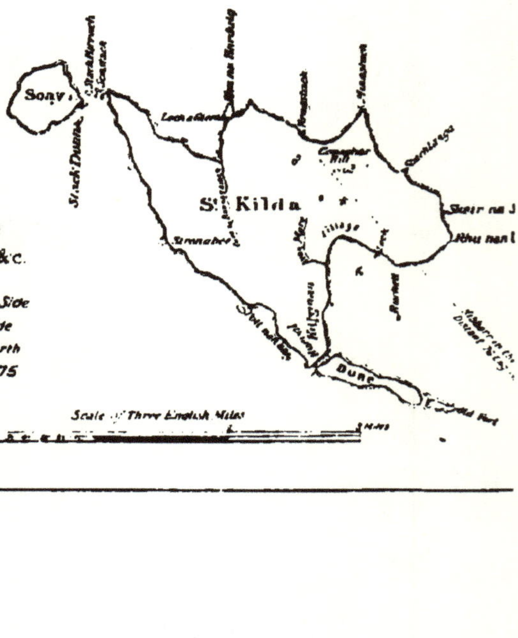

A PLAN OF THE ISLANDS
ST KILDA, BORRERAY &c.

St Kilda Landing place on the North Side of the Bay at the Leck lies in Latitude 57° 48' North and bears from the North of Barra North 44 West Distance 75 English Miles.

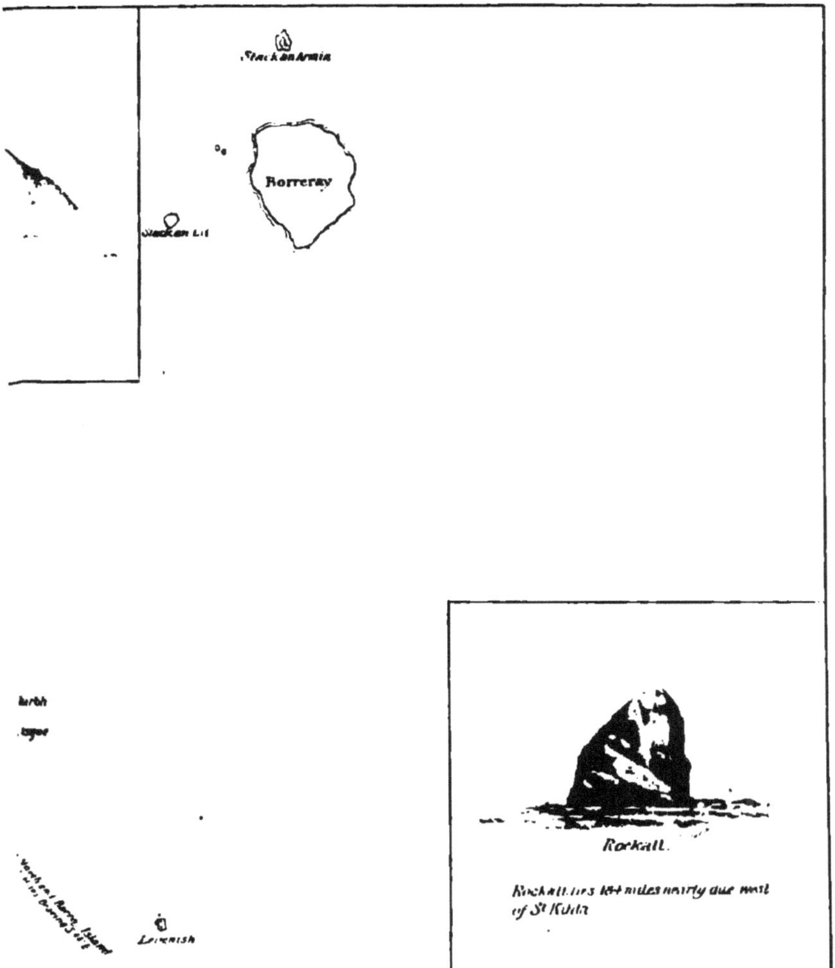
Stack an Armin
Borreray
Levenish
Rockall.
Rockall lies 184 miles nearly due west
of St Kilda

ROCKALL.[1]

But of all these sea-girt isles Rockall is the most oceanic. There is little known from published accounts of the banks and skerries of Rockall. Captain Thomas A. Swinburne, of Eilean Shona, Argyllshire, writes Harvie-Brown that he was fishing in sight of Rockall for nearly a month, during a part of May and June, and there was bad weather all the time, with south-west winds and a heavy swell, so that they did not even anchor; but occasionally, in finer weather, fishing smacks do anchor in deep water. Captain Swinburne and the crew of his yacht were usually fishing in forty to sixty fathoms, with hand-lines, and they did not find that the long lines answered, as it was seldom they could hoist out the boats. They made no attempt to land, and, in fact, seemed to deem it almost impossible, from the smallness and steepness of the rock and constant swell. The rock was constantly covered with birds, but it was not thought that they bred there. "The rock which," says Captain Swinburne, "forms the nucleus of the bank, is shaped like a hay-stack, and is about forty feet high. The upper part is covered with the dung of birds, the lower part constantly washed by the sea, and there are other sunken rocks near. The soundings near the rock are very irregular, with foul ground, and the tide runs strong with the flood." Captain Swinburne did not observe any strange birds. Gulls of sorts, Shearwaters and Guillemots are mentioned. Of fish, Cod, Halibut, Skate, and Blue Sharks were abundant—the offal thrown overboard soon attracting the latter. Captain Swinburne's son, Spearman Swinburne, supplements the above account *in lit.*, as follows: "Rockall is a quarter-mile long, and has high steep sides, and from a distance appears like a ship at sea. It has never been landed upon, and it has been thought impossible to do so. It is crowded with birds in the fishing season,

[1] For actual position, see Captain Basil Hall's *Fragments of Travel*.

but it is not known that they breed there. A rock lies east by north ½ north, 1¾ miles from it, but I do not know if it is awash or not."

Major Feilden (*in lit.*) tells us he looked into the subject of Rockall several years ago (about 1883), with the idea that it might be made a most valuable meteorological station, if connected with the mainland. On inquiries at the Admiralty, he found it to be as Captain and Mr. S. Swinburne describe it, a mere rock, the summit being about sixty feet above high-water mark. The Admiralty chart is very meagre, and the latest survey appears to have been that undertaken by H.M.S. *Chanticleer*, about 1823. A note on this chart says it is extremely difficult to land upon, but possible in very fine weather on one side. Its summit is white with sea fowl. In winter and during gales, it appears that the Atlantic waves must break right over it. "The botany," continues Major Feilden, "would perhaps prove the most interesting part, showing the distribution of plants in a purely oceanic islet. It would be well worth a visit, when every scrap of vegetation and lichens and mosses should be collected."

Mr. Sands also advocates its position for a meteorological station, in a newspaper article.

One other account of Rockall we reproduce here, kindly obtained for us by Mr. William Henderson, Burravoe, Shetland, in answer to an application making inquiry regarding a vessel said to have put in at Burravoe lately from "Rockall" fishery. Mr. Henderson writes: "No such vessel was here; but as Burravoe is not an uncommon name in this country where Pictish broughs are, it is more than likely it was on the west side. As there is a man here who fished at Rockall, I have thought it as well to get from him a true statement as far as he knows about the place, which I enclose." The said narrative is as follows:—

"Thomas Blanche, Burravoe, who fished at Rockall in the smack *Lily of the Valley*, in May 1869, gives the following description of the rock and its surroundings: 'We had been fishing at Faröe. We left Faröe and made the Butt of Lewis, and thence to Rockall. When we thought we were nearing the rock, we sounded, and when we found the bottom we knew we were not far off. We had hooks attached to the lead-dips, and caught two saithe of a large size before we sighted the rock. Four or five hours after we first sounded, the rock appeared, then distant about nine or ten miles. The first appearance was like a smack under full sail. We could not come within three miles of it, on account of the swell and the breakers. We fished here five or six weeks. On a fine day the sunk rocks show themselves, and appear like a number of small boats round a smack.

"The great reef goes straight out from the rock for about three-quarters of a mile to the north-east. The sea breaks from the rock to the point of the reef all at the same moment. Fishing vessels stand off the rock and the reef at night, and during thick weather. The fishing-ground which surrounds the rock is from two and a half to three 'boughts' (*i.e.* from 100 to 120 fathoms) deep.

"When the weather is heavy the whole place is in one white foam. The rock, which is thick at the base, gradually tapers into a cone. It is shaped very much like the 'Opy' (a conspicuous rock lying about three miles N.W. side of Rona's Hill) but is higher; indeed I should say it is larger every way. With bad weather the sea goes clean over the rock. On a fine day, when the sun is shining, the rock is white, the reason of which is, the hot sun drying up the salt water.[1]

"There is no grass nor moss on the rock. No bird is seen near it. I could not say if there is any landing-place on it, but should think not. I don't believe there is ever a day when a boat could come near it. The fish caught there are very large cod and small blue lings; also some common haddocks and saithe. I saw no wild-fowl, unless the mollymak, which is a North Sea bird, and is commonest at the Straits and at Greenland. They are of a greyish-white colour. We saw no seals, but there was a great quantity of sharks, which were large and fierce. We had to leave these fishing-grounds, on account of their destroying the fish while being hauled into the vessel. We saw no whales.

"While there, vessels were daily passing. My opinion is that many a vessel is cast away on that rock. One day a vessel under

[1] But see further on (p. xc.) Captain B. Hall's *Fragments of Travel*.

full stunsails, heading about east or north-east, appeared to keep clear of the rock but not of the reef. When two miles off the reef the sea broke over it. She then hauled to the west, and stood clear of the reef. They either did not know the reef, or did not think it was so far out from the rock. Had this affair occurred in the night-time or in thick weather, probably the vessel would have been lost."

So far the very succinct and clear account of Thomas Blanche of Burravoe.

It remains, however, for us to clear up one statement of Mr. Blanche's—the cause of the white appearance of the upper portion of the island,—and that can best be done by once more quoting fully, passages from the only printed account, so far as we are aware, of a visit paid to Rockall in Captain Basil Hall's *Fragments of Voyages and Travels*.[1] Adding the remark, that this landing upon Rockall, if one reads the whole account, appears to have been one of considerable difficulty and subsequent danger, or great physical discomfort, yet not worse than what might be experienced on other remote islands out of the track of ships, we proceed to quote what appears to us to be the present points of greatest interest:—

"In [*sic*] a fine autumnal morning, just a week after we had sailed from Loch Swilly, to cruise off the north of Ireland, a sail was reported on the lee beam. We bore up instantly, but no one could make out what the chase was, nor which way she was standing. . . . The general opinion was that it must be a brig with very white sails aloft, while those below were quite dark, as if the royals were made of cotton, and the courses of tarpaulin. . . . A short time served to dispel these fancies; for we discovered, on running close to our mysterious vessel, that we had actually been chasing a rock. . . . This mere speck on the surface of the waters, for it seems to float on the sea, is only 70 feet high, and not more than 100 yards in circumference. The smallest point of a pencil could scarcely give it place on any map which should not exaggerate its proportions to the rest of the islands in that stormy ocean. It lies at the distance of no fewer than 184 miles very nearly due

[1] *Op. cit.*, new edition, 1856, pp. 143-146.

west of St. Kilda, the remotest of the Hebrides, 290 from the nearest part of the main coast of Scotland, and 260 from the north of Ireland. Its name is Rockall, and is well known to those Baltic traders that go north about. The stone of which this curious peak is composed is a dark-coloured granite; but the top being covered with a coating as white as snow, from having been for ages the resting-place of myriads of sea-fowl, it is constantly mistaken for a vessel under all sail." On attempting to land they found "one side of the rock was perpendicular, and as smooth as a wall. The others, though steep and slippery, were sufficiently varied in their surface to admit of our crawling up when once out of the boat." A regular scientific expedition seems to have landed, "with hammers, sketch-books, and chronometers inclusive," and while some were set "to chip off specimens," others were appointed to "measure the girth by means of a cord." On leaving the island, the swell on the rocks having increased, "it required the "greater part of half-an-hour to tumble our whole party back again." Finally, after losing sight of their ship, *The Endymion*, and landing again one of the more active of the crew to obtain a look-out from the top of the island, and at last discovering her just before the approach of a fog-bank, the look-out crying, "I see the ship," which announcement "was answered by a simultaneous shout from the two boats' crews, which sent the flocks of Gannets (*sic*) and Sea-mews screaming to the right and left far into the bosom of the fog," and after taking "our shivering scout off the rock," and after again losing sight of the ship and finding her, our adventurous party boarded at last in safety.

In the above account of Captain Hall it will be noticed he refers to the whiteness of the rock being caused by birds resting on it, and he makes no contradiction of the statement after landing. Blanche's account therefore as regards an "incrustation of salt," as he did not land, may be dismissed. Captain Hall makes no mention of the dangerous nature of the seas around, nor of the innumerable sunken rocks or reefs. Captain Hall mentions "Gannets" as part of the throng of sea-birds, doubtless only resting colonies from St. Kilda and Sulisgeir.

In the remarks in the corner of Captain Vidal's map is the following useful note: "The height of Rockall is 70 feet, and the circumference of its base about 250 feet. The summit is made white by birds. The compass variations very near the rock were such as to induce the supposition that it is highly magnetic. Its composition is of coarse granite. Its summit may be reached on the N.E. side, but landing is at all times difficult. Helen's Reef may occasionally be seen in the trough of the wave at low-water springs, and has about 12 feet on it at high water. In fine weather, and easterly winds, it only breaks at considerable intervals, and with neap tides may probably not be observable. The existence of the Leonidas and Guide dangers is very doubtful. They are probably the same as Helen's Reef, but erroneously placed with respect to Rockall. They are here inserted, however, for future observation to confirm or disprove. It is said that a reef extends 3 miles S.E. from Rockall, but nothing of the kind was seen by Captain Vidal, and requires confirmation."

We ourselves had hoped to reach this far-off rock in 1887, but the lateness of the season, and uncertainty of calm weather after the middle of June, decided against the attempt; but we are pleased to find that the landing is probably not more abrupt nor difficult than on many other ocean islets of our archipelago on which a landing has been effected.

Were we able in any way to supplement our account of this mysterious islet by any personal experiences we would not quote so freely from others; but, considering how little seems to be known about it, we have thought it best to reproduce pretty freely all we have been able to gather, which must be our excuse for occupying so many pages.

The following later information regarding this lone rock in mid-ocean was communicated by Mr. W. Cursiter of Kirkwall to Buckley, under date of 7/10/88:—"When we were in Pierawall Bay, there was a Grimsby smack lying astern on the starboard quarter called the *Great Surprise*, Captain Leo. In course of conversation, he told Captain Herrison of having landed on Rockall, and taken about ten dozen eggs, and that another smack had done the same. Leo had just returned to Westray, and gave Herrison one of the eggs as a curiosity. Herrison noted down at the time the following :—'Leo, with a line, climbed, and estimated height of the rock at 78 feet. The top forms a ridge running from E. to W. The west side is almost perpendicular, and the east side slopes more gradually to the sea.'" The egg, which was forwarded to Buckley for identification, is that of a Guillemot.

MINGULAY VILLAGE

THE FAUNA OF THE OUTER HEBRIDES.

THE FAUNA OF THE OUTER HEBRIDES.

FAUNAL POSITION AND IMPORTANCE

OF THE

OUTER HEBRIDES.

In this section we desire to draw attention, as indicated by its title, to the Faunal position and importance or value of this group of islands: *first*, as regards its Mammalian and Reptilian Life; *next*, as regards its Bird Life, and the influences which these islands exert upon distribution, extension of range, and migration. Anything, however, that we may have to say regarding the Fish is freely acknowledged to be tentative, as will appear later in the text.

As regards the Mammals, nothing better has been written, nor probably can be written now, than the admirable article contributed by our deeply lamented friend the late Mr. Edward R. Alston to the series of the *Fauna of Scotland*, published by the Natural History Society of Glasgow, entitled "Mammalia"; and specially that introductory portion which occupies pages 2 to 7 inclusive. It covers all the ground up to date, in remarkably clear and concise language, and places before us a standard of efficiency of the highest order. With this consideration firmly impressed upon us, we make no apology, because we consider none is required, for reprinting those pages of Mr. Alston's article already alluded to, which we now proceed to do:—

"The following Catalogue of Scottish Mammals has been drawn up at the request of the Council of the Natural History Society of Glasgow. In its preparation I have been careful to confine my notes strictly to the department of geographical distribution, and have not entered into any details of description or economy. In the nomenclature I have endeavoured to reconcile the spirit and the letter of the British Association rules, to select the first 'clearly defined' name for each species, and at the same time to avoid all *unnecessary* changes of well-known and generally accepted names.

"The numerous scattered memoirs of previous writers have been collated—notably those of Walker, Low, Fleming, Selby and Jardine, the two MacGillivrays, J. Wilson, Baikie and Heddle, etc. etc.,—and their observations have been compared with my own field notes, and with those of many kind friends and correspondents. Amongst those to whom I am indebted for help are Professors A. Leith Adams, W. Boyd Dawkins, W. H. Flower, and A. Newton, the Rev. G. Gordon, Dr. J. Murie, and Messrs. J. M. Campbell, A. Heneage Cocks, J. G. Gordon, R. Gray, J. E. Harting, J. Kirsop, H. Saunders, and J. R. Tudor. More especially are my thanks due to Professor Turner, of Edinburgh, for many valuable notes on the Seals and Cetaceans, and to Dr. John Alexander Smith, of the same city, for corrections and additions to the account of the extinct forms. Also to Mr. Lumsden, of Arden, who obtained for me lists of the species found in Islay from Mr. M'Kenzie, and of those of Mull from Mr. Cameron; and to Mr. J. A. Harvie-Brown, of Dunipace, who, besides supplying observations from the Outer Hebrides, and rendering much other kind assistance, has procured information as to the Gaelic names from the Rev. Dr. Thomas M'Lauchlan, of Edinburgh, from Mr. D. Campbell, of Callander, and from other sources.

"The number of well-ascertained recent Scottish Mammals recognised in the following pages is *fifty-one*, the proportions of the different orders, as compared with the faunas of England and Ireland, being shown in the following Table:[1]—

[1] "In these Tables I have rejected several species of so-called British Bats and Cetaceans as not being well ascertained. Cf. Bell's *British Quadrupeds*, 2d ed. (1874)."

Distribution of British Mammals.	England.	Scotland.	Ireland.
I. Chiroptera, .	12	3	7
II. Insectivora, . .	5	5	2
III. Carnivora, .	13	13	9
IV. Cetacea, . .	17	16	11
V. Ungulata, .	2	2	1
VI. Glires, .	12	12	7
	61	51	7

"The *fifteen* English species not hitherto recorded as having occurred north of the Tweed are the following:—

1. *Rhinolophus hipposideros.*
2. *Rh. ferrum-equinum.*
3. *Vesperugo serotinus.*
4. *V. noctula.*
5. *V. leisleri.*
6. *Vespertilio nattereri.*
7. *V. bechsteini.*
8. *V. mystacinus.*
9. *Synotus barbastellus.*
10. *Phoca hispida.*
11. *Balænoptera laticeps.*
12. *Grampus griseus.*
13. *Delphinus delphis.*
14. *D. albirostris.*
15. *Muscardinus avellanarius.*

"The *six* Scottish species not yet included in the English fauna are—

1. *Trichechus rosmarus.*
2. *Ziphius cavirostris.*
3. *Mesoplodon bidens.*
4. *Delphinapterus leucas.*
5. *Delphinus acutus.*
6. *Lepus variabilis.*

"The *four* species found in Ireland, but not in Scotland, are *all* Bats, namely—

1. *Rhinolophus hipposideros.*
2. *Vesperugo leisleri.*
3. *Vespertilio nattereri.*
4. *V. mystacinus.*

"Whereas no less than *nineteen* Mammals are on the Scottish list, whose presence is not yet authenticated in the sister isle:[1]—

1. *Sorex tetragonurus.*
2. *Crossopus fodiens.*
3. *Talpa europæa.*
4. *Felis catus.*
5. *Mustela vulgaris.*
6. *M. putorius.*
7. *Trichechus rosmarus.*
8. *Megaptera longimana.*
9. *Balænoptera sibbaldi.*
10. *Hyperoodon latioeps.*
11. *Ziphius cavirostris.*
12. *Monodon monoceros.*
13. *Delphinapterus leucas.*
14. *Capreolus capræa.*
15. *Mus minutus.*
16. *Arvicola agrestis.*
17. *A. glareolus.*
18. *A. amphibius.*
19. *Lepus europæus.*

"From the above tables it is clear that the principal distinctions between the Mammal-faunas of England and Scotland are to be found in the aerial order of Chiroptera and among the marine Fissipedes and Cetaceans, while that of Ireland differs in the absence of no less than twelve species of land animals. My friend Professor A. Leith Adams has recently shown[2] that both the recent and the extinct Hibernian Mammals agree with those of Scotland rather than of England, and has given strong reasons for believing that Ireland received this part of its fauna from the south of Scotland, after its separation from Wales and western England. It therefore becomes a point of some interest to compare the fauna of the Scottish Islands with those of the mainland and of Ireland. In such an investigation it is most convenient to restrict our attention to the indigenous terrestrial Mammals, dismissing entirely the Bats, Seals, and Cetaceans, and also the introduced and cosmopolitan Rats and House-Mouse, which may almost be regarded as domestic animals. Taking the Scottish Islands in four principal groups—(I.) the Inner Islands (Skye, Mull, Islay, etc.); (II.) the Outer Hebrides (the Lews, Harris, the

[1] "On the somewhat vexed subject of Irish Mammals I have followed the list given by Professor Leith Adams, on the authority of Mr. A. G. More, *Proc. R. Dublin Society*, 1878, pp. 40, 41. The complete absence from Ireland of some species, as the Common Shrew, Weasel, Harvest Mouse, etc., has been disputed by some writers."

[2] "*Proc. R. Irish Academy*, 2d Ser., III. pp. 99, 100; *Proc. R. Dublin Society*, 1878, p. 42."

Uists, Benbecula, etc.); (III.) Orkney; and (IV.) Shetland—we find the distribution of Mammals, as far as I have been able to ascertain the facts, to be as shown in the following Table, which represents the known range of twenty-four species of land quadrupeds:—

Distribution of Scottish Mammals.	Scotland.					Ireland.
	Mainland.	Inner Islands.	Outer Islands.	Orkney.	Shetland.	
1. *Erinaceus europæus*,	+					+
2. *Sorex tetragonurus*,	+	?		?		
3. *S. minutus*, . .	+	?	+			+
4. *Crossopus fodiens*, .	+	+		+		
5. *Talpa europæa*, .	+					
6. *Felis catus*, . .	+					
7. *Canis vulpes*, . .	+	?				+
8. *Martes sylvestris*, .	+	?	+			+
9. *Mustela vulgaris*, .	+					
10. *M. erminea*, . .	+	+			intro.	+
11. *M. putorius*, . .	+					
12. *Meles taxus*, . .	+					+
13. *Lutra vulgaris*, .	+	+	+	+	+	+
14. *Cervus elaphus*, .	+	+	+	+ (ext.)		+
15. *Capreolus caprœa*, .	+	intro.				
16. *Sciurus vulgaris*, .	+					+
17. *Mus sylvaticus*, .	+	+	?	?		+
18. *M. minutus*, . .	+					
19. *Arvicola agrestris*, .	+	+	+	+		
20. *A. glareolus*, . .	+					
21. *A. amphibius*, .	+	?		+		
22. *Lepus europæus*, .	+	intro.	intro.	intro.		
23. *L. variabilis*, . .	+	intro.	intro.	+ (ext.)		+
24. *L. cuniculus*, . .	–	+	+	+		+
	24	7	6	7	1	12

"The facts indicated by the above Table are, at first sight, somewhat contradictory. As Ireland possesses the greatest number of species in common with the mainland of Scotland, it might well be supposed to have been in connection with it up to a later date than even the Inner Islands. On the other hand, we have the presence of other forms, as of the Field Vole in the Hebrides, and of the same species with the Water Shrew and Water Vole in Orkney, which are conspicuous by their absence from the Irish fauna. It appears to me, however, that this apparent contradiction may be explained, if we remember the more northern position of the Scottish Islands and the nature of the country lying between them and the south-western source from which our Mammalian fauna was undoubtedly derived.

"A consideration of the relative depths of the channels which respectively divide Ireland and the Islands from the mainland of Scotland would lead us to the conclusion that the severance of the former took place first, and that the Orkneys remained longest uninsulated. An upheaval of about 240-270 feet would bring the latter again into communication with Caithness, while it would require a rise of about 300-320 feet to reunite the Hebrides with Skye, and of from 700 to 900 feet to restore land communication between the various parts of south-western Scotland and north-eastern Ireland. Nor does the distribution of Mammal life seem to me to contradict such a hypothesis. The absence from the known fossil fauna of Scotland and Ireland of most of the characteristic postpliocene English animals shows that the northward migration of these forms was slow, gradually advancing as the glacial conditions of the northern parts of our islands decreased in intensity. Thus it is not difficult to suppose that the Hedgehog, Ermine, Badger, Squirrel, and Mountain Hare may have found their way through southern Scotland into Ireland long before they were able to penetrate into the still sub-arctic regions of the Highlands. Subsequently, when the continued depression of the land had isolated Ireland, and the improvement of the climate had continued, the Shrews and Voles may well have found their way northwards along the comparatively genial coasts, before the larger beasts of prey could find a sufficient stock of game. When they reached Orkney, however, they appear to have found it a veritable *Ultima Thule*, for the absence from Shetland of any land animal (except the half-aquatic Otter) seems to indicate that those islands were already separated before the arrival of any form of Mammalian life.

"Such a hypothesis of the dispersal of English Mammals through Scotland and Ireland appears to me to be the only one which explains the peculiarities of their present distribution, and is likewise in accord with the facts of physical geography. Should it be accepted, the recent and extinct Mammals of Scotland may be arranged in five categories, in the order of the dates of their immigration. This I have attempted to show in the following list, in which the extinct species are marked with an asterisk :—

"LIST OF EXTINCT AND RECENT SCOTTISH MAMMALS, ARRANGED IN THE PROBABLE ORDER OF THEIR ARRIVAL FROM THE SOUTHWARD.

I. Before deposit of boulder-clay :—

*1. *Elephas primigenius.*
*2. *Rangifer tarandus.*

II. Before separation of Ireland :—

3. *Erinaceus europæus.*
4. *Sorex minutus.*
*5. *Canis lupus.*
6. *C. vulpes.*
*(*Ursus fossilis.*)[1]
7. *Martes sylvestris.*
8. *Mustela erminea.*
9. *Meles taxus.*
10. *Lutra vulgaris.*
*11. *Equus caballus.*
*12. *Sus scrofa.*
*13. *Megaceros giganteus.*
14. *Cervus elaphus.*
15. *Sciurus vulgaris.*
16. *Mus sylvestris.*
17. *Lepus variabilis.*
18. *L. cuniculus.*

III. Before separation of Hebrides :—

19. *Sorex tetragonurus* (?)
20. *Arvicola agrestis.*
*21. *Bos longifrons.*

IV. Before separation of the Orkneys :—

22. *Crossopus fodiens.*
*23. *Bos primigenius.*
24. *Arvicola amphibius.*

V. Since separation of Orkneys :—

25. *Talpa europæa.*
26. *Felis catus.*
*27. *Ursus arctos.*
28. *Mustela vulgaris.*
29. *M. putorius.*
*30. *Alces machlis.*
31. *Capreolus capræa.*
*32. *Castor fiber.*
33. *Mus minutus.*
34. *Arvicola glareolus.*
35. *Lepus europæus.*"

[1] "Remains of the Cave Bear have not yet been found in Scotland, but its former existence is rendered probable by their presence in Irish deposits.—*Cf.* A. Leith Adams, *loc. cit.*"

Mr. Alston then proceeds to the consideration of the details of the distribution of the species, taking first the recent, and second the fossil and extinct forms. We do not follow him into this part of the subject at this time, but refer our readers to the able paper we have so far quoted.

The Mammalia of the Outer Hebrides may be said to be pretty fully and accurately worked out, up to date, except the distribution in some instances, where a few minor and minute details are still wanting, and more definite records of the rarer oceanic forms.

The short list of Reptilia seems also fairly complete, unless a few facts regarding their distribution, and the history of their extension of range, be still wanting.

We first wrote, ourselves, a slight account of the Mammals of the Outer Hebrides in 1879,[1] having at that time kept notes regarding them for some six years previously. When we now come to revise and go carefully over what was then recorded, we find little to alter or amend, but a good deal of new material to add to these earlier notes.

Many of our own bird-notes have already appeared elsewhere, *e.g.* in Mr. Robert Gray's *Birds of the West of Scotland*; these were the results of our visit in 1870; and we only purpose alluding to such of them as require correction or amplification, or which appear useful in the context of a distributional handbook, such as this volume aims to be. Of the more interesting, or indigenous and peculiar species, we may be tempted occasionally to speak at greater length. Our insistence upon chronological sequence in observations, which means, to a considerable extent, progress of distribution and increase of range, we desire to put forward, still with a deep sense of the circumstances which militate against their positive usefulness, such as the often-cried "want of observation," "absence of resident naturalists," "overlooking of facts," etc. But this chronological method of treatment often carries with it its own criticism, as it gives every person who has written anything upon the subject the opportunity of future correction, emendation, or corroboration, and also allows others to judge of the same, with "critics' pointed arrows not too coarsely barbed."

[1] *Proc. Glasg. Nat. Hist. Soc.*, April 1879.

Of the direct relationship existing between the Outer Hebrides and Skye it may be desirable to say a few words.

Skye, lying as a projecting spur, as it were, outwards across the Minch, separated by narrow water-ways from the mainland, and within easy bird-view of the Long Island, forms a catchment area for many migratory species, and is therefore not only itself a resting-place for several rarities which do not occur, or occur only rarely, in most parts of Scotland, but it also serves as a land-communication for our smaller land- and wood- birds towards the northern portions of the Hebrides. Thus the Grasshopper Warbler of late years has become a breeding species in the extreme north-west of Skye, and we find the Nuthatch turning up on migration. Thus, too, we account for the breeding of the Willow Warbler, Whitethroat, and other woodland species in comparatively recent years amongst the plantations around Stornoway, and, to a less extent, in Rodel Glen. The first appearance of the Willow Warbler in North Uist, and its extreme rarity elsewhere in the Outer Hebrides, is quite within our own recollection. Now it is fairly abundant and increasing. It has been argued with us, that Skye is not more necessarily a factor in the population of the Outer Hebrides than the further off coast of Sutherland and Ross-shire, but we still think Skye *does* offer a more natural *spring* land-route for weak-winged woodland species, whereas the Pentland Firth, Caithness, and the Butt of Lewis is a more natural autumn route.

We have several times before spoken of the great migration stream which pours east and west, according to the season, through the tempestuous Pentland Firth, the latter forming, as it does, one of the main ways of migrants in spring and autumn. The further-reaching wavelets of this stream impinge upon the shores of the Outer Hebrides; and occasionally birds of rarity and far-eastern origin and distribution are borne along by fierce easterly gales to the lantern lights of the lighthouses of the uttermost isles. But this is comparatively of rare occurrence, the larger number of migrants being able either to "catch up" on the mainland, on the north coast of Scotland, or else somewhere in the more sheltered Minch between Scotland and the Hebrides, although, should continuous and severe gales last for weeks together, as was the case in 1880, many, many weary ones are cast into the Atlantic, hundreds of miles west of The Lews.[1] The route which birds follow in autumn, catching up at the Hebrides, is down along the land communication, throwing off large detachments at the Little Minch, across to Skye, and others which, with great regularity, cross the island of Tiree and the promontory of the Ross of Mull, whilst yet others, but in gradually lessening flights, reach as far as Monach Isles in the west, but after this, "hark back" on a more easterly course. In the autumn of 1878 few of the birds which are ordinarily resident in the islands migrated, or were much affected by the severe weather of 1878-79, Twites and Wrens remaining all winter.

We have already referred to the peculiar direction taken by migrants after reaching Monach Isles, turning again eastward from a previous N. and S. direction.

We ought also to take some short notice of the return wave of *spring*-migrants in this connection, and point out that migrants usually follow very much the same lines, coming the reverse way in spring from that followed in the autumn migration, and it is *in this respect* we advocate the importance of the peninsular form of Skye as a "leader" or permanent land-route towards the northern portion of the Outer Hebrides, thus throwing, as it were, a "side-light" of some importance upon the present and future of the bird-population of Harris and Lewis.

Of the more remote rocks and skerries, and isles of all sizes, and of varying interest, which we have visited, we have little to say in evidence of their faunal positions and values except what appears from their breeding species. Their influence cannot be very great except in one or two interesting instances, where they distinctly act as stepping-stones or resting-places and landmarks to the migrants. Thus it is curious to find the Whimbrel evidently nesting on North Ronay, and the Tree Sparrow (*auct.* R. M. Barrington) occurring there in the summer of 1886.

Could we imagine North Ronay and St. Kilda clothed in rich luxuriance of forest growth, and suitable for the reception of Warblers and sylvan species, we cannot but believe that in course

[1] See Migration Report, 1880. General Remarks, p. 92.

of time tempest-tossed wanderers amongst such bird-migrants might take permanent possession, though we admit that such outposts of colonisation would, even under most favourable circumstances, take a longer time to become populated by migratory species than those lying closer to the more normal route of the migration.

EXPLANATORY OF THE FAUNAL LISTS.

(*a*) The following Lists of species contain all the Mammals, Birds, Reptiles, Amphibians, and Fishes, whatsoever, which are admitted to places in the British Fauna.

But, such as have not any notes attached to them are entered merely for purposes of comparisons and future additions, and have nothing directly to do with the present state of the Fauna of the Outer Hebridean Area. Our plan of work being distinctly chronological and progressive, will, we trust, do away with the idea that this is simply padding with useless names.

(*b*) *Square brackets enclosing ordinary type.*—This signifies that the information contained is still considered as more or less doubtful; and therefore that the species is not perfectly deserving of admission. Examples of this occur principally in the Mammal and Bird lists.

(*c*) When the information is only of extra-limital importance regarding the species, the notes are given in italics, and are only intended for purposes of comparison; as also are footnotes in brevier type.

Class 1. MAMMALIA.

Sub-class *MONODELPHIA.*

Order CHIROPTERA.

Sub-order *MICROCHIROPTERA.*

Family RHINOLOPHIDÆ.

Gaelic—*Ialt: Ialtag*=a Bat.

Rhinolophus hipposideros (*Bechst.*). **Lesser Horseshoe Bat.**

Rhinolophus ferrum-equinum (*Schreb.*). **Greater Horseshoe Bat.**

Family VESPERTILIONIDÆ.

Synotus barbastellus (*Schreb.*). **Barbastelle.**

Plecotus auritus (*L.*). **Long-eared Bat.**

Vesperugo serotinus (*Schreb.*). **Serotine.**

Vesperugo discolor (*Natterer*). **Particoloured Bat.**

Vesperugo noctula (*Schreb.*). **Noctule.**

Vesperugo leisleri (*Kuhl*). **Hairy-armed Bat.**

Vesperugo pipistrellus (*Schreb.*). **Pipistrelle.**

This common British species does not appear to be plentiful in the Long Island. Captain M'Donald told Harvie-Brown in 1870 that he had only once seen a Bat at Rodel, which is a well-sheltered spot, with a few trees around it, and there is also a plantation of considerable extent in the neighbourhood. Professor Duns includes "Bats" in his paper (*loc. cit.*), but gives no actual proofs of their existence, and previous writers are unanimous in excluding them. We may, however, with safety include, under the present species, all actual records of Bats, as this is the only one likely to occur. Since the above was written, and while staying at Rodel in June 1879, Harvie-Brown was repeatedly assured that "bats" had become much commoner there of late years, and that they are now far from rare, though he failed to observe any himself. Mr. D. Mackenzie, who has sent us many valuable notes on Mammals as well as on Birds, writes in an interleaved copy of a former paper of Harvie-Brown's in 1882: "Last summer, and the preceding one, I have seen bats in an arch near Stornoway."[1]

Vespertilio dasycneme, *Boie.*

Vespertilio daubentoni, *Leisler.* **Daubenton's Bat.**

Vespertilio nattereri, *Kuhl.* **Reddish-grey Bat.**

Vespertilio bechsteini, *Leisl.* **Bechstein's Bat.**

Vespertilio murinus, *Schreb.* **Mouse-coloured Bat.**

Vespertilio mystacinus, *Leisl.* **Whiskered Bat.**

Order INSECTIVORA.

Family ERINACEIDÆ.

Erinaceus europæus, *L.* **Hedgehog.**

Gaelic—*Graineag.*

Family TALPIDÆ.

Talpa europæa, *L.* **Mole.**

Gaelic—*Famh: Uireach.*

Family SORICIDÆ.

Sorex tetragonurus, *Herman.* **Common Shrew.**

Sorex minutus, *L.* **Lesser Shrew.**

Gaelic—*Beothachan-feoir* = little beast of the grass.

This is the species probably meant by MacGillivray as being "found in the Outer Hebrides on sandy pastures, where it is named 'Luch-feoir'" (*Edin. Journ. Nat. and Geog. Science*, vol. ii.), a name,

[1] "On the Mammalia of the Outer Hebrides" (*Proc. Nat. Hist. Soc. of Glasgow*, 29th April 1879).

however, properly belonging to the Field Vole (*vide* Alston, *Fauna of Scotland*, "Mammalia," p. 28). Harvie-Brown was fortunate in obtaining a single specimen of this species in North Uist, in June 1879, which he preserved in spirits for identification. They are very rarely seen except in harvest-time, so that his getting one in summer may be considered very fortunate. It was the only one he saw; but Mr. D. Mackenzie informs us that it is likewise found in The Lews, and this is corroborated by Mr. Greenwood.

Beothachan-feoir, given as the Gaelic name, is mentioned by Dean Munro so long ago as 1540, and the superstition of its being poisonous to horses is also taken notice of by him.

In 1831, MacGillivray gives a place to the Common Shrew, as inhabiting the Outer Hebrides. Little doubt, however, remains that the present is the species intended.

Crossopus fodiens (*Pall.*). **Water Shrew.**

Gaelic—*Lamhalan*.

NOTE.—"*This name is evidently a transformation of* Famh, *a mole, and* aln *or* alan = *water*, i.e. *the Water Mole*."—A. C.

Order CARNIVORA.

Sub-Order *FISSIPEDIA*.

Section ÆLUROIDEA.

Family FELIDÆ.

Felis catus, *L.* **Wild Cat.**

Gaelic—*Catfiadhaich*.

Obs.—Of many entries of "Wild Cats" in the very complete list of vermin killed between 1876 and 1885 inclusive (furnished by the courtesy of the Chamberlain of The Lews, Mr. William Mackay), not one can be held as applicable to the true wild species; and we mention this here in order to set up a guide-post to others in all future collections of similar statistics. The said records present a steadily increasing crop of Cats, averaging 28·7 (*sic*) for ten years, there being 30 killed in 1876, and 41 in 1885. Now, from all the records of true Wild Cats that we possess (and of these we have a large number from many parts of Scotland), such an increase is most unlikely, if not actually impossible; and, besides our own observations, we have the still higher authority of the late Mr. E. R. Alston, all pointing to the fact that "Felis catus" has not existed in the Outer Hebrides within historic times.

Section CYNOIDEA.

Family CANIDÆ.

Canis lupus, *L.* **Wolf.**

Gaelic—*Faol-chu* : *Madadh-allu.*

Canis vulpes, *L.* **Fox.**

Gaelic—*Sionnach* : *Madadh-rua* : *Balgaire.*

Section ARCTOIDEA.

Family MUSTELIDÆ.

Martes sylvestris, *Nilss.* **Marten.**

Gaelic—*Taghan* : *Taoghan.*

Even so long ago as 1813 the Marten is taken notice of by Professor Walker (*Essays on Natural History, etc.*, p. 483), as inhabiting the Hebrides, thus:—"*Locus*:—Habitat insulis Haraia et Lodhusa. Montibus scopulosis desertis sylvarum vacuis." And again:—"In Hebridibus prehensæ sunt, capsis viminois carne inescatis, apertura qua ingredi possunt, at reverti nequeunt."

About the year 1860 Martens were still abundant. In 1870[1] they were reported as being present in Harris, but not abundant. There being but few trees in Harris, any Martens which are procured are usually found in cairns of stones amongst long heather on the hill-sides. The Marten occurred in the Mhorsgail deer-forest, Lewis, at the time Professor Duns wrote, and it was recorded so long ago as 1777, by Pennant, as occurring in Harris (Preface to Lightfoot's *Flora Scotica, etc.*). The Polecats of the

[1] *Vide* a paper "On the Past and Present Distribution of some of the rarer British Animals," by Harvie-Brown (*Zool.* 1881-82, "Marten").

Old Statistical Account of Lewis are undoubtedly Martens. By 1879 the Marten had become very rare in Harris, and Mr. H. Greenwood tells us: "There seem to be no Martens left in The Lews. Certainly they were here at one time, but have been destroyed" (*in lit.* Dec. 27, 1879).

Mr. D. Mackenzie sends us the following note: "A shepherd in the Park district of The Lews says that about eight years ago—*i.e.* about 1873—he saw two Martens leave a cairn before the dogs and make their escape, and that he never heard of any having been taken there since. So," adds Mr. Mackenzie, "there may possibly be some still alive" (*in lit.* 1881).

The late Sir James Matheson's piper, Mackay, who came to The Lews many years ago, trapped many Martens, but *no* Pole-cats.

Mr. Notman writes me that the last seen on Park Lodge shootings was in 1869; in 1868, one; in 1867, one; in 1866, three; in 1865, three. Thus in five years, between 1865-1869 inclusive, nine Martens were seen or killed.

In the whole of the returns of vermin, as will be seen in the Appendix, not one Marten was obtained in Lewis between 1876 and 1885.

In 1881 Harvie-Brown finds this note in his Journal: "The Marten is not yet extinct in Harris, it is believed. The largest number killed in any one year was fourteen, by Mr. MacAulay, head-forester at Fin Castle (Abhuinsuidh). But by 1886 it appears to have entirely succumbed in South Harris, as we are informed, *viva voce*, by Mr. Finlayson, Lord Dunmore's head-keeper at Rodel."

In or about the year 1875, Mr. Osgood Mackenzie of Inverewe, whilst deer-stalking in the Mhorsgail Forest, found a dead Marten, which had met with its death in a curious way. We quote Mr. Mackenzie's own words: "I suddenly came to a place where there was a lot of wool scattered about, and there was every appearance of there having been a struggle between a sheep and some other animal. I continued my stalk. About 100 yards from where we found the wool, I came upon a large Cheviot wedder lying dead, with its head down-hill, and its shoulder jammed up against a stone which was sticking out of a bank. In passing, I gave the forequarters of the sheep a kick, and under its shoulder and neck lay a dead Marten. The wedder, in rushing madly down-hill with the Marten at its throat, had dashed itself against the sharp stone, which killed the Marten. The sheep's throat being cut, it had not strength to get upon its legs, but bled to death where it lay" (*in lit.* 1881).

The above data afford an interesting example of the time required, in these days of strict game-preserving, for the extermination of our rarer carnivora.

Mustela vulgaris, *Erxl.* **Common Weasel.**

Gaelic—*Neas: Nios.*

Mustela erminea, *L.* **Stoat, Ermine.**

Gaelic—*Neas-gheal* = white weasel.

Mustela putoria, *L.* **Pole-cat, Foumart.**

Gaelic—*Fumaire: Focalan: Feocullan.*

Lutra vulgaris, *Erxl.* **Common Otter.**

Gaelic—*Balgaire* (elsewhere in Scotland applied to the fox.) Also called *Biast-dubh* (*i.e.* black beast), *Biast-donn*, *Dobhar-chu*, *Dobhran.*—A. C.

The Otter is plentiful in some localities, generally frequenting the sea-shore, until the salmon and sea-trout begin to "run" in July, when it follows them up the streams, and frequents the fresh-water lochs. Exciting sport may sometimes be obtained, when men and dogs succeed in hemming one in upon any restricted area; as, for instance, one of the smaller lochs or burns. "Sixty-One" gives an interesting account of such a hunt in his *Reminiscences.*[1] A forester in Harris showed Harvie-Brown a small rock in Loch Resort where he once killed two at a shot. The recently frequented resting-place of an Otter is readily recognisable by the freshness of the grass; but the droppings themselves, which cause the greenness, rapidly dry up.

A total of ninety-five Otters is given for ten years, as paid for in The Lews, from 1876 to 1885 inclusive. In 1883 there were none; in 1876, only one. The greatest numbers were obtained in 1877, viz., seventeen; in 1884, eighteen; and 1885, nineteen; but these numbers may not necessarily be held to indicate any

[1] *Reminiscences of The Lews*, by "Sixty-One," p. 129.

actual increase, but only, perhaps, extra exertion or peculiar aptitude on the trappers' part.

Otters are found more or less plentifully all along the coast-line, through South Uist, and even to Barray Head.

A shepherd in North Uist, on his beat alone, had shot over seventy Otters during a residence of twenty-five years; and in one winter Mr. D. Mackenzie trapped twelve in Lewis.

By 1886, at least in Harris, in the Sound of Harris, and in North Uist, Otters had become much scarcer, owing to the incessant trapping, shooting, and persecution they were, and still are, subjected to, at the hands of both keepers and natives, the latter having lately obtained many of the traps left behind by the keepers. Skins of fine fur often bring as high a price as 15s. each.

John MacGillivray mentions the fact of the Otters of the Hebrides being of the dark-coloured type, but we have not had an opportunity of comparing them with those of the mainland.

Meles taxus (*Schreb.*). **Badger.**

Gaelic—*Broc: Stiallaire: Stiall-chu.*

Family **URSIDÆ.**

Ursus Arctos, *L.* **Brown Bear.**

Gaelic—*Math-Ghamhainn: Uraisg.*

Sub-order *PINNIPEDIA.*

Family **TRICHECIDÆ.**

Trichechus rosmarus, *L.* **Walrus.**

Instances of the occurrence of the Walrus in Scotland are given in Bell's *British Quadrupeds*, by which it will be seen that one killed in December 1817, at Caolas Stocnis,[1] on the east coast of Harris, was examined by MacGillivray,[2] who gave an account of it in vol. xvii. of the *Naturalists' Library*: and another was killed in April 1841, on the East Haskeir, near Harris, by Captain MacDonald, R.N., as mentioned by Dr. R. Brown in the *Annals and*

[1] Caolas Stocnis was a herring-curing establishment in 1879, when we saw it, but as such has since been abandoned. It lies at the entrance of Loch Stocnis.

[2] An early account by MacGillivray, who examined the specimen obtained at Caolas Stocnis, is given in the *Edin. Phil. Journ.*, vol. ii. p. 389 (1820).

Magazine of Natural History, 1871. In 1879, Harvie-Brown met Mr. Kenneth Mackenzie—over forty-five years a tenant on Lord Dunmore's property in Harris. He "remembered well" about the Walrus captured at Caolas Stocnis, but could not recall what had become of the carcase.

Captain MacDonald of Waternish, Skye—perhaps one of our most experienced seal-shooters and otter-hunters in Scotland—told Harvie-Brown that a Walrus was distinctly seen two years ago, close to the point of rock near Stein. It lifted its head quite out of the water, and the tusks were distinctly seen. It was afterwards seen off the coast of Sleat, in Skye, and fired at by a keeper, who correctly described the animal. (See also *Land and Water*, 20th December 1879.)

Fleming gives the measurements of the Caolas Stocnis specimen in his *History of British Animals* (1828, p. 19), as "upwards of ten feet in length."

Family PHOCIDÆ.

Phoca vitulina, *L.* Common Seal.

Gaelic—*Ron*=seal: *Biast* ♀: *Brinnal*=male seal ♂: *Ronbeag*= the little seal: *Ron-caolais*=seal of the Sound or Strait: *Moineis*=the female seal.

The Common Seal was, until quite lately, very abundant on these coasts, perhaps nowhere more so than in the Sound of Harris, and in Loch Maddy, North Uist, in which latter place Harvie-Brown has seen from twelve to twenty upon a small sloping rock, which was just sufficiently large to afford them resting room; and Mr. J. Henderson, formerly gamekeeper to Lady Gordon Cathcart in South Uist, has counted fifty of this species at one time in the Sound of Barray.

In the Loch Maddy district seals are now much rarer than when Harvie-Brown first knew North Uist, and indeed the Sound of Harris had become, to a large extent, depopulated by 1886. This is owing principally to the short-sighted policy of a "general merchant" in that district, and to the issue of some thirty gun licences now taken out in North Uist alone. The island of Berneray, too, at the present time swarms with guns, and under these circumstances it is not rash to predict the speedy and utter extermination of the seals. In face of these facts, would it not be desirable to establish a close-time, as has been done even in the more remote Arctic regions? There is an interesting account of the seal-fisheries of the Outer Hebrides given by Mr. John Knox in his descriptions of the "Fisheries of Scotland," which is deserving of consultation in this place (*vide* "Bibliography," 1785, p. 351 *et seq.*). He mentions that as many as 320 seals have been killed in one place, and at one time: but this may well apply to the Grey, and not to the Common Seal, the former from its haunts and habits giving greater facilities for such a slaughter than the latter species.

Martin (*op. cit.* 2d ed., p. 11) mentions that the inhabitants of the Long Island use the flesh of seals for food, and "find it as nourishing as beef and mutton."

Contrary to their usual habit elsewhere, the Common Seals of the Sound of Harris frequent rocks instead of sandbanks; and Mr. J. MacDonald informs us that he has never seen one in the latter situation. A northerly wind, however light, takes most of the seals in the Sound of Harris across to the Harris side, but otherwise the shoals, rapids, and rocks of the Uist side are preferred.

Captain MacDonald of Waternish, Skye, who has had a large experience of Seals on the west coast of that island, finds that there they are generally fatter, and consequently float longer after being shot than they do in other localities known to him. In the Sound of Harris those that were killed in the water were mostly lost, being at once swept away by the strong tides.

The habit that seals have of springing out of the water much after the manner, and with the actions, of a salmon has been noticed by almost every one who has written on these animals, and various reasons are given to account for it. John MacGillivray says: "*During a storm* I have seen them throwing themselves forward half out of the water, several times in succession." Mr. J. Henderson, formerly tenant of the shootings of The Ross of Mull and Tiree, has seen them doing so in calm weather *before a storm*, and this is also our own experience. Martin, again, in his ancient account, assigns this habit as specially exercised in cold weather, while other authors consider it as belonging to the *breeding season*, and assign it to the males chasing each other, but the habit is observed quite irrespective of the breeding season. With so many conflicting ideas, it seems impossible to say with certainty why seals indulge in this habit.

Harvie-Brown has several times witnessed similar efforts, often against a tide-race of great strength, the seals, upon reaching less troubled water, panting for breath after the unusual exertion.

It was a fine summer night in June 1887, with a light northerly "airie," as we lay at anchor close under the shore of Harris, that we witnessed a protracted battle, and subsequent long-drawn-out chase between two seals, evidently fighting for the possession of a silvery sea-trout which was held across the jaws of one of them. The combat continued for fully a quarter of an hour, within fifty yards of the side of the yacht, and was witnessed by those on board at the time, as well as by ourselves and part of our crew, from the ship's boat. These animals rolled over and over one another, making a great commotion and boiling of the water. Which of the two proved successful in the end we know not, but the battle at this place was followed by the flight of one of the combatants, and pursuit by the other, which was all followed with the eye for nearly a mile alongshore. They were springing out of the water in their eagerness, and with the speed at which they were travelling splashing the water high every bound they gave.

Phoca hispida, *Schreb.* **Ringed Seal.**

Evidence of the occurrence of this species in the seas of the Outer Hebrides is given in Bell's *Brit. Quad.*, 2d edit., p. 249, on the authority of Mr. M'Neil of Colonsay; beyond this we have no information.

Phoca grœnlandica, *Fabr.* **Greenland Seal.**

On the 2d May 1879 Harvie-Brown saw four of this species upon a rock in the Sound of Harris. Mr. MacDonald of Newton and Harvie-Brown had started down the Sound in a boat in search of seals, and after firing ineffectual shots at two in very rapid water, they spied several others lying on a rock some way out in the Sound. Getting well to leeward, they dropped down behind the rock and landed; but a slight scrape of an iron-shod boot upon the rough projection of the rock startled the seals, and before they got over the top the animals had slid into deep water. Running forward, a good view of them was obtained as they kept close in beside the rock, and in their sudden alarm and confusion, they several times rushed past within 10 feet of where Mr. MacDonald and Harvie-

Brown stood, and the large, splashy-looking dark marks on either side of the back were distinctly and unmistakably visible. Harvie-Brown fired into one close beside the rock, but before the boat could be brought round the animal sank, and the strong tide running made it vain to search for him.

As regards other evidence of this species occurring in the British seas, we think that that given by Mr. Henry D. Graham is well worthy of credence. Mr. Graham—who was well known in Scotland as a careful observer, keen sportsman, and a naturalist of ability—saw "three of these rare visitors to British waters in Loch Tarbert, Jura"; and, with the aid of a powerful telescope, "both he and his friends could distinctly make out the markings which characterise the Harp Seal; and again, "the animals remained in full view for three hours, constantly watched." Mr. Graham appears to have been well aware of the name "Tapvaist" being indiscriminately applied to several species of large seals (vide *Proc. Nat. Hist. Soc. Glasgow*, vol. i. p. 53; 1868).

Halichærus gryphus, *Fab.* Grey Seal.

Gaelic—*Ron mor*=the large seal; *Ron Haisgeir*=the Haskeir seal. *Cullach*=♂; *Cullach cuain*=the ocean bear, as applied to the male.—A. C.[1]

Haskeir Island, west of North Uist, has long been known as a resort of this species. Pennant mentions seals in "Hiskyr" in 1777. The late Captain MacDonald, R.N., Rodel, showed Harvie-Brown a very fine skin of one he had shot in the Sound of Harris, where, however, they are less abundant than in many other parts; and Mr. John MacDonald pointed out to him various favourite rocks in the Sound which the Grey Seal frequents in small numbers.

Some confusion exists amongst the natives as to the difference between the Grey and the Greenland Seals. The large dark marks on either side of the spine are very prominent in the adult Greenland Seal, and can hardly fail to arrest the attention. MacGillivray mentions that the Grey Seal seldom enters the shallow sounds, but Mr. J. Henderson, formerly the observant gamekeeper in South Uist, informs us of an isolated rock in such a locality; and in the

[1] For interesting remarks on this species as observed in Norway, *vide* Collett, Robt., "On Halichærus gryphus and its Breeding on the Fro Islands of Throndjem's-fjord in Norway" (*Proc. Zool. Soc. London*, vol. 1881, p. 380).

Sound of Harris there is a rock called Sgeir-nam-Tapbhaist which is still frequented by a pair of these animals. There, in June 1879, Harvie-Brown saw two, *very large* and *extremely* hoary in colour. This rock, Sgeir-nam-Tapbhaist, is out of gun- or almost rifle-shot of any other rocks in the Sound, and the seals rest and breed in security, the young having been found, even of late years, by Mr. MacDonald of Newton.

In the Sound of Harris, Grey Seals keep much more in pairs than the common species, but where there are large colonies this can hardly be said to be the case.

Up to the year 1858, it was the custom annually to have a *battue* at Haskeir off North Uist in November, as at this time the seals are found on the rocks with their young ones. The boats arrived usually about daybreak, and the men cut off the retreats of the seals lying on the rocks, and killed from 40 to 100, old and young. These used to be divided amongst the men: but the farms of Vallay, Scolpig, and Balilone were each entitled to a larger share. This annual *battue* was stopped by the late proprietor, Sir J. Orde, Bart. But the consequence of this, as stated often to us, has been that natives of other parts of the Long Island have benefited at the expense of those of North Uist, and the decrease of the seals of Haskeir remains unimpeded, if not accelerated. Martin gives some account of the catching of these seals at this locality, and apportioning them afterwards. A hundred years, however, previous to the date of Martin's work, six times the number stated above have been killed during one *battue*. On the rock of Easimuil 320 have been killed in one day (Martin's *Western Islands*, p. 62)—*i.e.* if we can accept the accuracy of those old records.

Besides these already mentioned, there are several other rocks and islets frequented by Grey Seals round the coast of Harris and The Lews, such as Gaskeir; Bo Molach, rocks lying inside Taransay; and Gloraig-su-Taransay, some rocks opposite the entrance of West Loch Tarbert; but perhaps the two most important localities are North Ronay and the rock of Sulisgeir, lying forty-three miles north-west of the Butt of Lewis. Expeditions to these two places are made almost annually by the fishermen of Ness for the purpose of killing seals, but it is very difficult to ascertain with any degree of certainty the numbers said to be killed there, all accounts varying in a remarkable degree. The accounts of the early writers, such as Martin, Knox, and others, were based upon, often, intentionally misleading statements, the tendency being with some to exaggerate for prospective purposes of their own; and with others, for similar reasons, to minimise the totals. We have heard of fourteen men killing 360 in 1882, and 230 in October 1883; but these figures cannot be relied upon. We ourselves saw between 30 and 40 on the Skerries off the South of Ronay in 1885.

Again, Mr. R. M. Barrington, who visited North Ronay in 1886, writes: "The seal statistics are most unsatisfactory. I asked the men who went there separately; then I asked John Morrison, to whom all the Ness skins are said to be sold; then I asked Anderson of Stornoway, the third party through whose hands the skins and blubber pass, and not two of them agreed. Anderson gets them from Morrison, and Morrison gets them from the Ness fishermen. The Ness fishermen give different numbers, and do not agree among themselves. John Morrison states the number as follows:—In 1885, 89 on Ronay; in 1884, 143 on Ronay; in 1883, 107 on Ronay; in 1882, 30 on Sulisgeir alone."

"I have received," writes Mr. J. N. Anderson, Danish Consul, Stornoway, *in lit.* July 24, 1886, to Mr. R. M. Barrington, "for the past few years most of the skins of the seals killed on Ronay or Sulisgeir, and their number has been from 120 to 150."

Mr. John MacDonald was present at the slaughter of fifty-three one day at Haskeir, where they breed. Donald MacLean, in his *Account of one of the Hebrides*, makes mention of hunting of seals with dogs, the services of which, however, could amount only to irritating them to resistance, and thus by detaining them, give time to the hunters to attack them with clubs. Donald MacLean was no doubt quoting from the original passage in Dean Munro's work, which however relates strictly to the island of Islay.

The Grey Seal, apart from his vastly superior size, is readily distinguished by the greater length of his nose and hoary grey appearance. When in the water, and looking straight towards one, the head looks very grey, and has a striking resemblance to that of a sleuth-hound, wanting only the pendent ears of the latter to complete the resemblance. The eyes appear to be deeply sunken in the sockets, this appearance being intensified by the lighter colour of the surrounding parts of the face, especially in old animals.

Cystophora cristata (*Erxl.*). Hooded Seal.

Order CETACEA.

Sub-order *MYSTACOCETI.*

Family BALÆNIDÆ.

Balæna biscayensis, *Eschricht.* **Atlantic Right Whale.**

Family BALÆNOPTERIDÆ.

Megaptera longimana (*Rudolphi*). **Hump-backed Whale.**

Balænoptera musculus (*L.*). **Common Rorqual.**

Balænoptera sibbaldi (*Gray*). **Sibbald's Rorqual.**

Balænoptera borealis, *Less.* **Rudolphi's Rorqual.**

Balænoptera rostrata (*Fab.*). **Lesser Rorqual.**

[*Obs.*—Mr. Southwell considers that doubtless both the Common and Lesser Rorqual occur off the Outer Hebrides, and also probably Sibbald's Rorqual at the herring season, but he has no record of either having been actually procured.]

Sub-order *ODONTOCETI.*

Family PHYSETERIDÆ.

Sub-family *PHYSETERINÆ.*

Physeter macrocephalus, *L.* **Sperm Whale.**

Sub-family *ZIPHIINÆ.*

Hyperoödon rostratum (*Chemnitz*). **Common Beaked Whale.**

Gaelic—*Muc-bhiorrach.*

Ziphius cavirostris, *Cuv.* **Cuvier's Whale.**

Mesoplodon sowerbiensis (*Blainville*). **Sowerby's Whale.**

[*Obs.*—Mr. James Wilson, during his *Voyage round Scotland*, saw whales "of the largest class" off the entrance of East Loch Tarbert. Mr.

E. R. Alston informed us that these were probably either *Balæno-ptera musculus* (Linn.), the Common Rorqual, or *B. Sibbaldi* (Gray); (cf. Bell, second edition.) A specimen of the former, as we are informed by Professor Turner, was brought into Stornoway in 1871. *B. borealis*, Rudolphi's Rorqual, has occurred off Islay. Other species of Cetaceans doubtless occur, but we know of no positive records. Large whales are often seen off Stornoway in the fishing season, and Harvie-Brown saw a large Finner Whale —either the Common or Sibbald's Rorqual—a little outside Loch Maddy, in 1881. Mr. Thomas Southwell, who pays much special attention to the records and occurrences of Cetacea on our coasts, writes to us that "the Hyperoödon on its northward passage prefers to pass along the east coast, and therefore would miss the Hebrides; and this appears to apply to its return journey also, although this species, as well as the White-beaked Dolphin, do occasionally pass on the west side of the kingdom. . . . Doubtless both *B. musculus* and *B. rostratus* are found in the seas off the Hebrides, and probably *B. Sibbaldi* also, at the herring season, but I have no record of any of either having been procured."

Even after whales are captured and brought ashore to be flensed, all record or notes of value are usually lost, unless there be any resident naturalist near at hand who can inspect them. Thus Mr. Southwell refers to the Whale taken by the Kerabost (Kirkabost?) fishermen, and towed into Stornoway in August 1886. Although the fish was purchased by Mr. J. N. Anderson, and though Mr. Southwell was in correspondence with him regarding it, final identification of the species has not been obtained. If Mr. J. N. Anderson, or any other person obtaining specimens of whales in this way, would forward one of the *cervical vertebræ* to Mr. Southwell, or some other capable naturalist, it would be valuable to science; and if he did this before selling it, he might often get a better price from a Museum for a complete set of bones than he does for the whole carcase. This, we think, is well worth his consideration, or that of any other person acquiring the carcases of the larger or even of any species of whale. It would serve a double purpose, mutually beneficial to them and to science.]

Family DELPHINIDÆ.

Sub-family *BELUGINÆ.*

Monodon monoceros, *L.* **Narwhal.**

Delphinapterus leucas (*Pall.*). **White Whale.**

Sub-family *DELPHINIDÆ.*

Orca gladiator (*Lacép.*). **Killer. Grampus.**

Gaelic—*Leumrach : Leumadair : Buallar*=the jumping whale.

Grampus griseus (*G. Cuv.*). **Risso's Grampus.**

Globicephalus melas (*Trail*). **Pilot Whale.**

The late Mr. E. R. Alston wrote to us: "Their occurrence among the Hebrides is rarer than at Shetland. More than 300 were taken in 1805, and 92 in 1882, at Stornoway (*Naturalists' Library*, xxvi. pp. 214, 215). Nearly 200 were taken there in 1869." Mr. MacDonald of Newton was at the killing of 100 of these animals some years previous to 1870. They were, as usual, pressed in confusion on to the shore of a small semicircular sandy bay not far from his house (Newton) on the sound of Harris. Professor Duns also informs us that he was present at the death of a large herd of Ca'ing Whales, numbering in all ninety old and young, and he had an opportunity of "cutting into" and examining them. Even as early as the days of Martin this species is mentioned, the fifty "young whales" spoken of so quaintly evidently belonging to this species. All Whales in the Outer Hebrides are called *Muc-mhara* (or sea-pigs), but at the same time several species are separately distinguished.

Phocæna communis, *F. Cuv.* **Porpoise.**

Gaelic—*Muc-mhara : Cana* or *Canach : Peileag : Puthag.*

Common in the seas around the Long Island, but does not so often approach close to land, nor can it be so easily induced to enter the sea-lochs as the last species. Still they are often found in quite narrow sea-lochs, and when food is abundant we have seen the entrance to these places crowded with them. They are occasionally caught in the herring-nets. Once, during a calm, Harvie-Brown saw a Porpoise apparently basking, but we have never with certainty seen them moving continuously along the surface of the water.

Delphinus delphis, *L.* **Common Dolphin.**

Delphinus tursio, *Fab.* **Bottle-nosed Dolphin.**

Gaelic—*Muc-bhiorach* = sharp-pointed sea-pig.

Delphinus acutus, *Gray.* **White-sided Dolphin.**

Delphinus albirostris, *Gray.* **White-beaked Dolphin.**

Order UNGULATA.

Sub-order *ARTIODACTYLA.*

Family SUIDÆ.

Sus scrofa, *L.* **Wild Boar.**

Gaelic—*Torc : Cullach : Torc-fiadhaich : Torc-nimhe.*

Family CERVIDÆ.

Rangifer tarandus, *L.* **Reindeer.**

Cervus elaphus, *L.* **Red-Deer.**

Gaelic—*Fiadh* = a deer. *Damh dearg* = the stag; *eild* = the hind. *Cabrach : Cabrach nan croc* = the antlered stag.

Martin in his time computed the number of deer in Harris at "at least 2000." "This was the computed number of deer in Harris when I left the Outer Hebrides in 1883."—*Note by Alexander Carmichael.*

John MacGillivray tells us that deer were so plentiful in Harris and Lewis thirty years ago (*i.e.* about A.D. 1800) that the poor had an abundant supply of food. A peasant is said to have killed five *at one shot*, and another to have killed eighteen in a season.[1] They rapidly decreased, however, when the local militia became instituted, and after Lord Seaforth's time, who had protected them. MacGillivray seems to have been under the impression that deer had become extinct in all the Outer Hebrides, except Lewis and Harris, but Lieut.-Col. Fielden and Harvie-Brown were told nothing in 1870 which would lead them to suppose that this had ever been the case in North Uist, although it was generally considered that it was, at that time, rapidly approaching

[1] In 1812 Dr. Walker wrote of the food of the Red Deer in the Hebrides: "Plantas submarinas hyeme in Hebridibus copiosè pascitur."

extinction there. [The normal number in North Uist is computed at thirty deer more or less.—A. C.]

In 1879 it was reported to Harvie-Brown that there was no increase in their numbers in North Uist since 1870, but at the same time no perceptible decrease.

One of the finest collections of stags' heads obtained in Lewis is that belonging to Mr. A. Williamson of Edinburgh, who rented the Aline and Soval shootings for several years. These Harvie-Brown had the pleasure of inspecting. The small but beautifully symmetrical horns are indeed a contrast to his grand trophies of Wapiti, etc., from Colorado, but they are none the less interesting and valuable. We are indebted to Mr. Williamson's courtesy for the following account of the deer of The Lews, and as these experiences cannot fail to prove valuable additions to our knowledge of the history of the Red Deer in Great Britain, we transcribe those parts of his letters in full which relate thereto.

"*26th March* 1879.—Stags on the Long Island I found rapidly deteriorating, as they are doing elsewhere. From a table I kept very carefully during the first three years I was at Aline I find the average weights were as follows:—

1872—16 Stags averaged 11½ stones (clean).
1873—18 " 11 st. 3 lb. "
1874—22 " 10¾ stones "

As I shot nothing under four years old, or any wretched old brutes, of which we had but too many, these averages give a correct idea of the size of the Lewis stags. I noticed one very striking peculiarity,—their immense craving for bones and old deer's horns. My predecessor shot an old horse a few days before he left in May, about two miles from the Lodge. When I arrived in August the deer were coming nightly to chew the bones, and all the latter had disappeared before I left in November of the same year.[1] I have often, when lying watching a herd, seen the hinds chewing the horns of a stag lying on the ground, and that this was a common practice was shown by the marks of their teeth upon the horns of almost every stag I killed late in the season. I never saw anything of the kind on the 50 stags I have since shot on the mainland. The heads of the Lewis deer are very pretty, though small, having generally more points than mainland deer. I generally

[1] This is a habit, however, of mainland deer as well as those of Lewis, though possibly not to the same extent.

killed two, sometimes three, Royals in a year, and '11-pointers' were very common.

"The cause of the deterioration in the Lewis and Harris deer I attribute to overstocking, not to their being overshot so much. Doubtless there, as elsewhere, though nothing like to the same extent, the killing off of the finest stags and hinds is telling; but I believe it is mainly owing to the poor feeding, on ground unable to carry the vast numbers of deer in the Harris and Lewis forests. The number of the hinds was far too great, as Sir James Matheson was opposed to their being shot down. If the severe winter of 1878-79 has killed off the half of them, it will have done great good."

From the foregoing remarks it will be seen that the best way to improve the breed would be to kill off a large number of hinds, and to import, from any good forest on the mainland, some good stags; as these, if preserved for three or four years, would infuse fresh blood through the whole herd. The greater appetite displayed by the deer of the Long Island for bones and cast horns may, we believe, be very simply accounted for by the almost total absence of bone-producing elements in the geology of the Hebrides. Possibly this too might, to some extent, be artificially supplied. Inbreeding must, in time, affect the stock, and so must deterioration of pasturage, and overstocking, as well as artificial selection by sportsmen, who follow the creed of taking the best stags and hinds, leaving only the younger animals to breed from. Too great stress can scarcely be laid upon the importance of a fuller regulation as to laying down the surface turfs after peat-cutting—a rich and even sward often being thus obtained, and grass growing where heather was before, though this does not invariably follow, each turf off a surface being again carefully and evenly laid down.

In Harris in 1870, and on many subsequent occasions, Harvie-Brown saw plenty of deer in Harris. As already shown by Mr. Williamson, the stags were not large, nor approaching in size those of certain forests of the mainland and Inner Hebrides. In one lot, however, there was one immense stag as compared with the others. The horns, though usually small, are seldom distorted: this Harvie-Brown learned from the inspection of a number of heads at Fin Castle[1] and other shooting-lodges, and also from the conversation of the foresters, with whom he spent many days when

[1] Now known by its Gaelic name, Abhuinsuidh (pronounced soft *Avonsuil*), a much prettier name.

in quest of eagles' eyries. In April, the hinds, as a rule, keep lower down the hills, and seldom associate with the stags. Only on one occasion did Harvie-Brown see stags and hinds together in that month. Several times a solitary hind was started from her lair low down amongst old rank heather, and on all such occasions she seemed to be an aged beast, greyer, more patchy and ragged than those which were going in herds. Probably these were old barren hinds. From 70 to 80 stags are killed in a season on the western half of the Harris forest, and about 60 on the eastern half —at that time leased by the Messrs. Milbank—"all by stalking, driving never being resorted to"—as Harvie-Brown was informed by the head-forester.

In addition to the weights supplied by Mr. Williamson, we have the following from the late Mr. Greenwood. He writes, 17th December 1879:—

"Deer here never require artificial feeding. As to weights after 'gralloching':—

1879,	Sept. 12.	1 Stag,	12 stones 11 lbs.
,,	Sept. 24.	2 Stags,	13 stones 3 lbs. 11 stones 12 lbs.
,,	Oct. 2.	1 Stag,	13 stones 3 lbs (Royal).
,,	Oct. 4.	1 Stag,	13 stones 9 lbs.
,,	Oct. 8.	1 Stag,	14 stones 10 lbs. (Royal).
,,	Oct. 13.	1 Stag,	12 stones 10 lbs.
,,	Oct. 20.	1 Stag,	13 stones (Royal)."

Mr. Greenwood added: "The horns of those stags, though small, are well formed, and are said to be larger on the east coast, diminishing in size towards the west coast of Lewis."

There is a fine collection also of heads and horns in Rodel House.

Captain MacDonald of Rodel, at that time factor to Lord Dunmore, informed Harvie-Brown that when fresh blood was introduced from the Athole forest, one of these fine large stags would not take up with the degenerate stock at all. This splendid animal wandered southward to Rodel, thence crossed the Sound of Harris, eight miles, going from island to island to North Uist. There the hinds did not please him either, and he travelled on and on till he reached Barray Head, and, as Captain MacDonald described it, he "smelt no longer the scent of land," and turning, retraced his steps, and attempted to land again in Harris. But, alas! two "sportsmen" who had taken the Borve shootings massacred the noble animal in the water before he even put foot on land. This occurred when Captain MacDonald happened to be away from home.

In North Uist, Red Deer are not so numerous as formerly, according to all the accounts Harvie-Brown has received; and this, although during many years none were killed at all. Up to the date of 1880, the only fresh blood we know of as having been introduced into that island was a single stag. This was three or four years previous to 1870. The former gamekeeper at Loch Maddy, Alan Maclean, informed us that the practice of *driving* had been given up, and that any deer which are now killed are killed by stalking only. The proprietor of North Uist, Sir John Campbell Orde, Bart., informs us that the heads of North Uist deer are *not now* so much deformed as they were. He wrote (*in lit.* 20th July 1880): "When I first went there—I think in 1856—there were more deer, and, I believe, I killed altogether, some six or seven with deformed heads in that and in another year. Since then I have never shot any such, and certainly never seen more than one or two at most. In 1874 I shot some four or five stags—royals and 'ten-pointers,' and, I think, in 1872, two royals. One of the former I believe to be superior to most, if not to all, of Mr. Williamson's heads from Lewis, and one of the most symmetrical heads I have ever seen. Indeed, in this respect, I consider them always superior to mainland heads. The horns have a curve backwards as well as outwards, and the brow antlers are generally of good development."

On the 10th May 1870, when our gillie, Robert Ross,—a Sutherland man,[1]—landed on an island, in a loch near Loch Maddy, for the purpose of digging out a nest of Shielduck's eggs, a stag jumped up from a hollow almost at his feet. Robert noticed that it was "lug-marked," and afterwards we were told that it was a solitary introduced stag. When in North Uist, in 1881, we learned that several new deer were just about to be introduced there by Sir John Orde; and the keeper was away to receive them the day we landed at Loch Maddy.

In South Uist, though once plentiful, deer are now extinct. In 1842 there was only a single hind in the whole parish, the rest

[1] Ross met his death, on the 10th January 1887, by slipping over a precipice during a night of hard frost.

having found their way northwards (*New Stat. Account of Inverness*, p. 165). They were extinct in Barray at that date, though many antlers found in the mosses testify to their former abundance there (*New. Stat. Account*, No. xxxi. p. 185).

The rough high island of Maoldonuich, at the entrance to Castlebay or Macneillton, is known to the old people as Eilean-nam-Fiadh, the island of the deer, because of deer kept there by the old Macneills of Barray. The better known name of Maoldonuich, Saint Dominus, is after one of the early Celtic saints who had his cell there.—*Note by Alexander Carmichael.*

In 1886, when on a visit to the Outer Hebrides,—as is our frequent custom in summer,—we learned that some Red Deer had been successfully introduced quite lately to the island of Pabbay, Harris; and in 1878, or 1879, we heard that an introduction of fresh deer had again been carried out in South Harris.

Cervus dama, *L.* Fallow Deer.

Gaelic—*Dais: Dathais.*—A. C.

Domesticated in part. Introduced. Lord Dunmore introduced some Fallow Deer to the island of Harmetray in 1878. By 1886, or even earlier, however, these became reduced to two or three in number. A year or so after their introduction they appeared to be thriving, and became so wild as almost to defy capture. Unfortunately Harvie-Brown had no opportunity of seeing them, though he knew Harmetray Island as a special haunt of Otters prior to the introduction of these deer. In 1887 he heard they had become extinct, or had been poached from the Uist side of the Sound, and that none now remain.

Capreolus capræa, *Gray.* Roe Deer.

Gaelic—*Earb*=roe: *Earbag, Ruadhag*=diminutive of roe. *Boc-earb*=roebuck: *Boc-ruadh, Ruadh-bhoc*=russet buck. Roe is evidently the Gaelic *ruo, ruadh*, red, russet; hence roebuck is simply the Gaelic *Ruadh-bhoc*=russet buck.—A. C.

Family BOVIDÆ.

Bos taurus, *L.* Wild White Cattle.

Gaelic—*Crodh: Spreidh: Greidh: Ni*=cattle; flocks.—A. C.

Order RODENTIA.

Sub-order *SIMPLICIDENTATA*.

Section SCIUROMORPHA.

Family SCIURIDÆ.

Sciurus vulgaris, *L.* **Squirrel.**

Gaelic—*Feorag: Earag: Easag.*[1]—A. C.

Family CASTORIDÆ.

Castor fiber, *L.* **European Beaver.**

Gaelic—*Leas-leathann.*—A. C.

Section MYOMORPHA.

Family MYOXIDÆ.

Muscardinus avellanarius (*L.*). **Dormouse.**

Gaelic—*Dallag: Dallag-fheoir.*—A. C.

Family MURIDÆ.

Sub-family *MURINÆ*.

Mus minutus, *Pall.* **Harvest Mouse.**

Gaelic—*Luch-fheoir.*—A. C.

Mus sylvaticus, *L.* **Long-tailed Field Mouse.**

Mus musculus, *L.* **Common House Mouse.**

Gaelic—*Luch.*—A. C.

Common. It would appear that this species also frequents the corn-fields, because in harvest-time, after the crops are cut and housed, they become much more abundant in the houses.

[1] See "The Squirrel in Scotland." Harvie-Brown.

Mus rattus, *L.* **Black Rat.**

Gaelic—*Rodan-dubh* : *Radan-dubh* = black-rat.—A. C.

[*Obs.*—Mr. Alexander Carmichael informs us that this species still exists in Benbeculay. He says: "I saw them about our house at Creägorry several times, and also at Gramsdail, on the opposite side of Benbeculay. I liked to see these pretty creatures, so glossy and black. I never felt that repugnance to the presence of this native rat that I have always felt to the presence of the brown Hanoverian rat."]

Mus decumanus, *Pall.* **Brown Rat.**

Gaelic—*Radan* : *Rodan* = a rat.—A. C.

Brown Rats are abundant on most of the islands, and greatly frequent the sea-shore, where they live upon shell-fish and dead animals thrown up by the sea, thus to some extent acting as scavengers.

Martin in his "Description" tells us that "about fourteen years ago" (*i.e.* previous to the date of the first edition of his writings, which would be about 1689) "a swarm of rats, but none knows how, came into Rona, and in a short time ate up all the corn in the island." He also tells us they were very abundant at Rodel, where numbers of cats were used for the purpose of exterminating them, and after a severe struggle, "succeeded so well, that they left not one rat alive." Rats in the Hebrides frequent the inland moors, far from houses, subsisting upon dead sheep, and also, to a considerable extent, on birds' eggs. Their burrows may be seen by the sides of the inland lochs and tarns quite commonly; and they are also frequently found upon the islands of the Sound of Harris, Loch Maddy, and other sea-lochs.

Evidently the determination to keep the plague of rats within restricted limits, and not to risk a recurrence of the fatalities mentioned by Martin, is evinced by the returns of rats killed and paid at the charges of the estate of The Lews. In the returns afforded us, we find a regular war carried on, and a total of 3820 rats are accounted for between 1876 and 1885 inclusive. The largest number recorded in any one year is 1016, viz., in 1884, and the next largest in 1885, viz., 525; but, as we indicated elsewhere, it is difficult to know whether such returns always point to extra

Peter Smit del. et lith. Mintern Bros. imp.

MUS HIBERNICUS, Thompson.

[illegible] Irish Rat.

[illegible] mammal [illegible] rat [illegible]

[illegible] the [illegible] views of the [illegible] 6 and 7.

The [illegible] Rat [illegible] to Great [illegible] gentleman reported the [illegible] Rat as [illegible] 21st [illegible] Aug[illegible] last, after the [illegible] portion [illegible] had been [illegible]. A [illegible] was [illegible] express [illegible] referred to the [illegible] for each [illegible] receiving [illegible] specimens [illegible] inspection to be [illegible]

It is not [illegible] Hebridean [illegible] also [illegible] Rat [illegible] following extracts from letters [illegible]

[illegible] between [illegible] the latter [illegible] dominates [illegible] into streets. They appear to [illegible] portions of the [illegible] upon it

Mus hibernicus *Thompson.* **Irish Rat.**

The discovery of this almost forgotten and neglected mammal in the Outer Hebrides is not only important, since a significant extension is thereby added to its limited geographical range, but also because it reopens for consideration the history and status of the creature itself, which it is thought most desirable should be undertaken.

To the fauna of these Islands the presence of this peculiar Irish quadruped must be regarded as an important link in the chain of evidence bearing upon the general zoological relationship of the archipelago—a link in strict consonance with the views of the late Mr. E. R. Alston, already expressed at pp. 6 and 7.

The occurrence of a Black Rat in the Islands had been known many years to Mr. Alexander Carmichael, but to Dr. John MacRury our thanks are due for reminding us of the fact, and enabling us to make an addition to the fauna of Scotland and to Great Britain. This gentleman reported the Black Rat as occurring there in a letter dated the 21st of August last, after the mammalian portion of our book had been printed. A desire was at once expressed to see specimens, and a reward offered to the inhabitants for their capture. This resulted in our receiving two specimens in spirit, and these proved on inspection to be what was described by Thompson as *Mus hibernicus*, which had not hitherto, we believe, been recorded out of Ireland.

It is not deemed advisable to enter into detail on the Hebridean range of this animal, but it is only withheld in the interests of a species limited not only in its distribution, but also in its numbers. Regarding its habits, etc., we give the following extracts from notes, kindly communicated to us by Dr. MacRury. This gentleman says:—there seems to be but little difference between their habits and those of the Brown Rat; the latter predominates, but it is thought the black species is holding its own; although they are not very numerous, they seem to be more so than they were within the recollection of the inhabitants. They appear to be most numerous on the sandy portions of the island, though found elsewhere upon it;

and they affect barns and outhouses, but Dr. MacRury never heard of their being seen in dwelling-houses.

As to the claims of this mammal to specific rank, and as regards its characters, general description, and history, we quote the words of our friend, Mr. W. Eagle Clarke, of the Natural History Department of the Edinburgh Museum of Science and Art, who has most kindly undertaken a thorough examination of *Mus hibernicus* and its relationship to the larger species of the British Muridæ—a service which it affords us much pleasure to acknowledge.

Mr. Eagle Clarke reports as follows :—

On the 13th of June 1837, Mr. William Thompson, the well-known author of the *Natural History of Ireland*, in a communication[1] to the Zoological Society of London, described and exhibited a new species of rat under the name of *Mus hibernicus* = Irish Rat. In this communication Mr. Thompson,—although he had heard of the animal before,—tells us that "until April last, when a specimen was sent from Rathfriland, county of Down, to the Belfast Museum, I had not an opportunity of either seeing or examining the animal. This individual differs from the *M. Rattus* . . . in the relative proportion of the tail to that of the head and body ; in having shorter ears, and in their being better clothed with hair, as is the tail likewise ; and in the fur of the body being of a softer texture. The difference in colour between the *M. Rattus* and the present specimen is, that the latter exhibits a somewhat triangular spot of pure white extending about nine lines below the breast, and the fore-feet being of the same colour. . . . These differences incline me to consider this animal distinct from *M. Rattus*, and being unable to find any species described with which it accords, I propose to name it provisionally *M. Hibernicus*. . . ."

It is not a little remarkable that after the careful examination made by Thompson, as is evidenced by his published detailed account, that this excellent naturalist should have associated this animal with *Mus rattus*, an error of judgment which is repeated in his *Natural History of Ireland* (vol. iv. p. 16), published in 1856, where it is obviously considered to be a mere variety of the Black Rat—an error perpetuated up to the date of issue of the second edition of Bell's *British Quadrupeds*, where the animal is merely alluded to as a variety of *Mus rattus*, and down to the present time. The primary fact of its colour being black seems to have exercised not only a misleading but a lasting influence on our naturalists.

Since Thompson's investigations *Mus hibernicus* appears to have received practically no attention at the hands of zoologists ; at least, endeavours to procure further published observations on it have failed. Its discovery, however, in the Outer Hebrides has reopened the question, and the writer has to express his obligations to the authors of this work for the opportunity afforded him of examining and reporting upon their specimens received in the flesh, as well as a series of skins furnished by their obliging correspondents.

[1] For full account see *Proceedings of the Zoological Society*, 1837, Pt. v. pp. 52, 53.

Before proceeding to the consideration of the true status of *Mus hibernicus*, it is desirable to institute a comparative examination of the British species of rats. This is conveniently and sufficiently afforded by the tabulated information below :—

	Mus rattus. (From Bell.)		Mus alexandrinus.[1]		Mus decumanus.		Mus hibernicus.	
	Ins.	Lines.	Ins.	Lines.	Ins.	Lines.	Ins.	Lines.
Length of Head and Body, .	7	0	7	4½	9	1	8	6
Length of Head,	1	9	2	0	2	0½	2	4
Length of Ears, .	0	10	1	1¼	0	9	0	8½
Width of Ears, .	...	...	0	9	0	7	0	7
Length of Tail, . . .	7	0	8	11	7	1	7	6
Length of Fore-feet and Claws,	0	8¼	0	9	0	10	0	8
Length of Hind-feet and Claws,	1	4¼			1	8	1	7¼

The four species readily fall into two groups :—

1. *The Long-tailed* and *Large-eared species*, in which the tail is longer than the head and body, and the ears comparatively large. *Mus rattus* and *Mus alexandrinus* belong to this section.

2. *The Short-tailed and Small-eared species*, in which the tail is shorter than the head and body, and the ears comparatively small. *Mus decumanus* and *Mus hibernicus* belong to this section.

This results in *M. hibernicus* being more nearly allied to *Mus decumanus* than to *Mus rattus*, with which it has hitherto been associated. This is most undoubtedly the case, and is borne out by a careful examination of the other characters of the two animals. Indeed, if specific rank be not conceded to *Mus hibernicus*, then it must be regarded as a variety of *Mus decumanus*.

It is now necessary to state the differences which lead to the belief that *Mus hibernicus* may be something more than a variety. Briefly stated, they are :—

1*st*, It is a smaller and more elegant animal than *M. decumanus*, which is a much coarser creature in build and other characters.[2]

2*d*, The fur is finer in texture, and silky to the touch. In this respect it is even finer than in *Mus rattus*, and affords marked contrast to the rough and somewhat harsh coat of *M. decumanus*.

[1] From H.M.S. *Devastation*. Forwarded in the flesh by Captain J. R. N. Macfarlane, R.N.

[2] Since the above was written I have been much indebted to Mr. G. Barrett Hamilton, of Kilmanock, Co. Wexford, for many valuable notes. Mr. Hamilton informs me that he has had specimens of *M. hibernicus* equal in length to ordinary specimens of *M. decumanus*, but that the former were always lighter in weight ; he also tells me that the head and tail are proportionately longer in *M. hibernicus* than in *M. decumanus*. I think it is possible, however, that melanic varieties of *M. decumanus* may sometimes be confounded with *M. hibernicus*.

3d, In the general colour of the fur, and its constancy in shade.

4th, In its peculiar and circumscribed distribution. This singularly limited and isolated western range, in which it has been so long known to exist in some numbers, is most remarkable and important, and, taken together with the fact that it does not appear to have been recorded for the mainland of Great Britain, nor from Europe, affords weighty evidence against *Mus hibernicus* being regarded as of varietal value only.

The following table shows the comparative measurements of *Mus hibernicus* and *Mus decumanus*, taken from specimens while in the flesh :—

	Mus hibernicus.				Mus decumanus.			
	Male.		Female.		Male.		Female.	
	Ins.	Lines.	Ins.	Lines.	Ins.	Lines.	Ins.	Lines.
Length of Head and Body,	8	5	8	3	9	7½	9	1
Length of Head,	2	4	2	0	2	0½	2	0½
Length of Ears,	0	8½	0	9	0	10½	0	9
Length of Tail,	7	5	7	8	7	3	7	1
Length of Fore-feet and Claws,	0	8	0	9	0	10	0	10
Length of Hind-feet and Claws,	1	7	1	8	1	7	1	8

DESCRIPTION.—The fur is glossy. The hairs on the back are of two kinds—the longer kind being white at the roots and darkening gradually to the tips, which are black ; and under this a shorter fur, of an ashy grey colour. The general colour of the upper surface is dark silvery grey, almost black. This shades into a paler tint on the sides. The under surface and limbs silvery mouse-grey. The head is slightly browner than the back, with the muzzle mouse-grey. The digits silvery white. The white stripe, regarded as *the* important diagnostic character by Thompson, does not possess that value. In both Hebridean and Irish examples examined by me the specimens wanting the stripe have been as numerous as those possessing it, so that it may be commoner to this than to *Mus rattus* and *Mus decumanus*, in both of which it is said occasionally to occur. When present in *Mus hibernicus* it forms a patch or stripe extending from between the fore limbs backwards, sometimes for a length of one-and-a-half inches.

I regret that owing to the scarcity of material it has not yet been possible to make a complete examination of the osteological characters, if any, of *Mus hibernicus*. The results of this and of further general research and comparisons are deferred for a future occasion.

A male having the white stripe, and a female, are figured, both Hebridean specimens, and have been kindly presented to the Museum of Science and Art, Edinburgh.

vigilance and activity on the keepers' parts, and others employed in their capture, or is due to actual increase of vermin; nor is it clear that these returns apply to the whole estate of Lewis, or only to a part. We are ourselves—all circumstances connected with the vermin list which will be found in the Appendix considered—inclined to the belief that actual increase of vermin occurred in 1884, owing to a propitious season and extra supplies of food, etc. In any case, such statistics are valuable and interesting, and cannot fail to mark off finger-posts, as it were, in our progressive knowledge of distribution and migration.

Sub-family *ARVICOLINÆ.*

Arvicola amphibia (*L.*). Water Vole.

Gaelic—*Lamhalan*, probably a corruption of the initial letter of *Famh-alan* = the water mole.—A. C.

Arvicola agrestis, *De Selys.* Common Field Vole.

In 1865 Professor Duns negatived the occurrence of any voles in Lewis (*op. cit.* p. 620). From a casual remark of MacGillivray's (*British Birds*, vol. iii. p. 331), it would appear that "*Field-mice*" were extremely rare in his time (1830-1850). Speaking of the Kestrel he says: "I have never seen it pursue a bird in open flight; but in such districts as the Outer Hebrides where, if *Field-mice exist, they are extremely rare* [the italics are ours], it can have no other prey during the winter." It seems uncertain, however, whether MacGillivray applied these remarks to a Sorex or to an Arvicola.

The Common Field Vole occurs in North and South Uist. No mention is made of it by MacGillivray, or by previous writers, save what we have given above, and considerable confusion existed concerning the voles and shrews of the Long Island, until, in June 1879, Harvie-Brown fortunately succeeded in obtaining specimens of both for identification.

This species is extremely abundant all over the hills, and in the machars or sandy meadows, and their runs in the grass are found to cross and intercross in a perfect network in certain favoured localities. The shepherds' dogs take great pleasure in hunting for them, and *eat them* with great relish. In autumn they are most abundant in the lower grounds and around the farms, but in summer they are equally numerous all over the more uninhabited parts.

Arvicola glareolus (*Schreb.*). **Red Field Vole.**

Sub-order *DUPLICIDENTATA*.

Family **LEPORIDÆ**.

Lepus europæus, *Pall.* **Common Hare.**

Gaelic—*Gearr: Giorr: Gearr-fhiadh: Maigheach: Moigheach.*—A. C.

First introduced at Rodel in Harris, where it became very numerous; and numbers could be seen at one time crossing and re-crossing the road between Rodel and Borve; but, prior to 1870 they had become much scarcer, and when Harvie-Brown was there in that year he did not meet with a single example. In 1879, however, they again increased, and, he was informed, had become much more numerous.

In Lewis, hares, according to Wilson, occurred as early as 1842 (*Voyage, etc.*, vol. ii. p. 120). They were also introduced into Barray, according to John MacGillivray, and W. MacGillivray confirms his son's statement (*Prize Essays of Highland Society of Scotland*, vol. viii. p. 275). Again, Buchanan puts on record the positive fact of the *absence* of hares from the Long Island in 1782-1790, and Pennant likewise in 1777 (Lightfoot's *Flora Scotica*). On Aline and Soval grounds, the Brown Hare was unknown to Mr. Williamson, the next species only being met with. But we get a more exact date than any of the above from that most useful old book the *Old Stat. Account of Scotland*, where we are informed that there were no hares in Barvas, in Lewis, "till of late a few have made their appearance from a breed introduced into the island by Lord Seaforth" (*op. cit.* vol. xix. (1797) p. 21). Mr. H. Greenwood, however, includes both Brown and White Hares in his List of the Mammals of The Lews sent to us, and adds: "White Hares are most plentiful towards the West Coast of Lewis and Harris, where the ground is more mountainous. In 1879, at which time there were five gamekeepers in South Harris, there was a very considerable increase in the number of hares there, and we ourselves saw many whilst passing between Rodel and Obb. They have since decreased again, owing to less perfect preservation.

Lepus variabilis, *Pall.* **Mountain Hare, Blue Hare, White Hare.**

Gaelic—*Gearra-gheal ; Maigheach-gheal.*—A. C.

This species, like the last, is not indigenous, but was introduced to the island of Harris at Rodel about 1859, and has thence spread rapidly all over Harris and Lewis. Now, in the forest of North Harris, they are killed down as vermin at all seasons, so rapidly did they increase, and we can remember this to have been the case in 1879, twenty years after their introduction, and doubtless for many years before. They come down from the hills to the hollows and mosses, even close to the shore, where Harvie-Brown has often seen them ; and they may be said to be generally distributed at all altitudes. They become white in winter like those of the mainland, and sometimes retain the white coat a long time. We saw some as pure white as they could well be at the end of April, whilst others were bluer. In all of them, however, at that season, we believe that bluish patches can be seen if the fur be raised, as we have often had opportunity of observing. When in Harris in 1870, we could not say that we observed any marked difference in the size of the White Hares from those of the same species on the mainland. Mr. A. Williamson is of the opinion, however, that grouse and even snipe are smaller there, as well as hares. This deterioration in size, which, usually accompanied with darker plumage or fur, is incidental to insular positions, in many cases, is what may be expected. Deterioration from an introduced stock may take some time to become very apparent, but a succession of extremely wet seasons, and late springs, may, and very probably does, hasten the process. The visible deterioration appears to us to be only one link in the chain of the destruction, decay, and final death of a species.

Lepus cuniculus, *L.* **Rabbit.**

Gaelic—*Coinean : Coineanach : Rabaid.*—A. C.

Rabbits were introduced to South Uist prior to 1842 (*New Stat. Account*, No. xxi. p. 176), also into Barray and Vatersay (Wilson's

Voyage),[1] where they have been abundant since that date. Several attempts to introduce rabbits into Lewis had failed before 1865 (*auct.* Prof. Duns). They have since, however—*i.e.* since 1865—been introduced successfully on an island of Loch Seaforth by Mr. Milbank.

The sandy downs or machars, grassy islands, and inland meadows of the west coasts of the Uists, and the islands to the southward, are in every way suitable to the species, but the rockier ground and moors of the northern portions, and the east coast northward, are likely to prevent it from gaining ground in that direction. The same deterioration already mentioned is distinctly noticeable in the rabbits upon the island of Loch Seaforth, and doubtless elsewhere, where they are confined to too restricted areas.

In N. Uist rabbits have increased from an introduction in the south of the island about 1870, and are slowly spreading northward along the west coast. One was shot at Scolpig in 1878, as Harvie-Brown was informed by Mr. J. MacDonald of Newton.

In the machars of N. Uist are many holes scraped in the sand, having every appearance of rabbits' work (1880). But these were, and are, made by the women and children of the island, who carry away fresh white sand to sprinkle on the floors of their houses. When the rabbits *do* reach the machars, they will no doubt take possession of these partially formed burrows. Since the above was originally penned, and by 1881, rabbits had increased, and are spreading north through N. Uist.

Rabbits have spread also into N. Harris (another separate introduction) at the back of the hills behind Tarbert, and they are not uncommon at several places between Stornoway and Soval Lodge.

By 1887 rabbits swarmed up as far as the extreme sandhills of Ardivorran Point in N. Uist. They are also plentiful at the back or south slope of Ben More and Ben-bhreac, on the farm of Newton, but at that time had not yet invaded the machars to the west, as the shepherd, so far, has been able to hold them in check.

[1] Quoted from W. MacGillivray (*Prize Essays*, vol. viii. p. 275).

DOMESTIC ANIMALS.

OF these it is only necessary to speak very shortly.

The DOG (Gaelic, *Cu*) appears to have been early domesticated in the Long Island, bones having been found in a "Pict's house" in Harris, carrying us back at least one thousand years; and so also with a small breed of SHEEP (Gaelic, *Caora*), and a small-sized HORSE (Gaelic, *Each*).[1] Bones of SWINE have also been found, but their age would appear to be doubtful, as Martin[2] mentions some facts which seem to cast doubt upon the asserted antiquity of shell mounds in the Long Island.[3]

Regarding the breed of HORSES (*Equus caballus*) in the Hebrides, the following note seems of sufficient interest to be reproduced here:—

"The native breed, or rather the early introduced breed of horse, is the Scandinavian, and appears to have been introduced in the eighth or ninth century, at the time of the Scandinavian invasion, and is the same as that found in Norway, Sweden, Switzerland, Tyrol, Hungary, Transylvania, and elsewhere, with slight local modifications."—(Walker's *Hebrides*, vol. i. p. 447.)

Arabian blood was also introduced,[4] and we ourselves well remember a very handsome cross-bred mare which we purchased at the Falkirk "Tryst" about the year 1856;[5] she came from North Uist, and we named her "Zilla." This little cream-coloured mare had many of the characteristics of the Arab, both in appearance

[1] Vide *Proc. Royal Phil. Soc. Edin.*, vol. i. pp. 141-207 (MacBain). Notice of various osteological remains found in a Pict's house in the island of Harris.

[2] Fielden, *Proc. Nat. Hist. Soc. Glasgow*, vol. ii. p. 58.

[3] *Op. cit.* p. 64.

[4] It does not appear that in the days in which Professor Walker wrote any mention was made of the introduction of Arab breed; nor that any special local variety was taken notice of in the Hebrides.

[5] Tryst = Anglicè, *market*, literally *meeting-place*.

and temper, and was a perfect beauty. All this fine blood has apparently disappeared, the moisture and fog of the Hebridean climate finally extinguishing it; but we understand it still lingers, or has been more recently revived, in one of the Inner Hebrides.

Of other domestic and semi-domestic animals it is only necessary to speak here of two. The Wild Goats (Gaelic, *Gobhar: Gabhar*) of Harris have very fine heads and horns, and are "as wild as the *fery deer themselves*," as Harvie-Brown has frequently been told, and has indeed realised. The rich clumps of grass sprouting deep in crevices of the gigantic cliffs, which scarcely any foot but that of the Goat can traverse, and which are enriched by the excreta of the wild birds, afford plentiful and luxuriant herbage to these creatures. Thus, although Red-deer in some measure fail to grow large horns, Goats do not, as they scramble into nooks and corners which are manured by birds of prey, and there find rich lime-reared grasses. We have seen them in profile majestic against the sky, at the top of the great precipices of Harris, daintily making their apparently perilous way far down, or high up amongst the ledges, where, indeed, it is not always easy to discern them. But there are few more interesting sights than to watch through a good glass the step-by-step advance of a Goat as he feeds down the face of what appears to the naked eye to be a sheer precipice. We have sometimes found it difficult to take the glass from our eyes, and, indeed, have only ceased to observe upon the animal lying down on a carpet of moss or grasses, which even through the glass seemed scarce big enough to hold a mouse. Walker says that cattle feed upon the submarine fuci or plants, so hardly pushed are they at times for more natural sustenance (*Essays*, p. 513). This still holds good at present, and also with deer.

The strange FOUR-HORNED SHEEP are to be seen upon several of the farms in Harris and North Uist. Mr. MacDonald of Newton usually keeps some of them. One which he gave Lord Dunmore some years ago had five horns, the fifth projecting from the centre of the forehead. Our friend Mr. J. J. Dalgleish has also had several, which he introduced into an island of Loch Sunart, one of which had no less than six horns, and this number is not by any means unusual. He has kindly presented us with one four-horned old ram, the head of which we have had preserved.

OF EXTINCT BRITISH SPECIES.

Possibly—but not very probably—we may trace the Wolf, *Canis lupus* (L.), in the name Loch Maddy (*Madaidh*) in North Uist, though it and the rocks at its entrance may only be named from a supposed resemblance to Watch *Dogs—Madaidh gruamach* (the grim dog), and *Madaidh mor* (the big dog).

Remains of the small fossil Ox, *Bos longifrons* (Owen), are referred to by Dr. J. A. Smith as having been found in a Pict's house in Harris, as noticed by Dr. James MacBain, R.N. [*op. cit.* in note, *supra*, or in "Ancient Underground Building at Nisebost in Harris," by Captain Thomas, of H.M. Surveying Ship *Woodlark* (*Proc. Antiq. of Scotland*, vol. ix. p. 634, "Notes on the Ancient Cattle of Scotland.")]

Class 2. AVES.

Sub-class *AVES CARINATÆ.*

Series ÆGITHOGNATHÆ.

Order 1. **PASSERES.**

Sub-order *OSCINES.*

Section 1. OSCINES DENTIROSTRES.

Family **TURDIDÆ.**

Sub-family *TURDINÆ.*

Turdus viscivorus, *L.* Missel Thrush.

Gaelic—*Smeorach*, a name applied to Thrushes generally, but to the Song Thrush (*T. musicus*) specially: *Smeorah-mhor*= large Thrush: *Sriochag*=the screecher: *Cullionag*=the holy bird.—A. C.

[There appears to be no record whatever of this species throughout the Outer Hebrides, until we find Prof. Duns's statement that it "appears in winter, and is commoner in summer than the Redwing." Although Prof. Duns correctly states that six species of *Turdidæ* occur in Lewis, there seems to be some confusion as to the occurrence of the Missel Thrush and Redwing. Through all our returns of Migration Schedules we cannot find a single entry which could be unhesitatingly put down to this species.]

Turdus musicus, *L.* Song Thrush.

Gaelic—*Lonag: Smeorach.*—A. C.

Has apparently always been recorded as common, from MacGillivray's time downwards, and by all observers as a breeding species. It is well known all over the Hebrides by its Gaelic name, and is marked as "permanently resident" by Gray. Elwes considered it extremely common in Harris and Lewis, but less so in Uist, with which remark we quite agree. The shelter and nature of the ground in Harris and Lewis suit them better, and, as Elwes rightly remarks, there is there more molluscous food. In 1881 we found them nesting among the heather, close to Loch Hamanaway, on the west of Lewis, often miles away from houses; and around Stornoway they were very abundant amongst the young woods, as certified by many earlier records. We find a note in our journal that on the 13th May 1870, near Loch Maddy, in North Uist, we heard a Song Thrush singing in a little dell, his mate probably nesting in the neighbourhood. We afterwards found nests newly formed near Loch Maddy, about the 20th May of the same year, and we believe this to be the first year in which the Song Thrush was known to nest in that island.

In the Shiant Isles, Harvie-Brown heard the Song Thrush almost constantly singing from its perch among the basaltic cliffs. There appeared to be only one or two pairs on the island, and these no doubt had their nests amongst the long tufts of grass and sorrel which crown all the broken columns in the cliff, and which owe their luxuriant freshness and greenness partly to the nature of the rock, and partly to the droppings of the sea-birds. In 1887 one Common Thrush was seen upon Eilean Mhuire.

The fact of the extremely dark plumage and small size of the Song Thrushes of the Long Island repeatedly forced itself upon our attention, sometimes in an almost startling way. The song seemed to us quite as rich and mellow as in sunnier climes.

The Song Thrush migrates regularly along the coasts of the whole line of the O. H.[1]

Turdus iliacus, *L.* Redwing.

Gaelic—*Deargan-sneac*=the ruddy bird of the snow.—A. C.

The report that this species had bred in the glen of Rodel was, no doubt, an early error of Bullock,[2] as no notice of it is taken by the MacGillivrays, but it appears commonly during the autumn migration.

[1] Hereafter the initials "O. H." may be understood as the contraction for *Outer Hebrides*.

[2] And see Fleming, *History of British Animals*, where he gives the date of Bullock's letter, viz., 23d April 1819, in which apparently the first announcement of the above was made.

Turdus pilaris, *L.* Fieldfare.

Gaelic—*Lia-truisg* : *Lia-triosg* : *Smeorach Lochlannach* = Scandinavian thrush.—A. C.

Bullock speaks of this species as nesting in Harris, but this is evidently an error. Gray speaks of it as well known in the O. H., but as not arriving till mid-winter; and it is principally observed on the pasture-lands of the west side of North Uist and Benbeculay. MacGillivray noticed it in 1837 in Harris as late as 25th May, and it is recorded on both migrations at Stornoway by Mr. D. Mackenzie, formerly head-gamekeeper there. It migrates southward throughout the whole length of the isles in vast numbers during some seasons, as reported, for instance, from Mingulay in November 1887, and simultaneously by Mr. Joseph Agnew at the Monach Islands, "rising and flying away to the N.W." (*sic*) in November 1887.[1] Mr. A. Carmichael has the note that he saw a flock of Fieldfares on the island of Haskeir, off North Uist, about the middle of October 1878.

Turdus merula, *L.* Blackbird.

Gaelic—*Lon* : *Londubh* : *Lonan* : *Lonag*.—A. C.

As early as 1821, MacGillivray included the Blackbird in his List of Birds in Harris, but we do not find it alluded to for a long time afterwards in local lists, until again MacGillivray in his greater work speaks of it as "not breeding" in the O. H.

At present it is common in the O. H., and breeds, but is not so abundant as the Song-Thrush. The woods at Stornoway are likely to afford room for an increase in their numbers. It is resident but rare in Harris and in N. Uist, and, so far as we can learn, absent in Benbeculay during summer, though a winter visitant there. In 1881 it was recorded as breeding near Stornoway by Mr. D. Mackenzie. As early as 1870, a blackbird with a white head frequented a small garden near Rodel, in Harris, and had done so for three years previously. We do not draw attention to this partial albinism as a rare occurrence, but only to instance the regularity with which the same individuals return to breeding haunts. By 1879 blackbirds had decidedly increased about

[1] It is as yet not always easy to understand the peculiar movements of birds at Monach Station, but see our earlier remarks on this head.

the Glen of Rodel, when Harvie-Brown saw several, and heard many more in June. In 1885 it was abundant around Stornoway Castle. Mr. Hubback was informed by the under-keeper that, along with Pheasants, Blackbirds were *introduced* at Stornoway, but we have not been able to verify the statement.

The Blackbird appears regularly on autumn migration from N. to S. along the whole chain of the isles, though not in great numbers.

Turdus torquatus, *L.* Ring-Ouzel.

Gaelic—*Lon-choilcireach: Lon-mhonaidh: Dubh-chreaige.*—A. C.

Dunn met with this species in Lewis, where, however, it appears to be rare, or very local; and in 1881 Harvie-Brown found it in one locality only in Harris. In 1871, however, Gray had not been able to trace this species in the O. H.

Harvie-Brown has certainly never met with the Ring-Ouzel since 1881, but then he has never had an opportunity of again visiting the locality where he formerly found it. It undoubtedly occurs during autumn migration.

Monticola saxatilis (*L.*). Rock Thrush.

Sub-family *CINCLINÆ*.

Cinclus aquaticus, *Bechst.* Dipper.

Gaelic—*Gobhachan: Gobhan Uisge*=little smith: little water-smith.—A. C.

Is considered rare by authors, and does not appear to have been included by MacGillivray; was occasionally seen near Stornoway, and breeds there, as recorded by Mr. Greenwood in 1879; and it is included in Mr. D. Mackenzie's List. In 1881, and again in 1886, Harvie-Brown found it not uncommon in Harris, when several were seen at the loch-side above Abhuinsuidh in June. It is also included in a list from Mhorsgail, in Lewis, by Mr. Heywood H. Jones in 1881.

Cinclus melanogaster, *C. L. Brehm.* Black-bellied Dipper.

Sub-family *SAXICOLINÆ.*

Saxicola œnanthe (*L.*). Wheatear.

Gaelic—*Clacharan: Cloichirean* = the stone bird.—A. C.

Common summer visitant,[1] arriving not earlier in most years than the 20th March, often not till considerably later. Is abundant throughout the whole range, and extends to many of the remoter islands. Near the top of Mulloch Mor, in St. Kilda, Harvie-Brown saw the bird perched on the stone houses which are built all over the hill-sides, and which are used by the natives for drying their fish. We have found it abundant also around Stornoway, at Gress, and in the north of the island, as well as in North Uist and other districts, and perched upon the heaps of stones gathered off the pasturage of North Ronay, where it was also seen by Swinburne and Barrington. Harvie-Brown observed only one pair in Haskeir, off North Uist, in 1881. In 1887 it was included as an inhabitant of the Shiants, having been observed sparingly on Garbheilean. It remains sometimes till late in October.

Saxicola albicollis, *Vieill.* Black-eared Chat.

Pratincola rubetra (*L.*). Whinchat.

Gaelic—*Conasgag: Conasgan: Conasag: Conasan: Gocan Conaisg.* All relating to whins, or the supposed cross temper of the bird. *Fraoichein* = the little heather chatterer.

First recorded from Harris, where, in 1841, J. MacGillivray found the nest, and saw one pair of birds. It is included in the Stornoway list of Mr. Greenwood, who obtained the eggs there in 1879, but it is not mentioned by Mr. D. Mackenzie. It is noted by Mr. H. H. Jones in the Mhorsgail list of 1881, and on June 5th of the same year it was identified by Harvie-Brown in Harris. It is evident that the species must be considered rare and local at the present time.

[1] A current belief in North Uist, and also in other islands of the archipelago, regarding this species, is, that it hibernates in winter in holes, under stones, and in moss-dykes.

Pratincola rubicola (*L.*). **Stonechat.**

Does not appear to have been recorded from any locality in the O. H., until Gray saw it in the old graveyard of Benbeculay in the autumn of 1870 or 1871. It is included in the Mhorsgail list by Mr. H. H. Jones, *along with* the previous species; and, in one instance, eggs have been taken near Stornoway by Mr. Greenwood, with a cuckoo's egg in the nest (1879). It is also included as breeding by Mr. D. Mackenzie.

It migrates regularly southward throughout the isles, some seasons in large numbers.

Harvie-Brown finds in his journals that two Stonechats were observed near the head of Loch Maddy on the 12th May 1870.

Ruticilla phœnicurus (*L.*). **Redstart.**

Gaelic—*Ceann-dearg: Ceann-deargan*=red-head: the little red-head. Names relate to the red head, rather than to the red breast, of the bird, and are wrongly applied.—A. C.

Ruticilla titys (*Scop.*). **Black Redstart.**

Sub-family *SYLVIINÆ.*

Cyanecula wolfi, *C. L. Brehm.* **White-spotted Bluethroat.**

Cyanecula suecica (*L.*). **Red-spotted Bluethroat.**

Erithacus rubecula (*L.*). **Redbreast.**

Gaelic—*Broidileag: Broinnileag: Bru-dhearg: Broinn-dearg: Nigidh*, etc. etc.—A. C.

In 1837 it is included as occasional by W. MacGillivray.

Recorded as only occasional in 1841 by John MacGillivray, and then only in the Glen of Rodel—"where is the only wood in any quantity in the country." Duns records it from Stornoway in 1865. Since MacGillivray wrote, Gray has been able to record this species from many other localities. He says: "On the north side of North Uist it is frequently seen at Paible, and is common in some parts of Lewis, especially around Stornoway." It is recorded as breeding in 1881 by Mr. D. Mackenzie, and as common around Stornoway in 1885 by Harvie-Brown.

We have never observed the Robin in any of the remoter or more barren isles, and we consider the negative statement as worthy of a place.

Daulias luscinia (*L.*). **Nightingale.**

Sylvia rufa (*Bodd.*). **Whitethroat.**

Gaelic—*Gealan-coille*: *Gealag-choille*=the white little bird of the wood: *Gealachag-choille*=the little white one of the wood. —A. C.

This is one of the species of more recent extension of range to the O. H. We find the statement by Gray, in 1871, that it was "wholly unknown in the Outer Hebrides." Mr. D. Mackenzie found it breeding at Stornoway in 1881, and Harvie-Brown, in 1885, found it "far from uncommon" in the woods around the Castle; but although he spent some time in the Rodel woods in 1885, he saw nothing of it there.

Sylvia curruca (*L.*). **Lesser Whitethroat.**

[*Obs.*—The only note of this species is one by Mr. Greenwood, who "distinctly remembered one seen singing every day on the same tree at Stornoway in 1879." Before then it had invariably been either omitted from all lists, or negatived. Gray says distinctly: "No trace of this species has yet been discovered in any of the Inner or Outer Hebrides." As Mr. Greenwood does not mention the common Whitethroat, we are obliged to bracket it for the present.]

Sylvia atricapilla (*L.*). **Blackcap.**

Gaelic—*Ceann-dubh*=black head: *Caillcachag Ceann dubh*=the little carlin of the black head: *Smutag*=snorter.—A. C.

Added to the List of the Birds of the O. H. by Mr. H. H. Jones, who picked up a male dead at Mhorsgail on Nov. 1st, 1879. It was blowing at the time a northerly gale, with heavy snow-storms. Thus the autumn extension *via* the Pentland Firth is distinctly indicated (see *The Vertebrate Fauna of Sutherland, etc.*, p. 108). That Blackcaps, as well as many other migrants, do not occur more frequently in the Long Island, is no doubt because they usually pass unobserved, and frequently during the night, and it is only under exceptional circumstances, or during exceptional weather, that they are forced to seek shelter in uncongenial districts. If they pass over in the day-time they carry on, not finding suitable localities beneath them. Thus whilst actual lines of migration of many birds may have existed for years and years, the birds themselves are seldom or never detected *en route*.

Sylvia salicaria (*L.*). **Garden Warbler.**

[*Obs.*—Professor Duns says he met with this bird in Lewis. This is possible, as it has occurred in Shetland, according to Dr. Saxby, but it would have been more satisfactory had he given fuller particulars of the occurrence of a bird so rare in Scotland.]

Sylvia nisoria, *Bechst.* **Barred Warbler.**[1]

Melizophilus undatus (*Bodd*). **Dartford Warbler.**

Sub-family *PHYLLOSCOPINÆ.*

Regulus cristatus, *Koch.* **Golden-crested Wren.**

Gaelic—*Drathain-donn Balloir: Ball-oir: Ballan-oir*=the wren with the gold spot. Again—*Crionan*=the mite-bird: *Crionag*=the mite-bird: *Crionan Ceann-bhuidhe*=the yellow-headed mite (male): *Crionag Ceann-bhuidhe*=the little yellow-headed mite (female).—A. C.

Professor Duns "found" the Goldcrest at Stornoway, but MacGillivray had never met with it in the O. H. Gray, "though failing to trace it to these islands, expected its appearance." Looking to the several Reports of the Migration Committee, we find that in the first report, 1879, Mr. Edgar, lighthouse-keeper

[1] It is worthy of note here that this species—new to the avifauna of Scotland—has been obtained near Broadford in Skye, as recorded in *The Field*, November 1st, 1884, by Mr. Dumville Lees. We mention it here simply as an illustration of the extensions and directions of migratory waves. This Skye bird and two others were obtained within a few days of one another, the latter two having been obtained respectively on the Yorkshire and Norfolk coasts. The dates of the three are Skye, 16th August 1884; Yorks., 28th August; Norfolk, 4th September. As the Skye bird dates earliest, Yorkshire next, and Norfolk last, it is not unreasonable to suppose that it was the E. to W. wave which touched Skye, and the N.W. to S.E. return of the same impulse that reached Yorkshire and Norfolk, though it is quite possible the migratory wave may have come direct from E. to W. on to the Norfolk coast.

at the Butt of Lewis, supplies the information that for many years previous to that date he had "never seen Goldcrests at the Butt of Lewis"—a negative fact of considerable value and interest. Again, in 1880, we have no returns of the species from the O. H. lighthouses on the W. coast of Scotland, except from localities south of Clyde: there was the same absence in 1881—our third report; nowhere north of Rhinns of Islay in 1882: in 1883, a few as far north as Duheartach Lighthouse. At last, in 1884, we have "one, at Monach Isles, *new* to Mr. Youngclause," but accurately described, therefore hitherto rare at that station. That year, another was recorded as far north as Ronay, Skye—showing an unusual extension to the northward. In 1885, again, there is an utter absence of records from the whole line of the O. H. Again, in 1886, only one spring record occurs—a single bird at Pentland Skerries,—but there is no mention of it anywhere in the O. H.

Since the above was penned, one, a male bird, was found dead at Monach Isles Lighthouse, by Mr. Agnew, 7th November 1887; this had arrived with a rush of many other land birds.

Regulus ignicapillus (*C. L. Brehm.*). **Fire-crested Wren.**

Phylloscopus supercillosus (*Gm.*). **Yellow-browed Warbler.**

[*Obs.*—The occurrence of this in North Unst, Shetland, as recorded in the Migration Report of 1886, may possibly point to its future occurrence still further west.]

Phylloscopus collybita (*Vieill.*). **Chiffchaff.**

Gaelic—*Caifein* = chafferer: *Caifein-coille* = chafferer of the wood.—A. C.

[*Obs.*—Gray, writing in 1871, says: "The Chiffchaff frequents Rodel, in Harris, as I have been informed by Mr. Elwes, who procured a specimen there in May 1868." But there is no notice taken of this unusual distribution in Elwes' article in the *Ibis* for 1869. On addressing a query to Captain Elwes in October 1887, he writes as follows: "I cannot tell what became of the Chiffchaff. I do not know now, but I expect I showed it to Gray with other birds." But later, on applying to Mr. William Evans, who drew up a catalogue of Gray's collections, he assured me there "is no Chiffchaff—at least nothing labelled—either from Rodel or anywhere else."

And again: "I never saw a Chiffchaff in Mr. Gray's possession." Under these circumstances the record must be included within brackets.]

Phylloscopus trochilus (*L.*). Willow Wren.

Gaelic—*Troicheilein*=trifler, dwarf: *Conan Conuisg*=Conon of the whins.—A. C.

There does not appear to be any record whatever of this now common summer visitant up to the date of Gray's work, in which he says he had "not been able to trace it on *any of the Hebrides.*" (*The italics are ours.*)

We first met with this species in North Uist, in the garden at Newton, in 1879—where, however, Mr. J. MacDonald observed them three years earlier; and it is just possible it may have occurred even further north a year or two earlier still. During the same summer season, Harvie-Brown, when passing Rodel Wood in June, heard a Willow Warbler, and so can record its occurrence and probable breeding in Harris in the summer of 1879. In 1881 the Willow Warbler was heard *in great numbers* in the young woods at Tarbert, in Harris;[1] showing the extension northward from Rodel so soon as suitable breeding-grounds are being opened up to them, their return spring flight being so arrested. In the same year it was included in the Mhorsgail list by Mr. H. H. Jones. In 1885 we found it *very common* in the woods around Stornoway Castle; and again, in 1887, we spent an hour or two listening to their notes in Rodel Glen, vainly hoping to hear a Chiffchaff or a Wood Warbler.

Phylloscopus sibilatrix (*Bechst.*). Wood-Wren.[2]

Gaelic—*Conan-coille*=Conan (*i.e.* Wren) of the wood.—A. C.

Sub-family *ACROCEPHALINÆ.*

Hypolais icterina (*Vieill.*). Icterine Warbler.

Ædon galactodes (*Temm.*). Rufous Warbler.

[1] These young woods are now, however, as before stated, almost extinct.

[2] There is no evidence as to the occurrence of this species as yet in the O. H., but we note it here for a possible future extension, because it is well known to be creeping northwards along the west coast of the mainland of Scotland, new localities becoming frequented almost annually.

Acrocephalus streperus (*Vieill.*). **Reed Warbler.**

Acrocephalus palustris (*Bechst.*). **Marsh Warbler.**

Acrocephalus arundinaceus (*L.*). **Great Reed Warbler.**

Acrocephalus aquaticus (*Gm.*). **Aquatic Warbler.**

Acrocephalus schœnobænus (*L.*). **Sedge Warbler.**[1]

Gaelic—*Ceolan*=the little bird-melodist: *Ceolan-cuilc*=the little melodist of the reeds: *Cuilcein*=the little reedling bird (male): *Cuilceag*=the little one of the reeds (female): *Loiliseag*=the little one of the sedge.—A. C.

Locustella nævia (*Bodd.*). **Grasshopper Warbler.**[2]

Locustella luscinioides (*Savi*). **Savi's Warbler.**

Family ACCENTORIDÆ.

Accentor collaris (*Scop.*). **Alpine Accentor.**

Accentor modularis (*L.*). **Hedge-Sparrow.**

Gaelic—*Donnan*: *Donnag*=the little brown bird.—A. C.

W. MacGillivray has no mention of the Hedge-Sparrow as occurring in the O. H. in his *British Birds*, 1839. John MacGillivray, however, says of it in 1841: "Occurs only in the Glen of Rodel, the only area of wood in the country."

No other record occurs until Gray (1871) says: "Well known on all the Hebrides, except the bleakest islands." It is included in the Mhorsgail list by Mr. H. H. Jones; and Harvie-Brown found it abundantly around Stornoway in 1887. It is curious, however, that not a single previous record of it occurs in all our journals, although Harvie-Brown visited all the most likely places

[1] A rare summer visitant in north-west Skye. Eggs taken at Talisker in 1884, and birds apparently breeding near Dunvegan in 1886.—(Rev. H. A. Macpherson, *Proc. Roy. Phys. Soc. Edin.*, 1886, p. 121.)

[2] All evidence is silent to date of 1866; but the possibility of its future occurrence exists, as it has been observed in the north-west of Skye every year for some time back by Captain MacDonald. It has been also noticed by Mr. Lees in the Broadford district, and is present at Uig.—(Rev. H. A. Macpherson, *Proc. Roy. Phys. Soc. Edin.*, 1886, p. 118.) Its extension along the west coast of the mainland is also well known to us as having taken place in recent years.

in Lewis, Harris, and North Uist, nor can he remember having met with it. This may have been an oversight, but, if so, it is a strange one, because our custom always has been, and is, to make *fresh* lists of species at *every locality each year* we visit them. In 1886 we found it at Rodel, and one was observed near Tarbert, in Harris, the same season.

Family **PANURIDÆ.**

Panurus biarmicus (*L.*). **Bearded Reedling.**

Family **PARIDÆ.**

Acredula rosea (*Blyth*). **Long-tailed Titmouse.**

Gaelic—*Cailleach bheag an Earbaill* = little carlin of the tail.—A. C.

Acredula caudata (*L.*). **Continental Long-tailed Titmouse.**

Parus major, *L.* **Great Titmouse.**

Parus ater, *L.* **European Coal Titmouse.**

Parus britannicus, *Sharpe and Dresser.* **English Coal Titmouse.**[1]

Gaelic—*Caillachag ceann-dubh* = little carlin blackhead: *Smutan: Smutag* = little snorter.—A. C.

Parus palustris, *L.* **Marsh Titmouse.**

Gaelic—*Ceann-dubh*: *Ceann-dubhag* = blackhead.—A. C.

Parus cæruleus, *L.* **Blue Titmouse.**

[*Obs.*—Professor Duns says he found the Blue Tit at Stornoway in 1865, but it is entered as "without record" in 1871, by Gray, notwithstanding the previous statement. We ourselves have utterly failed to meet with it anywhere in the O. H., and the entire negative evidence is infinitely stronger than the positive.]

Lophophanes cristatus (*L.*). **Crested Titmouse.**

[1] Gray says, "It is of course unknown in the O. H." It is likely to remain so, we believe, for some time, until our area becomes acceptable to the species for breeding purposes. At the same time, we have elsewhere spoken of its having bred and reared its young in the crack of a dry peat-bank in Sutherlandshire, far from the haunts where one would naturally expect to find it.

Family SITTIDÆ.

Sitta cæsia, *Wolf.* **Common Nuthatch.**[1]

Family CERTHIIDÆ.

Certhia familiaris, *L.* **Creeper.**

Family TROGLODYTIDÆ.

Troglodytes parvulus, *Koch.* **Wren.**

Gaelic—*Dreollan*: *Drethein*: *Dreathain*: *Drathain*: *Drathain-donn*: *Conan*: *Conan-crion*: *Fridein*=mite-bird: *Fridein Fionn*=the mite pale bird.—A. C.

"Everywhere common" is the edict even of the earliest recorders, and so it still remains. If we admit the local variety of St. Kilda as a species or sub-species under the name *T. hirtensis* (Seebohm), we add another bird to our British List.

Troglodytes "hirtensis," *Seebohm.*

[*Obs.*—This new species (?) appears to have become almost extinct in St. Kilda since Seebohm's record. Would it not have been better, if our friend had kept the new species to himself? Mr. R. W. Chase of Edgbaston, Birmingham, possesses one of these St. Kilda Wrens, a female, obtained in St. Kilda on 10th July 1886; and although he considers it very doubtful if it deserves specific rank, it is still interesting to record a date which may, at no very distant period, prove to be of considerable interest *to the last of the "race"*! The St. Kilda Wren is still likely to survive, however, on the more remote isles of the group, as these are less accessible and not so frequently visited.[2]]

[1] This bird is worthy of a note, owing to its occurrence in the spring of 1885 at Waternish, in N.W. Skye, as observed by Captain MacDonald, and recorded by the Rev. H. A. Macpherson (*Proc. Roy. Phys. Soc.*, 1886, p. 121), who adds: "I should have expected the form to have been *Sitta europæa*, but this is negatived by Capt. MacDonald's description of the *lower parts*." We have not ourselves had an opportunity of examining it.

[2] We in Scotland have heard of the short existence of the St. Kilda Wren as a species, its almost, if not quite, total extermination having been compassed in the

Family MOTACILLIDÆ.

Motacilla alba, *L.* White Wagtail.

W. MacGillivray includes this as an occasional visitant. This statement is not taken notice of by Gray in 1871, but Mr. E. T. Booth in his *Rough Notes* speaks of having seen numbers on the 6th May 1877, after a south-east gale—at least five pairs—on Loch Shell, in Lewis, and we shall certainly not be surprised if future observation shows it to be a regular spring-passing migrant.

Motacilla lugubris, *Temm.* Pied Wagtail.

Gaelic—*Bricein Bain-tighearn*=the Lady's speckled little bird.—A. C.

It is recorded as arriving in O. H. as early as the middle of March by W. MacGillivray in 1837. Not recorded in 1841 by J. MacGillivray; but noted at St. Kilda by Sir William E. Milner in 1848. Observed by Duns at Stornoway, but not marked as breeding by Mr. D. Mackenzie in his Stornoway Lists. Barrington, in the end of July 1886, found one specimen in North Ronay. It was catching flies upon the bodies of dead shags (*auct.* W. Williams, one of Mr. Barrington's party).

Motacilla melanope, *Pall.* Grey Wagtail.

Gaelic—*Breac an t-sil: Breacean-buidhe: Bricein-buidhe*=the yellow speckled little bird.—A. C.

Motacilla flava, *L.* Blue-headed Wagtail.

short space of time between the announcement of its supposed specific identity and the present period. Whether it be new or old matters little to it—poor little bird; it only serves to point a moral and adorn (?) a tale. There seems to be, alas! a somewhat loose idea current in certain collecting quarters as to the rights of property and of hospitality in Scotland, and especially in the Isles, which, however, curiously enough, becomes totally changed when it comes to be applied south of the Border. By some, Scotland seems to be looked upon as a savage waste, where collectors may do as they list, even as the winds do blow, and that no man hath a right to say them "nay"; yet when the same principle is applied south of the Border, a hue and cry is raised if even one clutch of eggs be taken in the *strictly preserved* places of England.

Motacilla raii, *Bonap.* **Yellow Wagtail.**

The only occurrence is that recorded by Barrington, who shot a single female in North Ronay in June, 1886 (*Proc. Roy. Phys. Soc.*, 1887).

Anthus pratensis (*L.*). **Meadow Pipit.**

Gaelic—*Bigean*: *Bigean-biag*=wee little bird: *Glaisein*=gray bird: *Riabhag*=the brindled. In Islay the *Riabhag*, or *Reamhag* rather, is applied to the Skylark.—A. C.

Always recorded as common by Sir W. E. Milner, and as occurring in St. Kilda in June 1848, and we have abundant records from all the other islands, which it is unnecessary to specify. Noted at North Ronay by Swinburne, but not seen by us there in 1885, nor by Barrington in 1886, nor do we find any note of it in our journals of 1887 at that locality.[1]

Anthus trivialis (*L.*). **Tree Pipit.**

Anthus campestris (*L.*). **Tawny Pipit.**

Anthus richardi, *Vieill.* **Richard's Pipit.**

Anthus spinoletta (*L.*). **Water Pipit.**

Anthus obscurus (*Lath.*). **Rock Pipit.**

Gaelic—*Uiseag-dubh*=the black shore-lark: *Bigean-mor*=big little bird.—A. C.

Recorded as common by earlier authors as well as by those of the present time. MacGillivray often found the nest at a distance from the sea, and we have frequently taken the eggs on level ground, near the summit of the outlying islands, on almost all of which it is very common, including St. Kilda. But at Stornoway it is said not to appear till autumn, and does not appear to be considered as with certainty a breeding species there. So said Mr. Greenwood, but, on the other hand, Mr. D. Mackenzie says, "it *may* breed here" (1881). (*The italics are ours.*) Gray speaks of it as adhering strictly to the sea-margin, but this does not invariably apply, although the general rule is upheld and proved by exceptions in certain localities.

[1] Negative and positive records in such a locality as North Ronay are of almost equal interest and importance.

Observed commonly in North Ronay by Swinburne, Barrington, and Harvie-Brown, and Barrington obtained one nest and eggs; found abundantly on the Shiant Isles in 1887, but without any previous note of it there by Harvie-Brown in 1881. Its rarity on the Shiants is difficult to understand, unless it be explained that the Puffins, by their vast population, make it unpleasant for other species.

Family **PYCNONOTIDÆ.**

Pycnonotus capensis (*L.*). Gold-vented Thrush.

Family **ORIOLIDÆ.**

Oriolus galbula, *L.* Golden Oriole.

Family **LANIIDÆ.**

Lanius excubitor, *L.* Great Grey Shrike.

Lanius minor, *Gmel.* Lesser Grey Shrike.

Lanius collurio, *L.* Red-backed Shrike.

Lanius auriculatus, *Müll.* Woodchat Shrike.

Family **AMPELIDÆ.**

Ampelis garrulus, *L.* Waxwing.[1]

Family **MUSCICAPIDÆ.**

Muscicapa grisola, *L.* Spotted Flycatcher.

Muscicapa atricapilla, *L.* Pied Flycatcher.

Muscicapa parva, *Bechst.* Red-breasted Flycatcher.

[1] Though not recorded from the O. H. it is well to note its occurrence in Skye in 1850, as recorded by Gray. Such a long time has elapsed, however, since then, that it must be only under very exceptional circumstances that their migration waves extend so far.

Section 2. OSCINES LATIROSTRES.

Family HIRUNDINIDÆ.

All the birds of the Swallow tribe, such as the Swallow proper, the House Martin, the Sand Martin, etc., are called *Na Famhlaich; Na Famhlagan*=implying the swift ones, the restless ones. Hence *Famhlag Tir* is the Sand Swallow, whilst *Famhlag Mhara* is a Sea-Swallow or Stormy Petrel. The Swift, though not a real Swallow, is also included.—A. C.

Hirundo rustica, *L.* Swallow.

Gaelic—*Gobhlan Gaoithe*=the forked bird of the wind.—A. C.

Not a very common passing migrant. Swallows are recorded by J. MacGillivray as not arriving till the end of June. He saw twelve or so at Pabbay, and one was caught at the school-house in Berneray a few days after. Gray says they do not appear to remain to breed in the O. H., but are seen every year. He observed them in North Uist, Benbeculay, and South Uist. Lieut.-Col. Feilden noticed swallows in Barray on 8th May 1870, and on the same day, Harvie-Brown met with others at Loch Maddy.

At North Ronay one pair was seen on the north side, about the cliff-face, by Barrington's party in 1886.

Chelidon urbica (*L.*). Martin.

One was shot at North Ronay in 1886 by a member of Barrington's party. A single bird was seen, on June 9th, 1887, in St. Kilda; and Mr. George Murray, who was teacher there from June 1886 to June 1887, adds the remark, "seldom seen here." (*Vide* Rev. H. A. Macpherson *in lit.*)

Cotile riparia (*L.*). Sand Martin.

Gaelic name—*Gobhlan Gaineacha*=the forked bird of the sand.—A. C.

Regular visitant to Lewis, Harris, and North Uist, breeding in sand-banks on the west side of these islands; also in South Uist and Barray, but not observed in Benbeculay, probably owing to the absence or scarcity of suitable haunts.

Progne purpurea (*L.*). Purple Martin.

Section 3. OSCINES CONIROSTRES.

Family FRINGILLIDÆ.

Sub-family *FRINGILLINÆ*.

Carduelis elegans, *Steph.* **Goldfinch.**

Chrysomitris spinus (*L.*). **Siskin.**

Serinus hortulanus, *Koch.* **Serin Finch.**

Ligurinus chloris (*L.*). **Greenfinch.**

In 1841 reported only from the Glen of Rodel, but in 1874 from North Uist and Harris. As a migrant the Greenfinch passes down the whole length of the Long Island, and occurs in passage abundantly at the Monach Isles, many entries of the species occurring in the Schedules sent to the Migration Committee by Mr. Joseph Agnew in the autumn of 1887, the migration commencing in October and continuing through November.

Coccothraustes vulgaris, *Pall.* **Hawfinch.**

Passer domesticus (*L.*). **Common Sparrow.**

Gaelic—*Gealbhonn : Glaisein* = gray bird.—A. C.

Only reported from one locality in 1841, viz., the ruins of Ormaclate Castle in South Uist. James Wilson, writing in 1842, says: "A pair built in 1833 in Stornoway, but we did not see descendants in 1841." Duns speaks of this species in 1865 as the "least common of the Fringillidæ." It is now spread over the north of the country, preferring to build in holes in rocks rather than to inhabit the thatched houses. By 1871 it had become very abundant, actually proving a scourge to the farmers, and by 1885 swarming around Stornoway. Duns says that the smoke in the thatch prevents them from building in it, and as it is taken off for manure, they prefer rocks to build in. MacGillivray says he found it among the ruins of Kilbar, Barray, where afterwards his brother, the late Dr. MacGillivray, lived. [May this not more likely have been the more sporadic Tree-sparrow?]

Passer montanus (*L.*). Tree Sparrow.

Gaelic—*Gealbhonn* : *Glaisrin*=the gray one ; also applied to the last-named species.—A. C.

This is recorded from North Ronay by Barrington. It does not appear to have been known as a Scotch species in MacGillivray's time. The curious sporadic extension of this interesting species receives fresh demonstration almost annually. Barrington and his party saw five birds at North Ronay. He himself thinks they were breeding, and shot three examples. This was in the end of June 1886. Mr. Dixon (*Ibis*, January 1885) mentions a pair at St. Kilda, one having been shot in 1884, and states that they breed in the holes of the rough stone dykes that enclose the fields. The question is raised—we think very correctly—by Mr. William Evans of Edinburgh (*in lit.*), whether it was not of the present species MacGillivray wrote in 1837, and not of the last, as first occurring in the O. H. in the ruined church of Kilbar, Barray (vide *Hist. of British Birds*, vol. i. p. 350).

Fringilla cœlebs, *L.* Chaffinch.

This species is spoken of by MacGillivray as confined to the glen of Rodel in 1841 ; Gray speaks of several localities in the O. H. in 1871. It is included, but not marked as breeding, at Stornoway, by Mr. Mackenzie in 1881. MacGillivray, in his *British Birds*, whilst speaking of it as permanently resident, "even in the bleakest parts of the North of Scotland," does not specially negative its occurrence in the Hebrides, though he was well acquainted with these isles : it seems doubtful if it occurred there at the time he wrote, but there cannot be any doubt as to its abundance now in suitable localities in summer, nor of its "swarms," in even the bleakest spots, during the autumn migration. At dusk of the 3d November 1887 the island of Monach (*i.e.* Shillay, the outermost of the group) was "swarming" with the species (*auct.* Jos. Agnew).

Fringilla montifringilla, *L.* Brambling.

Linota cannabina (*L.*). Linnet.

Gaelic—*Breacan-beithe* : *Bricein-beithe*=the speckled wee bird of the birch. Also applied to the Twite.—A. C.

[*Obs.*—Is noted as occurring in St. Kilda, but is probably an error for the Twite. Gray includes it as an inhabitant of the O. H., but, in 1870, Feilden and Harvie-Brown failed to find it there. Mr. Greenwood includes it in his list, and this is corroborated by Mr. D. Mackenzie, who also includes the Twite. We require however, more thorough evidence before admitting it.]

Linota linaria (*L.*). Mealy Redpoll.

Linota rufescens (*Vieill.*). Lesser Redpoll.

[*Obs.*—A good many notices occur, but they are, in almost all cases, as might be expected, distinctly in error for Twites. The earliest record, making the error self-evident, adds the information—*the only Hebridean species of the genus !*]

Linota hornemanni, *Holb.* Greenland Redpoll.

Linota flavirostris (*L.*). Twite.

Gaelic—*Bricein-beithe*=the wee little bird of the birch: *Bigean Bain-tighearna*=the Lady's little bird.—A. C. (See Linnet.)

The only known representative of the genus in the O. H., though perhaps others may occur on migration. The Twite has always been, as far back as records go, abundant in the O. H., and is, no doubt, the species represented as seen in St. Kilda by the Rev. —— Mackenzie. Several pairs of this species bred in the ivy on the side of Mhorsgail House, in 1880, but in 1881 very few were seen, and no nests were found at the same locality. In 1879 it is noted as breeding in the furze hedges near Stornoway. It breeds plentifully some years in currant-bushes on the garden wall at Newton, North Uist, and at other similarly suitable localities; but in other places, again, it appears almost entirely to desert these sites. Its numbers also, though always large, fluctuate considerably in different years. MacGillivray, speaking of its abundance, says it is often used by the natives for food, being caught in "riddles," or killed by throwing sticks among the flocks. It is exceedingly fond of nesting among the loose roots of the bent where the wind has shaken away the sand, and left long fringes hanging over the edges of the sandhills fronting the Atlantic. It is quite a characteristic bird of the Isles. Harvie-Brown has observed these birds going in flocks as late as the 11th May, and again as early in the autumn migration as the end of June (1887).

We found that nests placed amongst the roots of the bent, as above noted, were usually formed of dry grass outside, lined with sheep's wool, and with an edging of cows' hair; in another instance with a mixture of sheep's wool, cows' hair, and feathers. In another nest a considerable quantity of small, thin, reddish-coloured root was interwoven with the rest of the outside structure. This nest had the appearance of having been built on the top of an old one of the former year. Eggs of the Twite which we obtained prior to the 20th of May 1870, were for the most part fresh, but on that day we found a nest of five eggs in which the young were formed: thus many Twites would appear to have begun to lay about the 17th of May. On this date we found many nests at Newton containing only one egg.

Sub-family *LOXIINÆ*.

Carpodacus erythrinus (*Pall.*). **Scarlet Grosbeak.**

Pyrrhula europæa, *Vieill.* **Bullfinch.**

Gaelic—*Corcan-coille* = from *corcur*, red, purple, and *coille* = wood: *Deargan-coille* = the little red bird of the wood.—A. C.

Pinicola enucleator (*L.*). **Pine Grosbeak.**

Loxia pityopsittacus, *Bechst.* **Parrot Crossbill.**

Loxia curvirostra, *L.* **Common Crossbill.**

Gaelic—*Trastan* = the transverse (billed) bird: *Camaghob* = wry-bill: *An Deargan Giubhais* = the ruddy bird of the pine.—A. C.

[*Obs.*—The only mention of this species is a doubtful one, thus: "Crossbill seldom seen." This note is, however, by Mr. D. Mackenzie at Stornoway, a good observer.][1]

Loxia leucoptera, *Gmel.* **White-Winged Crossbill.**

Loxia bifasciata (*C. L. Brehm.*). **Two-barred Crossbill.**

1 Mr. D. Mackenzie left Stornoway in 1885 or 1886 for North America.

Sub-family *EMBERIZINÆ*.

Emberiza melanocephala, *Scop.* **Black-headed Bunting.**

Emberiza miliaria, *L.* **Common Bunting.**

Gaelic—*Ian Bollach a Ghort: Gealag-Bhuachair: Gola-Bhigein.*—A.C.

Always recorded as very abundant over the whole range, and increasing of late years where cultivation has gained upon the peat, as, for instance, at Stornoway, Loch Maddy, Newton, etc. Commoner along the west coast, where the greater amount of arable lands occur, but found here and there also where the merest patches of tillage exist. They are of course rarer in the islands which lack cultivation. Extends to, and breeds in St. Kilda, where Harvie-Brown found them far from uncommon, taking the extent of the cultivated area into consideration. They were only seen in the corn-fields and hay-stripes on the slopes below the town. Also not rare at Tarbert, in Harris, as observed in 1879 and 1881. In the west of The Lews, very abundant at Mhorsgail; seen in 1887 on the outermost of the Monach Isles (Shillay), near the lighthouse, and commonly in Mingulay. To a considerable extent the Common Bunting is migratory, leaving Mingulay in the end of August with its young, and others taking their place there as winter approaches.

Emberiza citrinella, *L.* **Yellow Bunting.**

Gaelic—*Bhuidheag* = the yellow one: *A Bhuidheag-Bhuachair* = the yellow one of the dung: *A Bhuidheag Bhealuidh* = the yellow one of the broom.—A. C.

Only recorded in 1841 as occurring at Rodel. Duns includes it at Stornoway in 1865. Seen at Rodel commonly in 1879 by Harvie-Brown, and in 1881 also at Tarbert. Mr. Hugh G. Barclay also found it in Uist in 1885, but not very abundant. Arrives on migration in the autumn in Mingulay, and most of the islands.

Emberiza cirlus, *L.* **Cirl Bunting.**

Emberiza hortulana, *L.* **Ortolan Bunting.**

Emberiza rustica, *Pall.* **Rustic Bunting.**

Emberiza pusilla, *Pall.* **Little Bunting.**

Emberiza schœniclus, *L.* Reed Bunting.

Gaelic—*Ceann-dubh Fraoich* = black-headed of the heather.—A. C.

This bird was included as early as 1831 in W. MacGillivray's List of the Birds of Harris, but was considered rare in 1841; now, a few pairs breed every year in the Long Island, and in North Uist. In 1879 it was recorded as "occasional" at Stornoway by Mr. Greenwood, and it is noted at Mhorsgail in 1881, in Mr. H. H. Jones's lists. We have observed it ourselves in several localities, and have obtained the eggs from the Vallay District of North Uist. It was also noted by Mr. Hugh G. Barclay in North Uist in 1885.

Plectrophanes lapponicus (*L.*). Lapland Bunting.

Plectrophanes nivalis (*L.*). Snow Bunting.

Gaelic—*Deargan Sneac* = ruddy of the snow: *Bigein Sneac* = little snow-bird.—A. C.

Was observed as late as the middle of May on Berneray, on the great sandhills, where John MacGillivray shot a pair at that season. Regular winter visitant, but in October 1882 came in *enormous* flocks to Mhorsgail. In 1879, however, only small flocks were recorded, and Mr. D. Mackenzie has a similar note that year from Stornoway.

Generally, it is considered strictly migratory in the O. H., remaining only a few weeks in the early part of winter, and Gray adds: "and pitch on the low grounds of Benbeculay and the Uists, but do not appear to be observed in the spring return migration." With north-east winds in the end of October, or early in November, these birds appear on migration throughout the archipelago, varying in numbers according to the weather, and becoming less abundant as the winter advances.[1]

[1] A report to the effect that it breeds, or has bred, in The Lews cannot be accepted, if we take into consideration the known altitude at which it breeds in our latitudes, and the corresponding temperature requisite. (See *Fauna of Sutherland, etc.*) Though it is known to breed in high latitudes, almost down to the sea-level, an altitude of 3000 feet—or to put it safely, say 2500 feet—would be necessary in the O. H. to afford it a suitable temperature, even if it would then obtain it. Exceptional cases of course are possible, but extremely unlikely.

Section 4. OSCINES SCUTELLIPLANTARES.

Family ALAUDIDÆ.

Galerita cristata (*L.*). Crested Lark.

Uiseag Mhoire=Mary's Lark.—A. C.

Alauda arvensis, *L.* Skylark.

Gaelic—*Uiseag. In Islay the Uiseag is called* An Reamhag, *i.e. Riabhag, the ordinary name of the* Pipit *elsewhere.*—A. C.

Duns, correctly perhaps, describes this species as "the bird of Lewis" everywhere, both inland and near the shore. A delightful sketch of the habits and history of the Skylark in North Uist is given by Gray, who speaks of it as occurring all over the barren moors. Included in all the lists received, and heard even at Rodel by Harvie-Brown, who found it common in most places, but rarest, or almost absent, on the rocky barrens, though plentiful on the mossy moors. Thus, he did not find it universal on the east coast of Harris, or the rockier portions of the west coast of The Lews; neither did it occur upon the sides of Ben Eabhal or Ben Leo, nor between Loch Maddy and Loch Obisary. Neither did he expect to find it universal on the east side of South Uist. He does not remember meeting with it in Barray, nor did he find it in Mingulay until 1887; and it did not appear to be present in the smaller isles. But on the lower-lying mosses of the interior of North Uist, and in the west, and in the pastured districts of all the islands, it was omnipresent. During a long summer day's drive from Tarbert to Stornoway, however, scarcely any bird-life was observed, and the Skylark was most conspicuous by its absence. Although nothing was seen of it in the Shiant Isles in 1879, and in but few localities amongst the wilder and outlying Hebrides, yet in 1887 it was found far from uncommon on Garbheilean, but scarce or absent on Eilean Mhuire. This is curious, as the pasturage and grasses upon the latter are deeper, more luxuriant, and apparently far more suited to the requirements of the species. We repeat here, what we have become more and more convinced of after successive years of observation in many localities in Scotland, that changes, rapid and marked, are annually going on in the increase and extension of range of species; observations at the present time cannot, therefore, be too carefully or too elaborately recorded.

The Skylark is migratory to a large extent in the O. H. as elsewhere, those arriving before winter taking the place of others which leave about the end of harvest, and many passing on—as Mr. Finlayson expresses it—"Northerners on their way south."

Alauda arborea, *L.* **Wood Lark.**

Calandrella brachydactyla (*Leisl.*). **Short-toed Lark.**

Melanocorypha sibirica (*Gm.*). **White-Winged Lark.**

Otocorys alpestris (*L.*). **Shore Lark.**

Section 5. OSCINES CULTRIROSTRES.

Family STURNIDÆ.

Agelæus phœniceus (*L.*). **Red-winged Starling.**

Sturnus vulgaris, *L.* **Common Starling.**

Gaelic—*An Druideag: An Druid* (St. Kilda): *Truideag: Truid.*—A. C.

As early as 1841 the Starling is recorded by John MacGillivray as abundant, beginning to flock in July, and resident in flocks till April. As far back as 1830 John MacGillivray speaks of it in St. Kilda, and gives its Gaelic name there as above; and in 1848 Sir W. E. Milner found them breeding there, the nest being situated under a large stone on the steep side of Conacher. Gray, in 1871, speaks of it as very common "under stones on the beach, in turf dykes, and deserted rat-holes." It is included in all the local lists received by Harvie-Brown. It is described by W. MacGillivray as inhabiting a cave on the west coast of one of the Hebrides in *vast* numbers about 1820, and, indeed, as is well known, it breeds in many similar localities, such as the cliffs at Loch Roag, and elsewhere. A locality where they are very abundant is amongst the rocks of the Luscantire[1] Hills, close to Tarbert; and in Mingulay it also breeds, assembling in vast flocks in winter. Local or general migratory movements are noted at many stations in autumn throughout the area.[2]

[1] *Losg-cinntir*=the headland of Lusk.—A. C.

[2] It is interesting, in connection with this, to find the following note in Charlesworth's *Magazine of Natural History*:—"About forty years ago (*i.e.* previous to 1837) people spoke of the Starling as being, in former times, a constant companion

Family CORVIDÆ.

Pyrrhocorax graculus (*L.*). Red-billed Chough.

Cathag = jackdaw: *Cathag nan Casa Dearg* = jackdaw of the red legs.—A. C.

[*Obs.*—EXTINCT. Though now unknown in the isle of Barray, the species was found there by W. MacGillivray, who states in his "Account of the Long Island" in the *Edinburgh Journal of N. H. and Geographical Science*, vol. ii. p. 323, 1830, that at that time it frequented the southern extremity of the range, but was not met with elsewhere. In the same way many of the inner isles have been deserted (*op. cit.* p. 165), and MacGillivray, in his *British Birds*, vol. i. p. 588, is equally circumstantial in his account of the above. The extinction of the Chough in the O. H. cannot, we think, therefore be ascribed to the influence of the more "fit" Jackdaw, as the latter is almost equally rare in the islands at the present time; though this accusation has, and perhaps not without some reason, been frequently launched at the head of this devoted bird at many localities on the Scottish coasts. The cause of extinction must be looked for somewhere else in this case at all events.]

Nucifraga caryocatactes (*L.*). Nutcracker.

Garrulus glandarius (*L.*). Common Jay.

Gaelic—*Sgriachag* = screamer: *Sgreuchag-choille* = the wood-screamer.—A. C.

Pica rustica (*Scop.*). Magpie.

Gaelic—*Pitheid* : *Piothaid* : *Pioghaid.*—A. C.

Corvus monedula *L.* Jackdaw.

Gaelic—*Cathag ghlas* = grey jackdaw: *cathag* = jackdaw.—A. C.

All previous authors have omitted mention of this species. Five wintered in 1877 near Stornoway, three in 1878, but none since (D. Mackenzie, 1881).

of the ruined towers, but which had so completely forsaken the southern counties (speaking of Scotland) that the first I ever saw were flying about the tower of the old Monastery of Rowdell, in the island of Harris, in 1804."—(W. L. *op. cit.* vol. i. p. 118.)

Corvus cornix, *L.* Hooded Crow.

Gaelic—*Feannag*; *Fionnag*: *Starrag* (Harris).—A. C.

This is the only species decidedly common in the O. H. besides the Raven, though both *C.* (var. *corone*) and *C. frugilegus* are met with in Skye.

The Hooded Crow is very common even in St. Kilda. Now it is omnipresent, and, as the Vermin Lists will clearly show, the united endeavours of the game-preservers scarcely succeed in reducing their numbers. W. MacGillivray speaks of the "Crows-cups" or "*Cragan-feannaig*"—shells of the Echini on the beach—as having been eaten by the Hooded Crows (*British Birds*, vol. i. p. 533), and known by that native name as having been so destroyed. (See *Oyster Catcher*.)

The Hooded Crow nests on the ground in the O. H., on sloping banks covered with long heather, and Harvie-Brown has often seen their nests "pointed" by a trained shepherd's dog. The reason for this, at least in North Uist, is self-evident, viz., the absence of more suitable nesting sites. (Vide *Kestrel* and *White-tailed Eagle*.)

On the farm of Newton the shepherd killed 180 "Grey Crows" in 1870; comparatively few were seen there the following season.

It is curious to find at times how Hooded Crows, and ducks and geese seem to hold a nesting site almost in common in these islands, their respective nests not unfrequently being placed in rank heather close to one another, *i.e.* say within two or three feet. Whether it is for purposes of mutual protection we cannot prove or disprove at present. But it can scarcely be that the Hooded Crow would rob that near neighbour's nest, as, if so, surely that neighbour would not again risk her valuables in the same spot: or can it be because the Hooded Crow in the Outer Hebrides and west coast of Scotland generally finds *more than sufficient* food on the sea-shore, comparatively near to which it breeds?[1]

[1] W. MacGillivray speaks of the Carrion Crow (*C. cornix*, var. *corone*) as utterly unknown in the O. H. Gray, in 1871, would seem to admit it, but from the text of his remarks we are inclined to think "black crows" are more probably rooks, as a perusal of the account of the occurrence of that species appears to indicate. We have utterly failed ourselves to obtain a single *reliable* record of its appearance, whereas we know that Rooks in the Migration Schedules are usually returned as "Black Crows." The question of the specific identity or otherwise of the Carrion Crow seems to us not yet definitely settled, though we have read the remarks by authors who take opposite views.

Corvus frugilegus, *L.* **Rook.**

Gaelic—*Rocuis.*

No authors speak of the Rook as occurring in the O. H. during the earlier chronology except W. MacGillivray, who, in 1830, speaks of it as "not a resident, but sometimes in hard weather," and acknowledges its recognition by a Gaelic name. Professor Duns includes it as "rarely seen in Lewis." Gray (1871) says it appears occasionally in large flocks, but stays only a short time. As instancing the irregularity of its occurrence or general scarcity, Mr. Greenwood's note, "never observed near Stornoway" (*i.e.* by him), is of considerable significance. Sir James Matheson tried to introduce them there, but failed.

By 1880-81 it is reported thus: "Several seen every winter about Stornoway; 100 in 1880-81; but all leave in spring." None are included in the Vermin Lists, thus indicating their scarcity even in winter, and their non-residence in summer; harmlessness in the eyes of keepers can scarcely be the ground for utter exclusion from mention.

Corvus corax, *L.* **Raven.**

Gaelic—(W. MacG.) "*Biadhlach*": (J. MacG.) "*Biadhach*" (St. Kilda) = glutton: *Fitheach*=feeder.—A. C.

In 1841 we find the statement, by one who ought to have known, —viz., John MacGillivray,—that "The Raven was generally distributed, but nowhere numerous." He instances a nest in a rock only 15 feet above the ground, but it was nearly inaccessible from below, and quite so from above. But, curiously enough, W. MacGillivray, as early as 1830, speaks of it as "astonishingly common in all parts," and instances a *white one* seen in Pabbay. He also repeats this in his more complete work, and refers to the "white-patched" variety as occurring in Harris, and to an example seen frequently on the sand-banks at Northhaven. This is the form more commonly observed in Faröe (*Corvus faroënsis* of J. Wolley), but which is unquestionably only a partial albinism, which no doubt has become hereditary. (Of course the question naturally occurs: "Was this Pabbay of Harris specimen a resident—or of resident stock—or a casual migrant from Faröe?")

MacGillivray then instances a vast gathering of Ravens upon Pabbay to feast on the carcases of Grampuses—*Orca gladiator*—

which had been driven on shore, and describes a device of the natives to get rid of them.[1]

Mr. H. H. Jones, in his notes in the list kindly given to us, says: "Ravens are more abundant near the Harris march than I have ever seen them anywhere else. I am afraid to say how many I have seen altogether at the same time." Of course they are included in all lists and all Vermin Lists received.

In MacGillivray's time the tameness of the Raven is made special mention of, as it is noted as sitting on hut-roofs and dung-hills (vol. i. p. 509), which recalls vividly to our memory the same fearlessness of the Magpie in Norway, which we have seen nesting in gooseberry-bushes and under the eaves of the farm-houses, and also of the same fearlessness of the Raven in Russia.

Young Ravens—well-grown—were seen on 26th April 1870 by Harvie-Brown, in North Uist, occupying a formerly inhabited eagle's eyrie. The Raven appears to have been more than usually abundant in 1870, judging from passages in our journals. In 1881, 100 were reported as occurring in MacAulay's Vermin Lists, in the west of Harris, and 637 was the total in ten years in the Lewis lists, running very close in numbers, all through the years, with Hooded Crows. The most were got in 1884. No trapping power appears to decrease their numbers, any more than it does of Hooded Crows. Kill every one in the O. H. and we believe their numbers will be replaced. In the Appendix of Vermin Lists, *q.v.*, it will be found that a superabundance of almost every species of vermin is apparent in the returns of 1884. How far it would be wise or expedient to theorise upon these figures it is difficult to say; but it would be curious to have similar statistics as to the supplies of shell-fish on the shore—the principal food of rats—and of the unusual general food supply for all species, and their consequent fertility, and success in the "struggle for existence." Every seven years, it is said, the razor-fish of the bays at Stornoway are cast up in *tons* on the shore, affording rich supplies of food to man and beast and bird alike.

Whatever the results of such inquiry may be, if we look at the

[1] "At length one Finlay Morrison devised a scheme which produced the desired effect. Having discovered their roosting-place, he and one or two others caught a considerable number of them alive. They then plucked off all their feathers excepting those of the wings and tail, and in the morning let loose amongst their companions these live scarecrows. The ravens terrified," etc.

list of vermin killed as given in the Appendix, we find an average of 63·7 per annum as against 69·8 of Hooded Crows. The smallest number was in 1876, possibly owing to less ground being preserved, or less energy exerted; but possibly also to lesser periodical rushes of immigrants having taken place. (See the full list appended.) But though large congregations of Ravens do take place, and especially large "rushes" do no doubt occur, the returns of the Lewis lists rather incline us to the supposition that *extra* vigilance had been expended, both upon rats and upon winged vermin.

But another reading may be—and probably is—the correct one, that an extra good breeding season for rats (consequent, it may be, on superabundance of shell-fish, *their* principal food in the O. H.) was followed by an extra influx of birds of prey—as is certainly known to be the case in Norway with regard to the Lemming years, and the southward extension of the breeding range of the Snowy Owl and other birds of prey—consequent upon the increase of food.

The above, however, would hardly be complete without our recording the fact that ravens and other birds of prey are not nearly so rigorously looked after on the more immediately adjoining properties—*i.e.* adjoining North Uist—or at least were not at the time of which we speak.

Three breeding-places are regularly occupied in Mingulay, year after year, by three pairs of birds, as we were assured by Mr. Finlayson, and by a pair on Bernoray, or Barray Head, in 1887.

Order 2. MACROCHIRES.

Family CYPSELIDÆ.

Gaelic—*Gobhlan Mor*=the big forked one: *Gobhlan Dubh*=the black forked one: *Clisgein*=the swift or rapid bird.—A. C.

Cypselus apus (*L.*). Swift.

No record appears till a very late date. Even in 1871 Gray says: "Totally awanting in the Outer Hebrides"; but in the Appendix to the same work he says, "A single Swift was seen on 27th May 1870 by Lieut.-Colonel Feilden [1] and Harvie-Brown."

[1] Then Captain H. W. Feilden.

On September 6, 1879, Mr. H. H. Jones saw six Swifts hawking over the loch at Mhorsgail House, and on the 8th a pair in the forest. [*N.B.*—The late stay of Swifts that year was taken notice of in *The Field*.] The Swift is included in the birds seen around Stornoway by Mr. D. Mackenzie during the migratory seasons. In 1887, upon the 15th July, Harvie-Brown saw two Swifts careering along the south side of Barray Head.

Cypselus melba (*L.*). **White-bellied Swift.**

Acanthyllis caudacuta (*Lath.*). **Needle-tailed Swift.**

Family CAPRIMULGIDÆ.

Caprimulgus europæus, *L.* **Common Nightjar.**

Caprimulgus ruficollis, *Temm.* **Russet-necked Nightjar.**

Order 3. PICI.

Family PICIDÆ.

Sub-family *PICINÆ.*

Dryocopus martius (*L.*). **Great Black Woodpecker.**

Picus major, *L.* **Great Spotted Woodpecker.**[1]

Picus minor, *L.* **Lesser Spotted Woodpecker.**

Gecinus viridis (*L.*). **Green Woodpecker.**

Sub-family *IYNGINÆ.*

Iynx torquilla, *L.* **Wryneck.**

[1] A. Carmichael has an interesting note, thus: "*A Chnag: A Chnagag-choille* = the little wood-rapper. MacGillivray says he never saw a British specimen of this rare bird; but Pennant saw it in August in the pine forests of Invercauld, Aberdeenshire. It was then thought to be peculiar to Sutherland." In this connection see Sir Robert Gordon's *Earldom of Sutherland*. As being *rare* in MacGillivray's time, Cnag (or "Knag") could hardly be applied to a Tree-creeper in Gaelic, and was therefore probably a true Woodpecker.

DESMOGNATHÆ.

Order 1. COCCYGES.

Sub-order *COCCYGES ANISODACTYLI.*

Family ALCEDINIDÆ.

Alcedo ispida, *L.* Common Kingfisher.

Gaelic—*Cruitein* = the crouched bird: *Biora Cruitein* = the crouched bird of the spit: *Bior an Iasgair* = the fisher and strong of the spit,—the spit being the bird's long spit-like bill.—A. C.

Family CORACIIDÆ.

Coracias garrula, *L.* Common Roller.

[A Roller is mentioned by Sir W. E. Milner as having been seen by the Rev. Mr. Mackenzie, during two or three days together, on St. Kilda "one winter."]

Family MEROPIDÆ.

Merops apiaster, *L.* Common Bee-Eater.

Family UPUPIDÆ.

Upupa epops, *L.* Hoopoe.

Gaelic—*Calman Cathaidh*: *Calman Cathaich*.—A. C.

Up to 1840 there appears to have been no record of this species, but it is designated as "occasional" in 1841 by J. MacGillivray, founded, most likely, upon a single specimen, concerning which there seems to be no reasonable doubt, which was found in an exhausted state at Balelone in North Uist (*auct.* Dr. M'Leod).

Gray also mentions two specimens as having been seen in North Uist in 1859, one of which, according to Mr. J. MacDonald of Newton, was shot.

Sub-order *COCCYGES ZYGODACTYLI.*

Family CUCULIDÆ.

Cuculus canorus, *L.* Cuckoo.

Gaelic—*Coi*: *Cuach*: *Cuachag* (poetical name): *Cuthag*.—A. C.

In 1841 we find the record that the Cuckoo is "seen and heard occasionally, especially about Loch Maddy, and elsewhere in North

Uist." W. MacGillivray relates in his great work some peculiar folk-lore regarding the Wheatear and Cuckoo in the Hebrides.[1]

It is not at all clear that the Cuckoo never remains in the Hebrides all winter, as authentic accounts are extant of its occurrence as early as March 25, 1882 (Alexander Carmichael, *in lit.* 27th March 1882), and we have similar accounts from other favoured inland localities.

Our own experience of the Cuckoo in the O. H. is that it is extremely common now, and perhaps more especially about Stornoway and the northern parts of the range. The 2d May was the earliest date in 1870 on which it was observed by Harvie-Brown.

Order 2. ACCIPITRES.

Sub-order *STRIGES*.

Family STRIGIDÆ.

Gaelic—*Cumhachag: Comhachag*=the lamenting one: the dolorous one.—A. C.

Strix flammea, *L.* Barn Owl.

Gaelic—*Chumhachag Bheag: A Chomhachag Bheag* = the little sorrowing one: *Sgriachag* = the little screecher: *Cailleach-oidhche Bheag*=the little night carlin: *Cailleach-oidhche Gheal* =the little white night carlin.—A. C.

Family BUBONIDÆ.

Asio otus (*L.*). Long-eared Owl.

Gaelic—*A Chumhachag Chluasach*=the long-eared lamenter: *A Chumhachag Adhairceach*=the long-horned lamenter.—A. C.

[*Obs.*—Mr. D. Mackenzie includes this species as occurring at Stornoway, but Mr. Greenwood never saw it there. Mr. Mackenzie

[1] Thus: "Should the wheatear be first seen on a stone, or the cuckoo first heard by one who has not broken his fast, some misfortune may be expected. Indeed, besides the danger, it is considered a reproach to one to have heard the cuckoo when hungry, and of such an one it continues to be said, that the bird has *muted* upon him, 'chac a chuaig air.' But should the wheatear be seen on a turf or on the grass, or should the cuckoo be heard when one has prepared himself by replenishing his stomach, all will go well. Such, at least," continues W. MacGillivray, "was the popular creed twenty years ago (*i.e.* say, 1820) when I began in earnest to look after birds." (*N.B.*—This is an interesting date to the bibliographer and to the admirers of William MacGillivray, of whom we Scottish naturalists are justly proud.)

likewise includes the Short-eared Owl (1881); so, having considerable measure of confidence in Mr. Mackenzie's powers of observation, we think the records worthy of place, at least in square brackets, all the more so that this species is said to have been rapidly increasing in the Dunvegan woods in Skye of late years, where it breeds, and is heard frequently in the summer nights by the officers of the s.s. *Dunvegan Castle.*]

Asio accipitrinus (*Pall.*). Short-eared Owl.

A species of owl—with little doubt this species—is noted by Major MacDonald of Aisgornis, in South Uist. The species is certainly common in Benbeculay, and still more so, perhaps, in North Uist. Harvie-Brown flushed the bird several times from high heather in the latter island, and has had eggs taken there, and H. G. Barclay has also obtained eggs, as well as many other collectors. It may be looked upon as quite a common species at the present time.

Syrnium aluco (*L.*). Tawny Owl.

Nyctea scandiaca (*L.*). Snowy Owl.

Gaelic—*Chomhachag Gheal* = white lamenter.—A. C.

This beautiful bird is occasionally shot and often seen in the O. H. Two are recorded by Professor Duns as having been shot at Butt of Lewis in 1855, and one at Uig in 1859. Gray regards it as a regular spring migrant to the O. H., especially in Lewis. It occurs also in Harris and Benbeculay. In the last-named island one was shot by Mr. J. Fergusson, surgeon. Several occurred in Lewis in 1868; one at Gress on 21st April; another at North Tolsta in April 1867. It is also spoken of by Mr. D. Mackenzie as a regular visitant, though he did not hear of any in 1880-81. It is not, however, mentioned by the MacGillivrays with any special references to the O. H.

Mr. John MacDonald of Newton also obtained a fine specimen in North Uist, which is now in Sir John Orde's collection; and in 1887, when we were at Mingulay, Mr. Finlayson informed us that one bird of this species, which he accurately described, was shot there last winter, "the only one remembered on the island." This was about January 1887. The specimen was not preserved, but was thrown out upon a dung-heap. All search, however, failed to reveal a single remaining feather.

Surnia ulula (*L.*). **Hawk Owl.**

Surnia funerea, *L.* **American Hawk Owl.**

Nyctale tengmalmi (*Gm.*). **Tengmalm's Owl.**

Scops giu (*Scop.*). **Scop's Owl.**

Bubo ignavus, *Forst.* **Eagle Owl.**

Athene noctua (*Retz.*). **Little Owl.**

Sub-order *ACCIPITRES.*

Family **VULTURIDÆ.**

Gyps fulvus (*Gm.*). **Griffon Vulture.**

Neophron percnopterus (*L.*). **Egyptian Vulture.**

Family **FALCONIDÆ.**

[*Obs.*—In the Lewis Vermin List all species of the smaller *Raptores* are included under "Hawks." By these returns it appears that 448 were paid for in the ten years between 1876 and 1885; and in this respect, as in others, 1884 bears the second heaviest returns, viz., 75, the largest being in 1878 with 91. What proportion of the following species are included in these figures cannot now be determined. Only their approximate proportions can be guessed at from the relative abundance of each species as recorded in the text.] [1]

Circus æruginosus (*L.*). **Marsh Harrier.**

Gaelic—*Clamhan Loin*: *An Spuillire Buidhe*=the yellow spoiler: *An Croman Loin*=the bog hunchback.

This last name is also applied in some places to the Snipe, and in some to the Woodcock (A. C.).

[*Obs.*—W. MacGillivray says (1830) that he once saw this species in Harris. Gray in 1871 says he saw it on the island of Benbeculay, and quotes MacGillivray as above. That it "may occur more commonly in North and South Uist," as Gray says, is not borne

[1] See Appendix.

out by subsequent research. Indeed its extreme rarity, anywhere in Scotland, if it really ever does occur, barely admits it to our list.]

Circus cineraceus (*Mont.*). Montagu's Harrier.

Circus cyaneus (*L.*). Hen Harrier.

Gaelic—The male is *Ian Fionn* = the pale bird: *Crom nan Cearc* = the hen hunchback: *Croman nan Cearc* = the hen huncher: *Clamhan Gobhlach nan Cearc* = the forked buzzard of the hens.—A. C.

In MacGillivray's time, recorded as "rather abundant, especially among the bogs of the two Uists (1841)."

Curiously, it is not admitted by Professor Duns, though he says: "Not fewer than ten species of *Falconidæ* occur in Lewis, or on some of the islands of the east or west coasts." Gray found the Hen Harrier "very common" in the O. H. when writing in 1871, and had seen twelve or fourteen in a day on Benbeculay and North and South Uist, and he notices its daily round of inspection, which Harvie-Brown has also observed, at Newton and elsewhere. The Hen Harrier is tolerably abundant near Stornoway (*auct.* Mr. Greenwood, 1879).

In North Uist fine old ash-coloured males are still seen, not uncommonly; these are comparatively rare in many other parts of Scotland. One such bird used to show himself every evening in the summer near Newton in 1879,[1] flying over the machar. Also in Lewis fine old birds are still to be seen in the more secluded deer-forests. A peculiar phase of flight of this species has not, we think, been generally remarked upon by authors, *i.e.* "mounting in the air, then diving downwards, closing and opening each wing alternately and uttering its kestrel-like cry all the time. The flight reminded us to some extent of the snipe 'drumming,' or even of the lapwing's flight in the breeding season. The bird exercising it came overhead at the time, and continued these peculiar movements as long as he remained in sight. The female bird was in sight at the same time."—(Journal, 1870, p. 24; J. A. H.-B.)

[1] In 1887, also at Newton, about the same hour in the evening, a Harrier, no doubt the same old bird again, showed itself hunting over the same ground.

Buteo vulgaris, *Leach*. **Common Buzzard.**

Gaelic—*Clamhan* : *Clamhan Luch* = the mice buzzard : *Am Bleidire* = the sorner : *Bleidire Tonach* = the large-hipped sorner : *An Gearra Chlamhan* = the broad buzzard.—A. C.

[*Obs.*—No records appear up to 1871, except that W. MacGillivray says :—

"Although it occurs in the larger Hebrides, it is rarely seen there." He probably refers to the larger Hebrides of the Inner group, as we cannot lay hold of any very tangible subsequent record. Harvie-Brown was assured that a pair of these birds bred, in 1870, on an island on the east side of Lewis, but examination into the statement did not result in any positive confirmation at that time, nor since.]

Archibuteo lagopus (*Gm.*). **Rough-legged Buzzard.**

Aquila clanga, *Pall.* **Large Spotted Eagle.**

Aquila chrysaetus (*L.*). **Golden Eagle.**

Gaelic—*Firein* : *Fior-eun* = the true bird, the bird *par excellence* = poetical name of eagle : *Iolair* = eagle : *Iolair Bhuidhe* = golden or yellow eagle : *Iolar Bhreac* = spotted eagle ; no doubt the young or immature bird : *Iolair Cladaich* = shore eagle (but is this not more correctly applied to the white-tailed eagle?) : *Iolair Dhubh* = black eagle.—A. C.

John MacGillivray did not, in his time, consider the Golden Eagle so common as the White-tailed Eagle, and indeed appears only to have known of a single breeding-place in 1841, viz., on the Hill of Northtown. W. MacGillivray, however, in 1830, considered it very abundant over all the range. Sir W. E. Milner saw one in the flesh which was trapped near Stornoway in 1848. Gray, in 1871, says : "Various eyries existed in 1867, from the Butt of Lewis to Barray Head ; and it is still a well-known bird in North Uist, there being two eyries last year (1870)," and one locality in South Uist every year.

The Golden Eagle, however, up to a certain time, may have been rarer in the O. H. than the White-tailed,—and Harvie-Brown has had considerable experience of the two species—at least in Harris, Lewis, and North Uist,—but now the reverse is the case.

In the Lewis Vermin Lists—a perfect store of valuable statistics—the total number of "Eagles" set down as paid for—*i.e.* including both species—is 30 in ten years, between 1876-86 (see Appendix), of which 7 were got in 1884, and 9 in 1879—the two largest returns. None were obtained (or recorded) in 1881, and only 1 in 1880, whilst only 2 are entered under 1876, 1878, 1882, and 1885 respectively.

In 1879 Harvie-Brown was informed that no eagles had bred in the North Uist range of mountains since he took the eggs in 1870, in which year that was the only eyrie on the island. The same year, 1870, there were some eight eyries of eagles in Harris, all of which Harvie-Brown visited. They were present in Mhorsgail deer-forest in 1881, and on an adjoining shooting in 1882. In 1879 Mr. Greenwood also knew of their breeding in the west of The Lews. The current tradition of their lifting children and bearing them away to their eyries, as related of the "*Mountain Eagle*," is also believed in Harris, in which version it is stated that a child was borne away across the Minch to the Isle of Skye, a distance of more than sixteen miles. How much is mythical, and how little is historical, does not appear, but it can scarcely be related as a solemn and proven fact.

Eagles constantly harass the wild-fowl on the loch at Rodel, but seldom succeed in carrying any away. They often hover for the most part of a day close overhead, frequently within easy range of the keeper's gun. Some of these are quite local birds, nesting in the vicinity of the loch, but others come, as certainly, from a distance.

There are specimens in many of the English collections—*few comparatively* in Scotch ones—of the birds shot in the Hebrides, as well as in other parts of the mainland.[1]

From the accounts in our journals of several eyries visited by us in 1870 we select the following :—

"At seven o'clock on the 29th April 1870, I[2] started with three assistants in a light gig for the distant eyrie. On arrival

[1] Mr. Chase has one shot in Lewis on 12th April 1879, an adult male, *therefore* shot in the breeding season, and he afterwards obtained eggs from the eyrie which had *previously* belonged to this bird, the female having obtained *another* mate.

[2] *i.e.* Harvie-Brown.

at the first known breeding-place we found that the rock projected fully 15 feet immediately above the eyrie, at a right angle with the rest of the cliff, and the nest was placed almost in the angle made thereby with the cliff below. We believe it would have been impossible to reach it with a rope, and indeed the presence of the long wooden ladder which we had brought part of our journey with us, by the advice of those who acted as our guides, proved that this was the general opinion. But the birds were not there. The old nest was in good preservation, and it projected considerably beyond the small rock-platform on which it rested. The keeper and I, after a careful examination of this eyrie with the glasses, proceeded to the other known nesting-place —an immense mass of rock detached from the rest of the cliff, and reminding me of another place where a few days before White-tailed Eagles' eggs were taken in my presence.

"The keeper and I ascended through a deep crack or crevice to the top of the rock, and, on looking over the edge, I could see into the old nest. This looked gloomy as regards our prospects; but fresh droppings which I observed on the summit led us to believe—or hope at least—that the birds were not far off. Shortly afterwards the cock eagle came flying round, dimly visible in the mist overhead. Two Rock Doves flew out of a crevice whilst we were ascending, but we found no nest of theirs.

"As the keeper and I retraced our steps, half-way between the two old eyries, we met another of our party, who had remained behind at the first breeding-rock. Just as we met, the hen came hurriedly off her nest, close above our heads. The eyrie was very simply chosen, the nest entirely new, and very large. We scrambled up a steep slope, overgrown by long coarse loosely-rooted heather and short dry moss, and stood on a green sheep-track some 20 or 22 feet below the bottom of the nest.

"I could have gone to the nest, I believe, without the rope, but the descent would have been difficult.

"The other men having joined us, R., one of the boatmen —a young active fellow—shouldered the ropes and went above. R. R. and the keeper remained under the rock. The other man stood lower down, and further out, in order to give directions or pass mine on to R. above. I climbed a few feet before I could reach the rope and fasten it round me, the rope being rather too short. I managed the rest without much difficulty, though one time I had a good swing for my carelessness, so eager was I to reach the edge of the nest, before the slack of the rope was quite taken in, or my feet fairly planted. But, luckily for me, there was a good man above, and I soon peeped in over the edge. Owing to the footing being bad, and to the projection of the nest itself, and the distance at which the eggs lay, I had to pull away a portion of the heather which composed it, and lean forward, resting on the nest itself. Then I could barely reach them. After a little straining, I reached the eggs with my right hand, and succeeded in packing them in a little tin box fastened under my arm, and made the descent easily; and we hurried off in order to catch the tide, having to row back part of the way.

"The interior portion of the nest was not more than 1½ feet in diameter, the eggs lay on a bed of dry grass, small quantities of the birds' own down, and withered blades of *Luzula sylvatica.* Two large fine bunches of the latter lay, roots inwards, and almost touching the eggs, as if the old birds had only lately dropped them there. The depression was very shallow, and just sufficiently deep to prevent the eggs from rolling. The whole nest was quite new from the foundation; and I was assured the same place had never before been used as an eyrie. Last year the second rock we visited held it, whence young had been taken with the assistance of a ladder. The fabric of the nest was entirely of heather stalks, none of which, that I examined, were thicker than a man's little finger. The depth of the structure was between 3½ and 4 feet, if not more, and the whole would have filled two ordinary wheel-barrows.

"The eggs are most beautiful. One is dark reddish-brown all over the small end, and has a broad band of the same rich colour stretching longitudinally over about two-thirds of the length, and the rest of the shell is well marked with both large and small blotches. The other is even finer in appearance, though more sparingly marked at one end. The small end is one mass of colour, having a purplish tinge, which is distinctly margined from the less highly coloured portions of the shell. Both eggs were deeply incubated, so I pasted several layers of gummed paper round the holes, and left the contents to decompose. It was not till the 21st June that I succeeded in emptying them at Dunipace. I preserved part of the upper mandible of the young bird showing the 'diamond.'"

Haliaetus albicilla (*L.*). **Sea Eagle.**

Gaelic—*Iolair Bhreac* = speckled eagle: *Iolair Cladaich* = shore eagle.—A. C.

Reported by our earlier authors as commoner than the Golden Eagle. Rewards were at that time offered for both species, and they were even then considered to be decreasing in numbers annually. One pair used to breed on a low, flat, deeply heather-covered island on a fresh-water loch, well known to Harvie-Brown, who has several times viewed the site. The nest was on a platform of sticks laid on the open ground on the top of the island. In May 1848 Sir W. E. Milner saw one trapped near its nest, not far from Stornoway. Gray mentions by name various eyries,[1] such as North Uist, Scalpay, Shiant Isles, Wiay, Benbeculay, and several in Harris and Lewis. But it is not necessary to particularise which of these are occupied at the present time, or were occupied in, say 1886, except the one on the Shiant Isles which is believed to be inaccessible. There is no doubt about the marked decrease in the number of inhabited eyries of the White-tailed Eagle throughout Scotland during the past ten or fifteen years. Harvie-Brown has also visited the site of an old nest of this species, now deserted, into which any child could walk. Dr. MacGillivray of Eoligary, Barray, had a tame White-tailed Eagle for some time, and this bird used to follow his sons in their rambles over the island.

W. MacGillivray, in 1840, speaks of a site somewhat similar to that in North Uist, above referred to, as existing at that time in Harris, "although there were lofty crags in the neighbourhood." They were still numerous in his time. Sea Eagles were reported also to Harvie-Brown to have bred on an island in the S.E. of Lewis in 1870, but he did not visit that eyrie. It is not yet out of the common to see a fine old Erne perched on a rock at the entrance to Stornoway harbour, as witnessed by W. Hubback in 1885 from the deck of the *Clansman*. In February 1879 a fine adult female was shot or trapped near Stornoway, and this, along with an immature male and female, killed respectively in March and June 1882, and also obtained near Stornoway, are now in the collection of Mr. Chase.

White-tailed Eagles have long since ceased to occupy the

[1] We think that their number may have been overstated to Gray even at that date, though, undoubtedly, they were infinitely more numerous than they are now throughout Scotland.

[redacted]

[illegible] May, 18[illegible] [illegible]

[illegible] to further [illegible] which [illegible] believed to [illegible]

[illegible] of this species, now [illegible] Dr. [illegible] and that [illegible]

[illegible] of a size somewhat [illegible] referred to, as existing at that time [illegible] were [illegible] in the neighbour[illegible] in his time [illegible] known to have bred on an island in the [illegible] but he did not visit the [illegible]. It is not [illegible] old [illegible] perched on a rock at [illegible] to Sto[illegible] by W. [illegible] in [illegible] the [illegible]. In February [illegible] fine [illegible] trapped near Stornoway, and [illegible] an immature [illegible] female, [illegible] respectively in March and June 1[illegible]2, and [illegible] near Stornoway, are now in the collection of M[illegible].

[illegible] the

[illegible] they were [illegible] that they [illegible]

Aonaig cliff of Mingulay: it is *forty years since they bred there*, as far as Mr. Finlayson could remember. Sea-Eagles, as migrants, used to frequent Mingulay much more numerously at one time.

In 1887 the Shiant Islands pair were still "to the fore," and gave our party a fine opportunity of watching all the phases of their flight. Long may they continue in their inaccessible retreat; and may the broken overhanging basalt columns, which project far beyond the giant ribs of similar structure down below, resist the tear and wear of time, and prove a sheltering roof to them. So far as we are concerned, we are as pleased with a feather ("tickled with a straw," if you like) which we picked up on the boulder-strewn beach below the eyrie, ay, and a great deal more than if we had shot the bird.

As still one of the characteristic birds of the Hebrides, we may be allowed to relate our personal experience there when procuring the eggs of this species so long ago as 1870, as follows:—

"After having taken two eggs from an eyrie of the Golden Eagle within a mile of the site of the White-tailed, and having failed to discover the exact position of the latter on the 10th April; having had also to contend with a terrific gale of wind, which came with irresistible force down the gullies of the Harris hills, and obliged us on several occasions to cast ourselves flat on the steeply sloping hill-side, almost overhanging the sea, we staggered home against the gale. Next day we started in a boat down the sea-loch in more moderate weather to renew our search. Arrived near the place, we landed, and shortly afterwards we saw the hen go on to the nest; but she would not move off again till we fired a shot, which was done in order to be perfectly sure of the exact spot. The men went above and Harvie-Brown went below. R. was let down. The place was much easier than yesterday's;[1] but the rope had to be worked considerably to one side of the eyrie. The nest was behind a corner, on the top of a green slope, which was directly overhung by the cliff, and so had to be approached horizontally for a space. The nest was two feet in diameter. R. brought down all the nest lining, which was composed of Luzula, dry grass, and moss, and a few sprays of eagle's down. At ten minutes past 8 A.M., R. took the only egg the nest contained, which, on blowing, proved fresh. The hen had returned four or five times close past the eyrie. After a time the cock came too,

[1] Another eyrie in Harris of the other species.

and both birds remained at no great distance all the time during our operations; and, whilst I was blowing the egg, the Golden Eagle, whose eggs we had taken the day before, also alighted within easy view on the summit of its own cliff. The two eyries and their birds, Golden and White-tailed, were thus visible from the same place simultaneously. . . .

"After taking the eggs of another White-tailed Eagle in another part of the island, and after a long tramp to and fro, and intervening searches for, and views of, other eyries, once more we started for the first-mentioned eyrie, hoping to take a second egg from the same pair of birds. This was on the 22d April, at 5 in the morning, and this time we rowed down again, as being more expeditious, because we desired to get on the same evening—a long drive over a very bleak country. We saw the cock come off a point of rock beyond the eyrie. We fired a shot, and the hen flapped hurriedly and clumsily off the nest. The men went above as before. R. descended with great ease, scarcely giving any directions. He had, however, to take off his boots when half-way down, as the rock was slippery after the morning rains. He slid quietly into the nest, and I could see him hold up the second egg above the edge of cliff which intervened between him and me. Both birds remained some time in the vicinity, but finally sheered off, and a Peregrine Falcon sailed swiftly past, and along the face of the cliff, within easy range of where I sat. The two Hooded Crows as usual tormented the eagles. R. told me, producing a leaf or two, that fresh Luzula lay plentifully in and around the nest. We now started to return, but almost a gale of wind with violent rain swept down the loch against us, and it was a hard united pull to reach the pier. The same night we started on a long drive about 10 P.M., and did not arrive at our destination till one in the morning of the next day."

In North Uist, during twenty-six years' experience in killing all kinds of vermin, the shepherd on Newton farm only succeeded in shooting two White-tailed Eagles, and these were the birds which bred in a low cliff in the north-east of the island, near Cheese Bay—a locality now entirely deserted.

We find a further note which seems to us to be of some interest, viz.: "Of eight eyries of Golden and Sea Eagles in the islands visited this year, whether occupied or unoccupied, seven were placed on rocks facing in a northerly direction. The eighth faced the east. In no instance was the cry of either species of eagle heard. In one case the female Sea Eagle returned and perched close to the nest, after having been disturbed by a shot, and in another, the same species *repeatedly* returned close past the nest, though the eggs, in both cases, proved to be fresh. But with the Golden Eagle circumstances were somewhat different, in one instance only showing for an instant or two to one of the party, and in another, that of the above narration, never re-appearing, although in both cases the eggs were partially incubated or hard-set. The mist, however, in the latter case was dense over our heads, and the view obscured.

Astur palumbarius (*L.*). Goshawk.

[Just as we go to press, we hear of a Goshawk obtained in the Hebrides, *now* in the possession of Mr. M'Leay, Inverness, but, up to date of leaving home—May 1, 1888—we have had no opportunity of examining it, so cannot say whether it is the European or American form. See Appendix.]

Accipiter nisus (*L.*). Sparrow Hawk.

Gaelic—*Seobhag* : *Seabhag* = a hawk : *An Speirag.*—A. C.

Professor Duns has the note that the Sparrow Hawk is common in Lewis, but the Kestrel is not so! Gray says that this species is probably confined to Lewis and Harris, except in winter, when a few range over the other islands. It was noted by Mr. Greenwood near Stornoway in 1879. As early as 1830, MacGillivray included it under its Gaelic name, as not rare in any of the larger islands. Harvie-Brown has not a single note of its occurrence, whereas he found the Kestrel far from rare. Harvie-Brown can hardly believe this species should not be bracketed.

Milvus ictinus, *Savigny.* Kite.

Gaelic—*Croman Luch* = the mouse huncher : *An Croman* = the huncher.—A. C.

[*Obs.*—According to W. MacGillivray, "very rare" in 1830. He includes it also in 1821 in a bare list of the birds of Harris; but see his *British Birds*, vol. iii. p. 269, in which he admits that he is not aware of its having been observed.]

Milvus migrans (*Bodd*). **Black Kite.**

Nauclerus furcatus (*L.*). **Swallow-tailed Kite.**

Gaelic—*Croman Gobhlach* = the forked huncher.—A. C.

Elanus cæruleus (*Desf.*). **Black-winged Kite.**

Pernis apivorus (*L.*). **Honey Buzzard.**

Falco candicans, *Gm.* **Greenland Falcon.**

Gaelic—*Seobhag Mhor* = large hawk : *Seobhag Mhor na Seilg* = the large hunting hawk : *Gearra Sheobhag* = the strong stout hawk.

Gray records one shot by Col. Gordon's keeper in South Uist (p. 21), and another in North Uist by Mr. John Macdonald in 1860.

MacGillivray says it breeds on St. Kilda—"several pairs," which it is needless to say is not borne out by subsequent inspection.

Sir John Orde records two Gyrfalcons (species ?) as having been obtained in North Uist (*in lit.* 30th Sept. 1886);[1] and Professor Duns records the Jer-falcon as "more frequently on the Flannen Isles"—presumably as a breeding species! This is quite a parallel paragraph to that of the Osprey (*infra*, q.v.).

Falco islandicus, *Gm.* **Iceland Falcon.**

Gray instances one shot at Vallay, North Uist, in 1865, by the late Major M'Rae, and afterwards in the collection of the late Dr. Dewar; a male, in October 1864, shot by Alan Maclean, at that time keeper there; and a third washed ashore about the same time in North Uist. It is spoken of as "rarely found" near Stornoway by Mr. D. Mackenzie. MacGillivray instances one, shot a few years previously, by Mr. D. Ardbuckle, and another seen in Pabbay by Mr. Nicholson of Berneray. From the description given by Mr. Findlay MacLeod—long time head forester in North Harris—to Harvie-Brown, we have little doubt that he shot an Iceland Falcon in Glen Meabhaig, in Harris. He described it as "nearly white, with black marks, and very big."

Falco peregrinus, *Tunstall.* **Peregrine Falcon.**

Gaelic—*Seobhag Ghorm* = blue hawk : *Seobhag Seilge* = hunting hawk : *An Lainnir* or *Lannair* = the gleamer : *Lainnir Sheilge* = the gleamer of the hunt.

[1] Vide *Ibis*, 1859, p. 469; and *op. cit.* 1861, p. 415.

NOTE.—*The root is* Lann, *a blade, a glaive, a sword.* Lainnir *might be translated 'the gleaming sword or lance,' in allusion to the colour of the bird, and its scimitar-like destructiveness.*—A. C.

Recorded as breeding in St. Kilda by John MacGillivray (who says nothing about the Greenland Falcon breeding there). "Seen," says Gray, "in all the islands of the O. H. (1871), and also in Haskeir; and several pairs on St. Kilda (*auct.* J. MacDonald of Newton)." Not very common in Mhorsgail, Lewis, where, however, Mr. H. H. Jones obtained eggs in 1881. Also recorded from the district of Stornoway by Mr. D. Mackenzie.

Many other localities occupied by the species of late years are noted in our books. The first record given is in 1830 by MacGillivray. W. MacGillivray names many previously tenanted eyries. Harvie-Brown has observed it in Barray and Mingulay, where there are at least three eyries; also in Harris, and records it as common throughout the O. H. Common on Carn shootings near Stornoway, and eggs obtained on the Flannen Islands in 1879 by Mr. Greenwood's gamekeeper. In North Ronay Swinburne found a pair near the south-west part of the island, where he considered they had a nest, "from the outcry they made when that portion of the island was approached." Harvie-Brown saw nothing of them in 1885, as time did not permit of that portion of the island being visited. Nor does Barrington appear to have met with them there either. But in 1887, as already mentioned in our chapter upon North Ronay, Harvie-Brown found them occupying the cliffs of the western horn of the island. This species holds its own still and is far from rare in Scotland, no doubt owing to its inaccessible retreats among these isles of the west and cliffs of the mainland.

Findlay MacLeod—long time head forester in North Harris—related to Harvie-Brown how, on one occasion, he saw a Peregrine Falcon swoop at a grouse, which, however, dodged under a rock, and "for more than two hours" the falcon remained on the watch for its re-appearance.

Falco subbuteo, *L.* Hobby.

Gaelic—*Obag: Gormag* = the little blue one.

[*Obs.*—There are many records, especially from north of Lewis, but *none* are trustworthy, the adult Merlins having been constantly mistaken for Hobbies. Mr. Greenwood, in all his letters and lists, was strong in the belief of the occurrence "in numbers" of the species, but all reason is against it, and reputed records *since then* are undoubtedly erroneous. We have had several opportunities of examining *female Merlins* sent as *Hobbies* from Greas in The Lews.]

Falco æsalon, *Tunstall.* Merlin.

Gaelic—*Speireag: Speireag Bheag: Speireag Bheag Bhuidhe* = the yellowish little hawk: *Speireag Bheag an Fhraoich* = the little heather hawk.—A. C.

From the earlier records apparently not very abundant. John MacGillivray speaks of "one or two seen" in 1841. On the other hand, another authority states, in 1832, that it is, "though once abundant, now rare." W. MacGillivray, as early as 1830, says: "Not very uncommon in some parts of the range," which we think is most descriptive. The Merlin has throughout Scotland held its own perhaps better than most of the persecuted Falconidæ. Gray includes it as inhabiting "all the Hebrides." Mr. H. H. Jones says: "A few seen in autumn on Mhorsgail, never in spring." Mr. D. Mackenzie includes it as breeding, but Mr. Greenwood *omits it altogether.* Even in MacGillivray's time it seems not to have been really abundant; but we suspect, from parallel circumstances in the counties of the north of Scotland, that it was much more abundant than authors have hitherto recorded. Certainly at the present time it holds its position well.

Harvie-Brown has himself taken their eggs in North Uist, and as an autumn migrant its occurrence throughout the group cannot be disputed.

Falco vespertinus, *L.* Red-legged Falcon.

Falco tinnunculus, *L.* Common Kestrel.

Gaelic—*Deargan Allt* = the ruddy of the burn.
(*See also* Red-necked Phalarope.)

Common or "not infrequent" in 1841, according to John MacGillivray. Is *not* common in Lewis, according to Professor Duns. In 1870, Lieut.-Col. Feilden took Kestrels' eggs on the ground—*not Merlins'*—amongst long heather in North Uist; and in 1879 the place was pointed out to Harvie-Brown by Stoddart,[1] who was then shepherd on the ground, and who accompanied Feilden. This habit is also shared by the Hooded Crow, which nests commonly in the heather, owing, in both cases, to a general absence of more suitable sites. In such a situation Harvie-Brown has *often* found the nest, and on one occasion called, by imitating their notes, five of the Hoodies close around him, to the amusement of our good friend Mr. J. MacDonald, who wished we had had a gun. He also has met with it very generally in the islands, and it is included in all the local lists received.

But it is curious to find W. MacGillivray record it as "by no means common" as early as 1830. It is certainly less common in the southern half of the O. H. than in the northern, and in 1830 MacGillivray's experiences were more intimately related with the southern than with the northern half, which no doubt accounts for the statement, at least in some measure.

Pandion haliaetus (*L.*). Osprey.

Gaelic—*Iolair Uisge* = water eagle: *Iolair Iasgaich* = fishing eagle: *Iolair Iasgair* = fisher eagle.—A. C.

MacGillivray never met with it. Gray considers it as a rare straggler to the O. H., but records one at Barray. Professor Duns considered it "more common in Lewis than either the Golden or White-tailed Eagle," and adds the information that it breeds in the Shiant Isles. The true Osprey is reported as a rare passing migrant, very irregular in its visits at Tarbert.

Order 3. STEGANOPODES.

Family PELICANIDÆ.

Phalacrocorax carbo (*L.*). Cormorant.

Gaelic—*Sgarbh an Uc-ghil* = white-breasted cormorant (*i.e.* immature): *Orag: oragann* (*i.e.* "Odharag" = the young scart): *Am Fleigire* = the flecked one: *Am Ballaire-bothain*: *An Sgarbh-buill* = the scart of the spot.—A. C.

[1] Mr. George Stoddart, sheep-manager, Newton, North Uist, an observant, intelligent, and trustworthy man.—A. C.

MacGillivray speaks of this species as resting in large flocks on a rock in the Sound of Harris during the winter months, where they were captured in considerable numbers by natives well acquainted with the place.[1]

Recorded also in 1841 as rather plentiful, breeding, along with the next species, at St. Kilda, Shillay, and Towhead. At the present time it is looked upon as rare in Loch Roag, Lewis, whatever may have been the case formerly. In 1830 considered not rare on some of the smaller islands, but less common on the main range, the next species taking its place there to a great extent. Both this and the Shag are common on Haskeir, off North Uist, keeping, however, in separate colonies (see description of Haskeir), or rather upon the more western group of islets known as Haskeir-Eagach.[2] In 1881, when Harvie-Brown visited these islands for the first time, a colony of this species occupied the rock *fourth from the west*, whilst a colony of about fifty pairs of the Shag occupied the rock next to them. It is evident from the earlier notes of Captain Elwes (1868) that this order was then reversed. We have often before observed the habit which Cormorants have of shifting their ground when disturbed, or their nests tampered with.

Several times we have sailed past the rocks known as the Gloraigs su Taransay, near the entrance of Loch Reasort, and have usually observed that the principal rock held a colony of the great Cormorant, but never made out distinctly whether it was a *nesting* or only a *resting* colony, on account of the difficulty of landing, or even approaching, the sea around bristling with sunken rocks and overfalls. Cormorantries are decidedly not abundant on our coasts.

Phalacrocorax graculus (*L.*). Shag.

Gaelic name. W. MacGillivray apparently did not distinguish the Scarts, but called the Common Shag *An Sgarbh*, and the summer plumage or Crested Shag *An Sgarbh Beag*= little scart: *Am Fitheach uisge*=the water-raven: *Sgarbh an Sgumain*=the crested scart.—A. C.

Much commoner than the last, according to most authors, and recorded as very abundant everywhere. Sir W. E. Milner found

[1] Up to date of 1886 we have had no further account of this locality.

[2] The notched rock of the ocean.—A. C.

it at St. Kilda in 1845, and took eggs. Fully noted by Gray, whose remarks are worth perusal (*Birds of West of Scotland*, p. 457).

The Scarbh or Shag is permanently resident, gregarious, and a cave-dweller, great numbers fishing in the sounds of Harris and the Uists. Not less than 200 to 300 Green Cormorants are present in the cave of Liuir, on the west side of Harris, in the breeding season, and it is very abundant around the west coasts of Lewis at Mhorsgail. MacGillivray also speaks of the cave on the western extremity of Capval, and mentions that he has seen and counted 105 in one flock, many others being under water at the time.

On the 24th May 1888 we had an opportunity of closely inspecting the whole rocky coast of the western extremity of Capval, on both sides of Towhead (Cape Difficulty of the charts), from the sea, and firing shots at the entrances of all the caves, of which there are at least six or eight in number. We found Uamh Liuir as marked, under the cliffs of Liuri (*sic*), on the Ordnance Survey map, below the cliff-edge contour of 300 feet. On firing a shot, only some five or six Rock Doves and a few pairs of Scarts flew out. During the whole course of our row along shore—about three miles—we did not see, certainly, as many Scarts as in MacGillivray's *lowest* total, *counted* by him; and Cormorants were *entirely* absent. Nor can we conceive that the cave of Liuir in its present condition could possibly afford foot-room for more than a few pairs. Unfortunately we could not land, owing to the Atlantic roll being too heavy, but we were assured by several different natives at Obb, that the cave penetrates very little further than it is seen to do from the sea. It is, however, quite possible that, since MacGillivray visited the cave, great changes may have occurred, both as regards its dimensions and its capabilities otherwise, as there are many fallen blocks at its entrance; and it may also be the case that the continuous netting of the birds by night, by torchlight, has exterminated the Cormorants' colony altogether, and decimated both Doves and Shags. We are particular in giving these minutiæ, as MacGillivray seems to have "set great store" by this cave in his time. MacGillivray also refers to *vast* numbers of Starlings frequenting "a cave on the west coast of one of the Hebrides about 1820." If this Uamh Liuir is the cave referred to, we can only say not one Starling was seen on any part of the coast by us in May 1888 [*vide* text, under Starling, p. 68].

[illegible] ante in an S[illegible], [illegible] ous parts, [illegible] superior [illegible] [illegible] towards [illegible]

[illegible] the [illegible] her [illegible] [illegible] the [illegible].—A. C.

[illegible] at dawn [illegible], and [illegible] (J. MacGillivray). The [illegible] viewed from [illegible] feet of [illegible]. [illegible] it is very interesting [illegible] ever [illegible] in S[illegible] of the [illegible] St. Kilda." They [illegible] off G[illegible] Head [illegible] in imitation of the [illegible] favorite food, though they [illegible] of the coast [illegible] Kilda, and [illegible] East [illegible], [illegible] than the [illegible]. It far surpassed all pictures of [illegible] the two Garnet [illegible] which were covered [illegible] with [illegible].

[illegible] was [illegible] are [illegible] in active year [illegible] Craig put together."

SULISGEIR

Abundant on North Ronay, as generally observed, to the exclusion of the last species, and fairly common on Suliagoir; but evidently crowded out of the likeliest situations purely and simply by the superabundance and moral superiority (?) of the other rock-haunting species. This appears to Harvie-Brown to be a well-ascertained fact.

Sula bassana (*L.*). Gannet.

Gaelic—*Amhasan: Amhasag: Asan: Sulaire*=the eyed or the eyer: *Ian Ban an Sgadan*=the white bird of the herring: *Ian Glas an Sgadan*=the grey bird of the herring: *Guga*=the young of the species. Also *Sulaiche*=the watchful-eyed, which may give the root *sula*.—A. C.

The two MacGillivrays speak of it in almost similar terms: breeding only in St. Kilda, entering the sea-lochs at dawn to fish, and returning to St. Kilda in the evening (J. MacGillivray). The Gannets of Boreray, St. Kilda, as viewed from the sea, give the rocks the appearance of a chalk cliff—800 feet of Gannets.

Gray's article on the Gannet is very interesting (p. 460). Fleming says: "No instance had ever occurred of a Solan Goose reposing on any of the shores of the Hebrides, but only seen fishing, or flying to and from St. Kilda." They frequented Loch Reasort in autumn, and are seen off Gallon Head in spring (*auct.* H. H. Jones). This is no doubt an indication of the movements of herring or herring-fry, their favourite food, though they also feed freely on "cuddies," or the young of the Coal-fish.

In 1879, when we visited St. Kilda, and sailed past Boreray, nothing struck us more forcibly than the marvellous colony of Gannets on the latter island. It far surpassed all pictures of our imagination. We passed close under the two Gannet stacks, Stack-an-Armin and Stack Lii, which were covered along their whole slopes and summits, and far down their sides, with countless and closely terraced rows of Gannets.

On Suliagoir, near North Ronay, *they swarm*. Swinburne speaks of "2000 to 2500, and in some years as many as 3000, birds being taken, more than double the number which are obtained annually from the Bass Rock in the Firth of Forth—where in some years not more than 800 are obtained—and Ailsa Craig put together."

[illegible] ant on North [illegible] [illegible] observed, to the [illegible] of the last species [illegible] is common on St [illegible] [illegible] [illegible] out of the [illegible] st situations [illegible] and [illegible] [illegible] ance and [illegible] superiority (?) of the [illegible] [illegible]. This appears to [illegible] to be [illegible]

[illegible]

[illegible] *Asseig*, *Amhasg* [illegible] the [illegible] or the [illegible] *ban* on St Kilda the white bird of the herring [illegible] the grey bird of the herring [illegible] the young of the species. [illegible] the [illegible] [illegible] give the root [illegible] — A. C.

[illegible] speak of it in almost similar terms [illegible] the [illegible] at dawn to fish, and [illegible] in the evening (J. MacGillivray). The [illegible] St Kilda, as viewed from the sea, give the [illegible] of a [illegible] feet of Gannets.

[illegible] the Gannet is very interesting (p. 460) [illegible] had ever occurred of a Solan Goose [illegible] of the [illegible] but only [illegible] from St Kilda." They frequented [illegible] off [illegible] Head in spring [illegible] an indication of the movements of [illegible] food, though they also feed [illegible] the young of the Coal fish.

In [illegible] St Kilda, and sailed past Boreray, nothing [illegible] than the marvellous colony of Gannets [illegible]. It far surpassed all pictures of our [illegible]. We passed close under the two Gannet stacks, Stack-an-Armin and Stack Lii, which were covered along their [illegible] and [illegible], and far down their sides, with countless and closely terraced rows of Gannets.

On Sulisgeir, near North Rona, *[illegible]*, Swinburne speaks of [illegible] to 2500, and in some years as many as [illegible] birds being taken, more than double the number which are obtained annually from the Bass Rock in the Firth of Forth, where in some years not more than [illegible] are obtained [illegible] and Ailsa Craig put together."

SULISGEIR

Our visit to Sulisgeir in 1887 will ever be remembered as one of the most remarkable incidents of our Hebridean peregrinations during many years, more especially if we look to the ghastly lonesomeness and geologically disintegrated nature of the whole place; almost pathetically sad in its collection of rough stone huts, the solitary wretched sheep, and the remains of another, and the heads of defunct Gannets strewn all over the surface, as already described, partly, in our introductory chapters.

Order 4. HERODII.

Family ARDEIDÆ.

Gaelic—*Corra-riabhach: Corra-ghlas.*

Ardea cinerea, *L.* **Common Heron.**

Gaelic—*Corra-sgriach* = the screeching heron: *Corra Ghribheach* = the billed heron: *Corr.: Corra-ghlas* = the grey-billed heron.—A. C.

From the earlier records downwards all authors are unanimous that this is not a breeding species in the O. H. Captain Elwes is at a loss to understand why it does not breed in these islands. Duns says, "Often seen in winter; very rare at any other season." Mr. H. H. Jones cannot hear of any breeding-place in Lewis, and they are never seen at Mhorsgail in spring. M. Greenwood says, "They are not uncommon near Stornoway, but no notice occurs of their breeding" (1879). Gray is silent as regards its occurrence *at any season.* Harvie-Brown has found them commonly on the rocky shores of the northern half of the group in summer.

Ardea purpurea, *L.* **Purple Heron.**

Ardea alba, *L.* **Great White Egret.**

Ardea garzetta, *L.* **Lesser Egret.**

Ardea bubulcus, *Audouin.* **Buff-backed Heron.**

Ardea ralloides, *Scop.* **Squacco Heron.**

Ardetta minuta (*L.*). **Little Bittern.**

Nycticorax griseus (*L.*). **Night Heron.**

Botaurus stellaris (*L.*). **Bittern.**

Gaelic—*Corra-Ghrain* : *Bubaire* : *Graineag* : *Am Buirein* = the lowing bird. *Am Buiriche* = the lower.—A. C.

[Gray mentions one shot in North Uist a few years previous to publishing his work (*Birds of West of Scotland*, p. 279), but had then no other record. At present we have no corroboration of it.]

Family **CICONIIDÆ.**

Ciconia alba, *Bechst.* **White Stork.**

Ciconia nigra (*L.*). **Black Stork.**

Family **PLATALEIDÆ.**

Platalea leucorodia. *L.* **Spoonbill.**

[*Obs.*—Gray has some rather vague remarks regarding birds of this species seen in the O. H., where, especially at Barray, they seem to occur occasionally, and five were said to have been seen together at one time. We find the record that three Spoonbills were killed in one day in Barray ("Admiralty Sailing Directions for the West Coast of Scotland, Part I. Hebrides,") and quoted from earlier editions (*op. cit.*, 1885, p. 3); but on whose authority it is stated we have been unable to find out.]

Family **IBIDÆ.**

Plegadis falcinellus (*L.*). **Glossy Ibis.**

Order 5. ANSERES.

Family **ANATIDÆ.**

Gaelic general names—*Giadh* = a goose : *Ganradh* = a gander : *Giadh-ban* = a white goose : *Giadh-glas* = grey goose (greylag ?).

Anser cinereus, *Meyer.* **Greylag Goose.**

Gaelic—*Lia-Ghiadh* : *Glas-Ghiadh* = grey goose.—A. C.

In the MacGillivrays' time (though mis-identified by them as the Pink-footed Goose) this species was found breeding in large numbers in the Sound of Harris, and also in the interior of North Uist. Observed in flocks in the beginning of May, in pairs in the middle of May, and the young were flying by the end of July, and by the middle of August had collected again into flocks. By 1848, when Sir W. E. Milnor found them on Loch Langabhat in Lewis, he reports them as fast decreasing and leaving the O. H., being much disturbed. This he relates also of the Sound of Harris, where only a few pairs were seen. On 5th June nests on Loch Scatavagh had been robbed and birds had left.

The oft-repeated error of MacGillivray (vol. iv. p. 592), arising from his mis-identification of this species, is remarkable. The changes were rung in Sutherland by Selby and Sir Wm. Jardine, only to be corrected by later writers. The whole chapter by MacGillivray must be read as referring to the Greylag in the O. H. and to the Pink-footed in the East of Scotland.

At the present time, however much reduced they may have become since then, they still breed in most of the O. H., are common in Benbeculay and South Uist; also in Lewis and Harris in large flocks. It is a resident species. Many have been tamed and are kept at the farm-houses; and a large flock of hybrids, and tame geese and Greylags may be seen at Rodel in Harris.

As instancing the marvellous regularity with which certain wild-geese visit the same grounds, there is a small patch of turf on the road between Stornoway and Garynahine in Lewis, upon which Greylag Geese are seen almost (if not quite) daily in many successive seasons, both in summer and winter (v. Mr. Hubback's notes). Besides flocking in spring and autumn, Greylag Geese are seen in flocks even in midsummer. These are either males, barren birds, or birds not old enough to breed, but it appears most probable that they are males and moulting birds, as we also constantly observe flocks of male Eiders and Red-breasted Mergansers at the same season.

At Rodel is a large flock of Greylag Geese, which represents the successive progeny of a few young birds reared in the garden by the late Lieut.-Col. MacDonald, late factor to Lord Dunmore.

Amongst them are several crosses with domestic geese. These are very much whiter about the tail and wings, but appear to retain the plumage of the Greylag in other parts, even in the legs and bill. They are larger birds, but fly equally well with the purer wild ones, and have a sprightlier carriage than domesticated geese. At times the whole flock repairs to the rich grazing of Killegray Island in the Sound, coming back to Rodel Bay and the vicinity of the houses in the evening.

About the 12th July, when the young geese have grown large and the old geese feel the moulting season approaching, old and young repair to certain well-known haunts to complete the moult in the old, and the growth of quills in the young, birds. These localities are *now* usually sea-girt isles with a certain amount of good grazing, though they do not require so much food, nor eat so abundantly at this season. Formerly they often betook themselves to the mosses of the larger islands, but *continuous* persecution, besides diminishing their numbers, has at last driven them to still more inaccessible shelters at the moulting-time. Many of these places, though known to us, we decline to make public. These facts concerning the decrease cannot be "rubbed in" too hard on those whose duty it is to protect them, whether Scotch lairds or English shooting tenants.

In winter, however, when the geese can usually take fairly good care of themselves, they have other and more extensive pasture-grounds, and are the cause of no inconsiderable loss to the farmer, cropping the short sweet grass of the machar, or the pasturage of the larger islands in the sounds of Harris and Barray. Favourite resorts at this season are Killegray and Lingay in the Sound of Harris. Almost any day during the winter a considerable body of Greylag Geese may be found on the seaward slope of Lingay. But now nearly all the crofters carry guns, and, in their short-sighted folly, are rapidly exterminating not only the geese, but also the seals, and all the other wild creatures which, to the southerner, make their cold grey rocks so charming, and bring an annual shower of English gold into their country. They will discover their error when at last too late.[1]

[1] Even as we now write, we hear of the senseless attacks upon the deer by the crofter population in Lewis.—(Daily papers, Nov. 22, 1887.) The short-sighted policy of these imperfectly educated, poor people, led by heartless agitators, for their own immediate and selfish purposes, is lamentable in the extreme.

Anser segetum (*Gm.*). **Bean Goose.**

Gaelic—*Muir-ghiadh* = sea goose.—A. C.

[There is still some uncertainty as regards the occurrence of this species in the islands.]

Anser brachyrhynchus, *Baill.* **Pink-footed Goose.**

[Included as breeding by J. MacGillivray, and the error perpetuated by W. MacGillivray; but it occurs sparingly in winter in October and November (Gray). Mr. MacDonald never found it.]

Anser albifrons (*Scop.*). **White-fronted Goose**

Gaelic—*Giadh bhlar* = white-faced goose.—A. C.

Reported always as rare. Rare in North Uist: one was obtained at Newton in 1856, and another in 1868. It is, however, common in Islay, and others of the Inner Hebrides. A pair were shot also at Loch Maddy, and presented to the Kelvingrove Museum by Mr. Ewen Maclean, Loch Maddy.

Bernicla brenta (*Pall.*). **Brent Goose.**

Gaelic—*Giadh-got*, or *Got-ghiadh*.—A. C.

The Brent is a less certain visitor to the O. H. than the Bernacle, and of more easterly distribution on migration. A few, however, appear at Loch Bee in South Uist, and also small flocks at Benbeculay and North Uist. To the *earlier* records it seems advisable not to attach too much importance, but they occur at Lingay shore in the proportion of about 7 to 10 of the Bernacle Geese.

Bernicla leucopsis (*Bechst.*). **Bernacle Goose.**

Gaelic—*Leadan: An Cathan*.—A. C.

We find this species is not noted by J. MacGillivray, so possibly the *earlier records* of the previous species are at fault. The Bernacle is very common in the O. H. They arrive in immense flocks in October, and pitch on the open sands of the fords before leaving. Gray gives a curious account of the order of their going, which we think it desirable to quote here in full, more especially as we have had the same account, over and over again certified to, by many intelligent natives of the isles, as well as by Mr. John MacDonald of Newton.

"Being a strictly migratory species, it takes its departure about the end of April or the beginning of May, by which time the Greylag Goose has commenced laying. Previous to leaving, the Bernacle Geese assemble in immense flocks on the open sands at low tide, in the Sounds of Benbeculay and South Uist; and as soon as one detachment is on the wing it is seen to be guided by a leader, who points the way with a strong flight northwards, maintaining a noisy bearing until he gets the flock into the right course. After an hour's interval, he is seen returning with noisy gabble alone, southwards, to the main body, and taking off another detachment as before, until the whole are gone. A notice of this singular habit was first communicated to me by Alexander Carmichael, and has since been corroborated by Mr. Norman MacDonald, who informs me that the inhabitants of the Long Island have been long familiar with it." How do exponents of the theory of "hereditary instinct" account for this?

MacGillivray speaks of the islands of Ensay and Killegray, in the Sound of Harris, as a long-established haunt of Bernacle Geese, and relates an amusing episode he experienced in the former, whilst employed in a chase after a flock (*History of British Birds*, vol. ii. p. 626).

Bernacle Geese assemble annually in vast numbers on the strand of Lingay just below Newton. This would appear to be their final halting-place before taking their departure in spring. They arrive at this place, *always* coming from the south,—*i.e.* not far from the gathering-place of the Greylag Geese before alluded to, but on the *strand*, not on the *grass slopes*.

Bernacle Geese frequent Mingulay to the number of 600 or 700—Mr. Finlayson says, "or indeed more"—between November and April, and then leave. In 1888 a flock of 100 were seen, however, flying over Monach in an almost north direction, or heading in the direction of St. Kilda, as late as middle of May.

Harvie-Brown has met with large flocks of Bernacle Geese in the Sound of Harris in the first week in May.

Bernicla ruficollis (*Pall.*). **Red-breasted Goose.**

Chen albatus (*Cassin*). **Cassin's Snow Goose.**

Cygnus olor (*Gm.*). **Mute Swan.**

[*Obs.*—Mute Swans have been introduced at Rodel in Harris, and Balelone in North Uist—the latter by the late parish minister—and they breed near there every year; eggs taken there in 1887 had been sent to Mr. F. P. Johnstone, Castlesteads, in Cumberland. In 1888 there was a nest also at Loch Hosta, but the eggs were taken by boys. The birds at times extend their flight as far as the branches of Loch Maddy, where they are occasionally seen by our friend Sheriff Webster.]

Cygnus immutabilis, *Yarr.* **Polish Swan.**

Cygnus musicus, *Bechst.* **Whooper Swan.**

Gaelic—*Eala*=a swan: *Eala-bhan*=white swan: *Eala-ghlas*=grey swan (cygnet).—A. C.

J. MacGillivray notes this species as occasionally visiting St. Kilda, on the authority of the Rev. —— Mackenzie. W. MacGillivray speaks of it as generally abundant; and later accounts testify to its regular appearance. Swans are noted at or near Stornoway by Mr. D. Mackenzie, and at many other localities on spring migration by Gray. Gray says they generally arrive on winter flight in November, though sometimes earlier. As many as 400 have been seen at one time on Loch Bee in South Uist, which never freezes over. As they fly low, they are often shot. "About the middle of April the noble congregation breaks up," continues Gray, "into detachments, as the Bernacle Geese are known to do, and after much sounding of bugles, summoning the feathered host into the air, they soon get into their line of flight, and are afterwards seen at a great height steering for their northern home."

In 1879, when Harvie-Brown was staying at Newton, North Uist, he was informed by Mr. John MacDonald that a pair of Wild Swans had been seen a week previously—*i.e.* previous to the 17th May—which was considered to be a very late date indeed—the latest known at that time. Between seventy and eighty passed over Newton towards the south-east, and eight wintered on the loch at Scolpig: some about the same time wintered at Tiree, but perhaps the larger portion of this flock passed on to Loch Bee in South Uist—their favourite haunt.

Cygnus bewicki, *Yarr.* **Bewick's Swan.**

This species is also common, and frequents the same localities with the last. Gray, however, appears to have been the first writer to take definite notice of it in the O. H. Since then it has been frequently observed and obtained. Mr. Hagan has met with it at Gress.

Tadorna cornuta (*Gm.*). **Common Sheldrake.**

Gaelic—In Uist is called *Uist nan cra-ghiadh: Cra-ghiadh.*—A. C.

The apparent increase of this species is somewhat interesting. In 1830 it is spoken of by MacGillivray as "not uncommon." In 1848 Sir W. E. Milner speaks of only a few inhabiting North Uist and the Sound of Harris; but, of course, he did not visit many of the southern islands of the O. H. Later, it is noted as common on all the islands of the Sound of Harris, except Berneray, Pabbay, and Shillay, and plentiful in many parts of Benbeculay and the two Uists. Gray records it as "very common, but only a summer visitant, over the whole O. H., and often kept tame at the farm-houses." An interesting cross took place at Scolpig in North Uist between a common domestic duck and a Sheldrake: it is a pity that none of the brood were saved. Mr. MacDonald described the duck as an uncommonly marked bird, being of a curious dun and yellow colour, the only one of the kind in the farmyard flock; hence doubtless the partiality shown to it by the Sheldrake.

In 1870, whilst excavating a nest of this species—the bird of which had been watched to its nesting-hole on an island the day before—it was found that the nest and ten eggs lay quite 9 feet from the entrance, and that the tunnel in the peaty soil in which they lay described an almost complete circle. The eggs, which were only slightly incubated on the 9th May, were laid upon a large mass of the bird's own down; they were quite warm when taken, and the bird itself must have retreated into one of the side branches of the tunnel, or escaped unobserved by another exit.

Tadorna casarca (*L.*). **Ruddy Sheldrake.**

Anas boscas, *L.* **Mallard. Wild Duck.**

Gaelic—*Lach, Lacha: Lacha-ghlas, glas-lacha: Lacha-riabhach* and *Lacha-ruadh* (Uist): from *riabhach*=brindled: *ruadh*=red, ruddy: *Lacha-chinn uaine*=the green-headed duck.—A. C.

In 1830 W. MacGillivray speaks of it as "not very common," and another author as "not common." Gray considers it "very abundant on all the inner and outer groups." Mr. H. H. Jones speaks of it as breeding abundantly at Mhorsgail in Lewis, but "all leave the ground in autumn."

Harvie-Brown did not find it common in either Harris or North Uist in the breeding season of 1870, nor upon any occasion since: single males were seen on May 2d.

Chaulelasmus streperus (*L.*). Gadwall.

Gray mentions one—a male—shot at Barray in the winter of 1863; two males and females were shot by Dr. MacRury in Benbeculay in March 1864; and twelve were also seen in a flock at Barray in 1868 (*op. cit.* p. 367). We are also aware of a route followed by this species, we believe regularly every winter, not far removed from the Long Island.

Spatula clypeata (*L.*). Shoveller.

[*Obs.*—There appears to have been no positive record of this species, but in 1871 Gray records the occurrences of "ducks with broad bills," frequently reported to him by some of his Hebridean correspondents, but these may have been Scaups.]

Querquedula crecca (*L.*). Teal.

Gaelic—*Crann-lach*, *Crion-lach*=stunted duck.—A. C.

In 1830 the Teal is recorded as having been shot in Lewis by the Rev. Alexander Simpson. Is said to have bred in South Uist as early as 1841, but W. MacGillivray did not get it there. Occurs sparingly in Lewis and Harris, but Gray never found it breeding. Seen at Mhorsgail in spring, "and," says Mr. H. H. Jones, "no doubt breeds there," but by 1881 he "had never found the nest."

MacGillivray speaks distinctly of the Teal as "extremely rare in the Hebrides," but it had been seen in Lewis, and more frequently in Skye. We have never ourselves met with it in summer, but there can be no reasonable doubt that it occurs sparingly in winter,—accepting the evidence of "Sixty-one" (*Reminiscences*, p. 188),—and Sheriff Webster, who has shot a few during most of the winters he has resided in North Uist. Mr. John MacDonald considers them numerous, and they are doubtless fonder of the west side than of the east side of North Uist.

Querquedula circia (*L.*). Garganey Teal.

Dafila acuta (*L.*). Pintail.

Gaelic—*Lacha-mhara*=sea duck: *Lacha-stuach*=wave duck: *Lacha-stiurach*=rudder-duck.—A. C.

Only, up to 1871, was one instance of the occurrence of the Pintail known to Gray in the O. H. The bird was shot on the farm of Milton, South Uist, by A. Carmichael, in the winter of 1869-70. In the *Proceedings of the Royal Physical Society of Edinburgh* for April 1880, Gray notes the occurrence of a flock in North Uist, and one was shot.

We have elsewhere recorded in detail the actual breeding of this species on one of the Inner Hebrides, where there seems every reason to believe it had both bred before, and has done so again, since Professor Heddle and Harvie-Brown discovered it there in 1880.[1]

Mareca penelope (*L.*). Widgeon.

Gaelic—*Glas-lach*=grey duck: *Lacha Lochlannach*=Scandinavian duck.—A. C.

W. MacGillivray wrote of this species first in 1852 (*British Birds*, vol. v. p. 87), and says, as he believed, that it never had occurred in the O. H. Whether in this he was correct, or that he was unaware of its advent in winter, is not now easy to say, but it is difficult to believe that such a conspicuous and boldly-marked

[1] That this may originally have been an escape is of course possible, but the record must stand, as it would now be almost impossible to ascertain that point. Certain it is that they were not escapes, at all events from Rodel, as Pintails were not introduced there till long after this discovery, nor are we aware of any private lochs or ponds elsewhere, at least in the west, where Pintails had, or have since, been introduced. The situation, an almost barren rock in the sea, is somewhat different from those in which Alston, Seebohm, and Harvie-Brown found their nests in North Russia in 1872 and 1875. It still remains to be seen if any appreciable extension from this centre of the species takes place. The paper in which the original account occurs is called "Haskeir, off Canna; and its Bird-Life," etc., and we possess in our Egg-book the *feather and down* which led to its positive identification. We may reproduce the whole evidence at another opportunity.

bird could have escaped all notice, even at that season. Be that as it may, even the history of past omissions, or of past errors, is worthy of record in such a chronologically-arranged account as this professes to be. At all events, in 1871 Gray disposes of the past by the distinct records: "that it is abundant in winter over the whole of the Long Island, crowding many of the shallower lakes of South Uist and Benbeculay, and it is likewise a very common bird in almost every district on the western mainland." Gray had never taken the nest of the Widgeon in the O. H., but had "little doubt of at least a few pairs nesting there regularly,"—pairs having been reported by correspondents, "at the season when all other water-fowl are breeding." Harvie-Brown has never in all his experience of the Long Island obtained any definite evidence of its breeding, although pairs have been reported as occurring in summer.

Fuligula ferina (*L.*). Pochard.

Gray gives no positive record of the Pochard in the O. H., but says it is common on many of the Inner Hebrides. Mr. D. Mackenzie, however, includes it from Stornoway. We are not aware of any place among the Inner Hebrides where at present it can be considered "common." Probably Gray thought of Mull when he wrote thus.

Fuligula rufina (*Pall.*). Red-crested Pochard.

Fuligula marila (*L.*). Scaup.

Not very abundant, but a few are found frequenting the likelier localities, especially the oozy and sandy shores between Benbeculay and North Uist; also between Grimisay on the east, and Baleshare on the west, and they occasionally take shelter in the Sound of Harris in very stormy weather.

Fuligula cristata (*Leach*). Tufted Duck.

Gaelic—*Lach an Sgumain: Sgumalach*=crested duck.—A. C.

Occurs sparingly in the O. H., and is sometimes seen as early as September: Gray himself saw it in North Uist. Considered by MacGillivray in 1852 as much rarer north of Clyde and Tay than south of these Firths. Now, in 1887, and for many years past, the account of its distribution in Scotland is very different indeed. It had not been recognised in North Uist by Mr. John MacDonald up to 1888.

Nyroca ferruginea (*Gm.*). **White-eyed Duck.**

Clangula albeola (*L.*). **Buffel-headed Duck.**

Clangula glaucion (*L.*). **Golden Eye.**

As early as 1830 was reported by MacGillivray as "visiting the lakes in winter, and is usually seen in pairs in Harris not after the beginning of May, and always on small fresh-water lochs.

Gray quotes MacGillivray, and adds that stray examples sometimes linger in Benbeculay and North Uist till the beginning of May.

Mr. Greenwood includes it in his list, but not as common, at or near Stornoway. The Rev. H. A. MacPherson, whilst on a short visit to North Uist in 1886, tells us that he had, on July 10th, an "excellent view of a Golden Eye, as it rose within easy shot from a small deep pool, and went off to a large loch. It appeared to be a male of the year. The edge of this loch was quite bare, and was not more than five miles from Loch Maddy." Mr. John MacDonald speaks of it as "common in winter and spring."

Harvie-Brown has never met with them so late as July, but on the 2d May 1870 he saw five or six on a small fresh-water loch in one of the islands of the Sound of Harris, when, along with Mr. John MacDonald of Newton, he was in quest of seals.

Cosmonetta histrionica (*L.*). **Harlequin Duck.**

Harelda glacialis (*L.*). **Long-tailed Duck.**

Gaelic—*Ian Bochainn: Ian Bachainn*=the ocean bird: *Ian Bochainn*=the melodious bird.—A. C.[1]

Very common in small flocks in the Sound of Harris in the early part of summer, as recorded in 1830 by J. MacGillivray. Gray

[1] "From a fancied resemblance in its voice to the name M'Candlay—a family name in Berneray—it is called 'M'Candlay-cun' (*i.e.* M'Candlay's bird). It is considered by the natives as a harbinger or herald of winter. (Compare 'Coal an' candle licht' of the East of Scotland.)" The Gaelic names in Berneray are often peculiar to that island.—J. A. H.-B.

says, "very common in the Sound of Harris from October to March." A few are seen near Stornoway, as noted by Mr. Greenwood in 1870.

Somateria mollissima (*L.*). Eider Duck.

Gaelic—*Lach mhor: Lacha Heisgeir* (so called in North Uist): *Colc: Colcach*.—A. C.

The following is the chronological sequence of the records of this species. At the time of the MacGillivrays, whilst there were astonishing numbers on Haskeir—giving to them a very local habitation *and a name*—only small numbers were found on the uninhabited islands of the Sound of Harris, and Sir W. E. Milner, in 1848, corroborates this remark.

Gray says in 1871, "very abundant in the Sound of Harris," and "sometimes" on the moors of the interior (p. 378). At that time their visits to the fresh-water lochs were *not* of long duration. At Mhorsgail Mr. H. H. Jones reports Eiders as breeding "plentifully by 1881" on many of the islands of the west coast of Lewis. Harvie-Brown's experience of the O. H. teaches him that the evidence of the increase is decided and undoubted, and is borne out by the authority of many witnesses now living. The increasing numbers have pushed through the Sound of Harris eastward, and are now invading the north-west and west coasts of Skye, and have even descended far down the coast of Ross-shire.[1]

In further evidence we may adduce the following: Mr. John MacDonald of Newton well remembers when Eider Ducks nested only by the seashore, and mostly upon islets of the salt water. Allan MacLean, who had resided in North Uist for forty-five years, and was gamekeeper there previous to 1879—and who died in 1887—distinctly remembers when *not one Eider-duck nested in the interior*, and indeed he equally well remembers when they were entirely confined to North Ronay and Haskeir.

In 1879 it was reported to Harvie-Brown that it was only twenty years previous to that date that they had *begun* to occupy the interior of North Uist. At the present time they breed all

[1] As regards the latter extension, it is quite possible, however, that this has received its impulse from a very old breeding centre on one of the Inner Hebrides, known long prior to 1852, the birds at that time being distinguished as "Colonsay Ducks."

over the interior in large numbers, often far from water, and frequently at a considerable elevation. By 1879, indeed, Eiders had quite populated the east coast of Harris.

At Haskeir off North Uist, where, in 1881, Harvie-Brown found this species very plentiful, the nests were generally placed in a warm corner, sheltered on the north and west, and facing the south and east; and this rule seems generally to hold good both here and on Shillay, and elsewhere among the Hebridean islets. He found some fifteen to twenty nests in all. One nest contained nine eggs, but most of the others from three to four. When startled off the nest the Duck has a very nasty habit, which entitles her—according to an old Scotch proverb—to have the epithet bestowed upon her of a "foul bird." As is well known to most collectors, the stench then raised is very pungent indeed.

According to old accounts, North Uist and Haskeir appear to have been the centres from which the population now occupying the O. H. has spread in comparatively recent years. Some idea may be formed of their numbers, at least in autumn, when it is stated that "not less than 3000 (*sic*) Eiders sheltered upon the west coast of Harris during east winds in the autumn of 1886 within four miles extent of coast-line, whereas the year previous there was not one" (Jas. Cowan, Esq., *in lit.*).

Very abundant in North Ronay, as mentioned by all observers. Barrington found very few males there, however, by the end of June.

A few Eiders even frequent that far-off "islet o'er the sea"—Sulisgeir—as we saw them there in June 1887, though we did not stumble on any of their nests.

Mr. Henry Evans of Jura deer-forest remarks upon the large size of the mussels swallowed, shells and all, by Eiders, many being quite 2½ inches in length (*in lit.*).

Somateria spectabilis (*L.*). King Eider.

[*Obs.*—We have no positive records of this species. There seems to be a possibility of its having occurred on St. Kilda. One day in June 1879, Mr. Boyd, late of Greenock (a good field ornithologist and experienced sportsman on the west coast), and Captain M'Ewan of the ss. *Dunara Castle* very accurately described to Harvie-Brown a male of this species seen by them at close quarters, that same day, when rowing round the shore of the Dune. But before absolute record can be registered more observation will be required. Dixon also records its occurrence *as seen by him*, but this and most of such records are quite unsatisfactory. Mr. Henry Evans, who has visited St. Kilda during *many* years, quite disregards such records, as do those who know the islands best, and who are most intimately acquainted with the nature of their shores, and whether they are boulder-strewn or otherwise.]

Somateria stelleri (*Pall.*). Steller's Duck.

Œdemia fusca (*L.*). Velvet Scoter.

Gaelic—*Tunnag ghleust*=the intelligent or cunning duck.—A. C.

No mention appears to have been made of this species, but there can be no doubt it is a regular winter visitor in small numbers to the western end of the Sound of Harris, where, at one locality, they are constantly observed—viz., in the Sound of Lingay. Curiously enough, they have not been observed anywhere else in the Sound of Harris by Mr. John MacDonald, who resides there all winter—and in whose opinion, when firmly expressed, we have never yet regretted placing confidence; and where he is uncertain he never makes an assertion, over which a practised ornithologist need ever stumble.

Œdemia nigra (*L.*). Black Scoter.

This cannot be looked upon as at all common. It is sometimes seen in the Sounds, as stated by Wm. MacGillivray. Gray makes no mention of it as occurring in the O. H., but it is included by Mr. D. Mackenzie in his Stornoway lists, and as being represented in the museum there.

Œdemia perspicillata (*L.*). Surf Scoter

The only specimen up to date of 1871, known to have occurred on the west coast of Scotland, is recorded by Gray as "shot in the winter of 1866 at Holm, near Stornoway, by Mr. MacGillivray of Stornoway." This bird is now in the collection of Lewis birds in the Museum at Stornoway Castle, and is last noted by Mr. D. Mackenzie in the list sent to me, which has already been alluded to.

Mergus merganser, *L.* Goosander.

[*Obs.*—MacGillivray appears only to have *seen* this species on some of the larger lakes in summer (*British Birds*, vol. v. p. 212). Dr. Dewar's record is extremely doubtful, all the more so that the egg, described by Mr. R. Gray, was slightly *darker* in shade than that of a Red-breasted Merganser. The set of eggs Harvie-Brown himself saw in Dr. Dewar's collection were certainly very like those of the Goosander. Captain Elwes, however, seems to us fully to dispose of all previous records; and even its subsequent occurrence in the O. H. appears to have been in very few and rare instances. In view, however, of its known and recognised spread northward through the mainland of Scotland, its future appearance may be looked for in the O. H. both in winter and in the breeding season.

At the time when Mr. Gray, trusting to the records of Dr. Dewar, said "that the statement that the Goosander undoubtedly breeds in North Uist, or has bred, can hardly be called in question," the Goosander had barely made its appearance as a breeding species in the mainland of Scotland. The increase of the species since then is as patent and incontrovertible, in our opinion, as its rarity and limited distribution were about the time mentioned—say, prior to 1871. Although all records of this species nesting in the O. H., and also even the accounts of its occurrence in winter, must, we consider, as yet remain within brackets, and though no additional evidence obtained of late years warrants their removal, still it seems to us desirable to mention one or two more observations, for what they are worth. In the winter of 1865-66 Sheriff Webster shot a bird, and accurately described it as a Goosander to Harvie-Brown, but unfortunately no portion of the bird was preserved. Sheriff Webster is perfectly acquainted with the Red-breasted Merganser, as indeed are all the natives, and he has the sportsman's eye, if not the naturalist's. In June following, Mr. John MacDonald reported a "stranger," which he thought "was like a Goosander," to be frequenting the upper armlets of Loch Maddy, and Harvie-Brown, to whom he related this, drove back with Mr. MacDonald as far as the bend of the road nearest to where it had been seen that same day. Harvie-Brown remained searching these creeks for some hours, but all in vain; and nothing more has been reported of the Goosander in North Uist since then. If the Goosander would take as kindly to the long tunnelled holes of the Sheldrake, and as readily, as the Red-breasted Merganser has done, for breeding purposes, some day may see the extension of its range to these bare islands. Meantime the cloven roots of ancient birch-trees on the shores and islands of the better-wooded lochs of the mainland will probably first come to be selected; and until all such are occupied by the expanding waves of pressure, then only need less likely habitats be expected to give them acceptable homes.]

Mergus serrator, *L.* Red-breasted Merganser.

Gaelic—*Siolta: Sioltain*, from *Siolag* = sand-eel, upon which the Merganser principally feeds.—A. C.

John MacGillivray, in 1830, said he had met with this species, but that he seldom found it inland; confusion, however, appears in the internal evidence of the passage, as he seems to have been wrong about the previous species. Mr. H. H. Jones writes that it is very common in spring. "On the 1st or 2d June I saw a flock of about twenty, of which half were males in full plumage; they were fishing in Loch Roag." He also, a few days later, got eggs from two nests on a fresh-water loch near there. Harvie-Brown found them numerous at Loch Maddy and elsewhere from 1871 down to date. They also occur, but more sparingly, at Carn, near Stornoway, but they are really common birds in all suitable localities in the O. H., breeding numerously in holes in the peaty islands of Loch Maddy, their eggs and those of the Sheldrake being not uncommonly found together in the same nest, as we have ourselves found. We have not, in our experience, found that they are so common amongst the sandy machar and fringing sandhills of the west coast, but of late years we know of their extension of range along certain portions of the Skye coast. Just as the impulse of the Eider Duck's extension appears to have moved eastward from its earlier settlements in the *west* of the O. H., so do the impulses of extension of the Sheldrake and Red-breasted Merganser appear also to have pressed eastward from *their* earlier centres in the *east* of the O. H. The machar of the west is eminently suited for the latter, yet there they are rare or absent, so far as we are aware. It certainly seems to us that waterfowl breed on the lines of their natural migrations, which lines are usually north and south of the centres of first population, just as extension of range of land-birds seems to take place east and west of their centres, or at least along the annually followed routes of *their* migrations. This is not altogether hypothetical, as we believe there is much solid material to back up the opinion, or theory, if it must still be so considered.

Mergus cucullatus, *L.* **Hooded Merganser.**

Mergus albellus, *L.* **Smew.**

There is a Smew in the museum at Stornoway, which, according to Mr. D. Mackenzie, was obtained in The Lews. Such statements from Mr. Mackenzie may, in our opinion, be fully relied upon. Although we know of the occurrences of a considerable number of birds of this species along the line of the Inner Hebrides and west coast of the mainland, authentic records from the line of the O. H. are decidedly scarce.

Series SCHIZOGNATHÆ.

Order I. COLUMBÆ.

Family COLUMBIDÆ.

Gaelic general names—*Fearan*: *Colman*, *Calman*=a pigeon: *Calman Taighe*=house pigeon: *Calman Fiadhaich*=wild pigeon: *Calman Gobhlach*=fantailed pigeon: *Calman Gorm*=blue pigeon, rock pigeon: *Calman Coille*=wood pigeon: *Guragug* =dove.—A. C.

Columba palumbus, *L.* **Ring Dove.**

Gaelic—*Calman Coille*=wood pigeon: *Smud*: *Smudan*: *Duradan*.—A. C.

Gray records this as only a straggler to Benbeculay and South Uist in spring and autumn; it is also included in Mr. D. Mackenzie's list of birds at Stornoway. It is nowhere common, owing to the limited area of woodland, but appears to be resident or breeding in the Glen of Rodel, where Harvie-Brown saw one bird on June 15th, 1879, and on one or two subsequent occasions several pairs were observed and heard there.

Mr. MacElfrish, Loch Maddy, shot one wood pigeon on the west side of the island of Barray. The bird was alone, and, though strong, permitted itself to be approached within gun-shot; it appeared as if it had only just alighted after a long flight. This was on the 15th October 1887, about the date of the great rush of other migrants which "showed up" so remarkably on the Monach Isles, as related by Mr. Joseph Agnew in his Migration Schedules from these islands.[1]

Columba livia, *Bonnat.* Rock Dove.

Gaelic—*Calman Gorm*=blue pigeon: *Calman Creaige*=rock pigeon.—A. C.

By all writers described as common, and "nowhere more abundant than in Pabbay." "In summer their food consists almost entirely of *Helix ericetorum* and *Bulimus arcticus*—shells very abundant among the sandy pastures"[2] (John MacGillivray, 1841). Appears generally plentiful wherever suitable caves occur, and on that account, no doubt, most so on the west coast of North Harris and parts of Lewis, and on the east side of North and South Uist. Harvie-Brown visited some caves near Loch Maddy, which had also been visited by Captain Elwes, and obtained eggs, not without some difficulty. Rock Doves are also very common on the east side of Lewis, south of Stornoway, where as many as twenty couple may be shot by expert hands in a day. Though usually less numerous upon islands at any distance from the main range, we found them not uncommon on Haskeir in 1881, and took eggs there. Indeed the Rock Dove is abundant in many places all along the more suitable parts of the coasts.

They are also seen, probably on partial migration for purposes of food supplies, even on the sandy low-lying islands of Monach in the autumn, as reported by Mr. Joseph Agnew.

[1] We are indebted to our friend Mr. J. E. Harting for drawing our attention to the above occurrence. It was communicated to him by Mr. Thomas Wilson, son of Mr. Peter Wilson, Officer for the Scotch Fishery Board at Girvan, Ayrshire.

[2] Not food surely, or not food necessarily; only grit wherewith to grind the corn in their crops, and assist in making their egg-shells.—J. A. H.-B.

Columba œnas, *L.* Stock Dove.

- **Turtur communis, *Selby.* Turtle Dove.**

Gaelic—*An Turtar.*—A. C.

[*Obs.*—Professor Duns says: "Some years ago one of Sir James Matheson's keepers shot a turtle dove." Gray ignores the record, and says, "none" have as yet turned up. Was this an Egyptian turtle dove (*T. senegalensis, L.*), one of which, "evidently an escape," says Gray (*op. cit.* p. 223), was caught at Scolpig, North Uist, on 20th October 1866. This species is not included in the British list.]

Ectopistes migratorius (*L.*). Passenger Pigeon.

Family PTEROCLIDÆ.

Syrrhaptes paradoxus (*Pall.*). Pallas' Sand-Grouse.

On October 13th, 1863, one was shot (a solitary example) on the island of Benbeculay, by C. W. MacRury, Surgeon; it afterwards passed into the collection of Dr. Dewar of Glasgow, since deceased. This specimen is not mentioned in Professor Newton's article,[1] nor have we been able to learn of any other occurrence at that time in the O. H. except this. But as Dr. Saxby notices that several appeared in Unst, Shetland, and if we bear in mind the lines of migration as lately worked out, we believe it is quite possible that others may have escaped observation in the Long Island during that year.

Of the second great "Tartar invasion," which took place in May 1888, we have much fuller records in the Long Island than that of 1863. Observing from a newspaper paragraph (*Scotsman*, June 14th, 1888, signed J. J. F.) that a detachment of Pallas' Sand-Grouse was recorded as occurring in Benbeculay, Harvie-Brown took the usual and necessary steps to have the record put on sufficiently strong evidence, and his letter of inquiry was replied to in course of post by John MacRury,

[1] "On the Irruption of Pallas' Sand-grouse" (*Syrrhaptes paradoxus*) in 1863. By Alfred Newton, M.A., F.L.S., F.Z.S., *Ibis*, 1864, p. 185.

M.B., of Dunruadh in that island,[1] who was the first person to see and identify the birds. The first he saw was a single bird on the evening of the 22d May, on the grassy moorland near his own house, and the next day he found it again at the same place. On the 24th May Dr. MacRury again saw this solitary bird get up near the same place, flying this time so close over his head that he distinctly made out the speckled breast and pointed tail. The ground which this bird frequented was grassy moor, where the soil is black and peaty, and which was reclaimed from the heather not many years ago. On the 25th Dr. MacRury went down to the shore, where there is abundance of sand, and there met with a flock of ten birds, which rose out of a recently ploughed field where the sand was loose, white, and dry. They alighted again, and he stalked up to within sixty yards of them, and watched them with much interest for some time. We here give his own words: "They travelled slowly over the ground, moving about like Red Grouse, and evidently picking beetles and other insects out of the sand. On putting them up, and looking over the ground where they had been feeding, I noticed it full of small holes in the sand, as if one had marked it with the point of a walking-stick, and this was always present wherever they had been feeding, so that travelling over the machar I could easily tell where they had been. Their mode of getting up is something like the partridge, rising suddenly and rather high, but their flight is more like that of the Golden Plover when travelling at a good speed. In flying, too, they utter a sharp cry, something like 'whirr-whirr,' quite unlike the note of any bird I know here. In flying I noticed that although there were ten in the flock each pair kept a little apart, but when they alighted they seemed to mix together, and they kept very close to each other."

Dr. MacRury here speaks of their apparently feeding principally on insects, but also goes on to say that "they might also be eating grass, and the young 'braird' which was just appearing above ground in the field where I saw them." All the birds Dr. MacRury saw were, he says, wild "for foreigners." "On approaching them at first they would try and hide themselves, lying down close,

[1] C. W. MacRury and J. MacRury are brothers. J. MacRury has only been back from New Zealand about eighteen months.

but before I could get within gun-shot they would rise suddenly and fly away quickly." We are glad also to continue Dr. MacRury's notes, as follows: "I was rather reluctant to shoot the poor strangers, and I don't think they were fired at either in Benbeculay or South Uist. They seemed to be travelling south, as in getting up they always turned in that direction, and I saw some cross over to South Uist late in the evening of the 25th May; and subsequently for a week or two after the above date some birds were seen by others about the locality where I saw the ten, but not so many, only five or six." Dr. MacRury now missed seeing any until the 10th June, when he saw a pair on a farm on the west side of South Uist, which were also observed at the same time by the tenant, Mr. R. Ferguson. They rose out of a sandy ploughed field, rather wild, and flew southwards. Again, on the 19th and 20th of June, Dr. MacRury saw one flying high over the west side of Benbeculay. He says: "I watched it cross the ford to South Uist. For some days after this I did not see or hear any more of them, and I was under the impression I had seen the last of them for a season, but two days ago (26th) I again saw three birds near the same place where I met the ten. These flew north, but alighted again not very far off. They may still be going about. I looked carefully over the ground for any appearance of nests, but saw nothing of the kind. There are, however, long stretches of sand covered with bent here and there along the west of North Uist, Benbeculay, and South Uist, where they might nest undisturbed."

Dr. MacRury's interesting notes reached us just in time for insertion in the text of first proof (viz., on the 4th June); had they been a day or two later we would probably have been obliged to place them in an Appendix.

Order 2. GALLINÆ.

Family PHASIANIDÆ.

Phasianus colchicus, *L.* Pheasant.

Gaelic—*Easag.*—A. C.

Professor Duns mentions the introduction of Pheasants at Stornoway. At first they roosted in trees, but later they are stated not to have done so. The Professor asks if they gave up this habit because there are no foxes nor ground vermin in the district. They have survived so far, as in 1879 Mr. Greenwood notes them as doing fairly well in the Castle grounds, and in 1881 Mr. Hubback observed "plenty" of Pheasants in the same enclosures.

Caccabis rufa (*L.*). Red-legged Partridge.

Caccabis petrosa (*Gm.*). Barbary Partridge.

Perdix cinerea, *Lath.* Partridge.

Gaelic—*Cearc Thomain*=the knoll hen: *Chearc Chruthach*=the horse-shoe hen, from the semicircle on the breast resembling a horse-shoe.—A. C.

Said to have been introduced into Lewis by MacKenzie of Seaforth, about eighty years previous to 1871. Gray understands they have since been introduced more than once into Lewis, but without good results. One account says that partridges, which had been introduced two or three times, had never lasted through the winter (Hubback—*fide* under-keeper at Stornoway Castle).

The Partridge was introduced to South Uist a few years ago, as we are informed by Mr. John MacDonald, and one was seen in the fields at Newton, North Uist, by a Lowland shepherd lad, who knows the bird perfectly. This was in March 1888.

Coturnix communis, *Bonnat.* Common Quail.

Gaelic—*Gearra-gort*, from "Gearr," a word of doubtful meaning, and of varied application in the names of birds and animals, and *gort*, famine, the famine—"gearr,"—possibly from the bird having been the food of the Israelites during their wanderings in the desert at the foot of Mount Horeb.—A. C.

Very rare in the Long Island. There is one in the collection of the late Sir James Matheson, Bart., at Stornoway, which was shot in Lewis, and where, in 1868, one pair was known to breed. Also a nest of twelve eggs was taken in North Uist by Mr. J. MacDonald of Newton, about the 20th July 1870, eight of which are now in the collection of Sir John Orde, the proprietor.[1]

[1] Vide *Ibis*, 1871, p. 112.

Mr. D. Mackenzie in his list also takes note of the specimen in the Castle museum above mentioned.

Family TETRAONIDÆ.

Lagopus mutus, *Leach*. Common Ptarmigan.

Gaelic—*Gealag bheinne: Ian Ban ant-Sneac: Sneacag: Tarmachan: Tarmachan Beinne*=mountain ptarmigan.—A. C.

In 1830 W. MacGillivray states that they occur on hills over 2000 feet in height, and that they are smaller than those on the mainland. In 1841 John MacGillivray speaks of them as occurring sparingly on Ben More, and Hekla in South Uist, and on Roniebhal in Harris, but more abundantly on the summits of the forest-hills.[1] It is also related that one was seen at St. Kilda a few years previously. Mr. J. MacDonald never met with it in North Uist.

By 1871 it was wholly confined to the rocky peaks of Harris and Lewis, as stated by Gray. It is still included in Mr. D. Mackenzie's list, but he, in this instance, no doubt, alludes to the whole of Lewis. As late as 1881 it was reported to Harvie-Brown as "not uncommon" near Tarbert, but in 1879 (two years before) Harvie-Brown did not see a single bird, though he traversed nearly every mountain-top in Harris, whilst in search of eagles eyries; but this negative evidence, as ornithologists are well aware, proves nothing, not even its rarity, and far less its entire absence.

Lagopus scoticus (*Lath.*). Red Grouse.

Gaelic—*Cearc-Fhraoich*=heather hen: *Coilleach-Ruadh*=red cock: *Coilleach-Fraoich*=heather cock: *Ian Fraoich*=heather bird. —A. C.

Abundant in 1830 and also in 1841; recorded as very common in all the larger islands; but noted as rare at Mhorsgail in Lewis in 1881. Tolerably abundant in Lewis, Harris, North and South Uist, and Barray, and are smaller and lighter in colour than those of the mainland (Gray, *Birds of the West of Scotland*, p. 234).

[1] These forest-hills were the North Harris hills, as no forest (*i.e.* deer-forest) existed south of Tarbert at that time.

In 1879 Harvie-Brown finds this note in his Journal: "Red Grouse are said to be increasing of late years around Rodel in Harris." Lord Dunmore at that time had five keepers in South Harris. But the old and rank heather, which we have ourselves seen behind Rodel, and upon Roniebhal's eastern exposure, certainly requires burning (1879). By 1886, heather which was in its prime had been burned by the crofters, and old heather had been just as often left, with the result that feeding-grounds have again become limited in extent, and birds again scarcer. Ignorance and prejudice among such matters are of course hard to battle with.

Sportsmen of experience are well aware of the marvellous results in increase of Grouse on an area of heather, where the latter is subjected to careful revision by the proprietor or his agent, during the burning season. There is often also a vast improvement—where old heather is too abundant and young heather scarce, as it is in a great many West Highland estates—where the whole beat is taken determinedly in hand, the old birds of each covey shot down, and the coveys broken up. From an annual bag of some thirty or forty brace, we have known such an estate, or portion of an estate, after two or three years' judicious treatment, yield from 200 to 300 brace, and that simply by killing old birds, breaking up coveys, and vigorously hunting down the vermin. When these conditions—judicious burning and judicious shooting—are combined with the active destruction of those worst of vermin, Carrion or Hooded Crows, the best results are sure to follow. Add to the above, where such is practicable, judicious driving both early and late in the season, by which means a large percentage of old birds is sure to be obtained, and coveys are mixed up and separated. This benefits the moor, and lessens also, we consider, the chances of disease.

We have already taken occasion, under *Lepus variabilis*, to speak of deterioration in size of grouse and other animals in The Lews. We feel sure that if the above methods were closely followed out all over the area of the Long Island, not only would grouse increase in number, but also in size of individuals, and if, in addition to this, fresh blood were liberally introduced, the results would be simply astonishing in a very short time. We speak from personal experience of such experiments, both as regards Grouse and Partridge. In one case, where unproductive partridge-ground yielded some miserable twenty or thirty brace in a season, by introducing fresh blood, the same ground in two years yielded as many brace in a day, and the birds were remarkable for size, strength, and beauty, whilst on adjoining ground, where fresh blood had never been introduced, the birds were miserably small, even when full-fledged. We could say much more in illustration, did space permit, of what are, or ought to be, now generally recognised facts, but we refrain, hoping however that the above notes may prove of interest and use to many a highland laird in the west. He has no surer way now-a-days of improving his property, and the cost is comparatively trifling.

We do not lose sight of the fact that Grouse in the Hebrides have other enemies to contend with—the short-sighted policy of the easily-influenced crofter population, dampness of the climate, and late frosts in spring, and "vermin"; but we believe *in-breeding* is the greatest of these enemies, especially where heather-ground is injudiciously treated.

With Grouse in the Long Island, the process of deterioration seems to be steadily advancing towards the final act. Mr. A. Williamson is also of this opinion. He writes: "After an experience of nine years, and giving much thought to the subject, I have come to the conclusion that the Grouse in Lewis are slowly dying out, for which I blame severe and late springs."

Tetrao tetrix, *L.* Black Grouse.

Gaelic—*Lia-chearc*: *Cearc-lia* = grey hen: *Coilleach dubh* = black cock.—A. C.

[*Obs.*—Professor Duns records that efforts had been made to introduce Black-game to Lewis, but these attempts had not proved successful. Gray takes no notice of the species as a Hebridean bird, either as a native or as an introduced species, nor have we been able to gather any information on the subject, though never losing sight of the above statement.[1]]

[1] In reply to inquiries, Professor Duns informs us that his Notes—the substance of which was given to the Royal Society of Edinburgh—will not be available till August; if ready by then, we hope to give the results in our Appendix. Mr. John MacDonald never heard of the introduction of Black Grouse anywhere in the Long Island.

Tetrao urogallus, *L.* **Capercaillie.**

Gaelic—*Caber-coille: Caprioc.*[1]—A. C.

Family TURNICIDÆ.

Turnix sylvatica (*Desf.*). **Andalusian Hemipode.**

Order 3. GRALLÆ.

Family RALLIDÆ.

Rallus aquaticus, *L.* **Water-Rail.**

Gaelic—*Snagaire nan Allt*=the creeper of the streams: *Gearra Dubh nan Allt*=black *gearr* of the streams.—A. C.

J. MacGillivray found it in North and South Uist. Rare in 1830 (*auct.* W. MacGillivray). "Not common, but found throughout the year" (Professor Duns). But the truer record is given later by Gray, who says, "Distributed over all Scotland, including the O. H.; plentiful in Lewis, also in Harris and North Uist." Mr. Greenwood had twice observed it near Stornoway by 1879. Mr. D. Mackenzie has the note, "Arrive towards the end of harvest, and leave in spring," near Stornoway; but we shall not be surprised yet to hear of its nesting. MacGillivray only met with it in winter in Lewis and Harris, but, owing to its skulking, silent habits, we strongly incline to the belief that it is a species much more common at all seasons in Scotland than is generally credited, and the absence of records is owing to its easy concealment under summer- and even winter-growth of reeds and marsh plants. We have received the species in the flesh, from Mr. Joseph Agnew, from the Monach Isles, dating 17th November 1887.

Porzana maruetta (*Leach*). **Spotted Crake.**

Porzana bailloni (*Vieill.*). **Baillon's Crake.**

Porzana carolina (*L.*). **Carolina Crake.**

Porzana parva (*Scop.*). **Little Crake.**

[1] For full treatment of the Gaelic names, see *The Capercaillie in Scotland.* David Douglas, Edinburgh. We had never met with the name *Caprioc* before.

Crex pratensis, *Bechst.* Land-Rail.

Gaelic—*An Treun* (St. Kilda): *Treubhua*: *Treona*: *Treun-ri-trean.*—A. C.

Very common in summer, frequenting, after their arrival, the dense patches of *Iris pseudacornis*, and then taking later to the corn-fields.

We have very late records, and also very early ones, of this bird in the O. H. Thus, whilst Mr. H. H. Jones relates that there are a few pairs at Mhorsgail, and that they are found "occasionally" very late in the autumn,—viz., October 13, 1879,—we have still later dates of its occurrence, one, for instance, at Tiree (Inner Hebrides) on 24th November 1880, the specimen being now in the Kelvingrove Museum, Glasgow.

The Corn-crake is said to arrive about the middle of May in the Hebrides generally, but quite ten days sooner in Berneray (Duns)! According to W. MacGillivray, it arrives in May and disappears in August. It is chiefly found in patches of *Iris pseudacornis* and *Spiræa ulmaria*, and was very abundant, according to Gray, in 1871—even extending to the rocks of Haskeir, and St. Kilda, and the Monach Isles. The earliest record we have of the arrival of the species is afforded us by Alex. Carmichael, whose acquaintance with the species is undeniable. It was heard on the 25th March 1882 close to Scolpig House, North Uist. Carmichael says: "This place is on the edge of the Atlantic, as you know, and as bare, bleak, and shelterless as any place in Scotland. . . . The weather is cold and stormy in the extreme, and as for vegetation there is none whatever, nor will there be for the next ten weeks. I have never previously seen it here before June." A. C.'s letter is dated 27th March 1882.

Gallinula chloropus (*L.*). Moorhen.

Gaelic—*Cearc-uisge* = small water hen.—A. C.

Considered by Gray as pretty common in suitable ground. It is spoken of by W. MacGillivray as common in the Uists and Benbeculay, and as also found in Harris. Has only been observed at one locality in some marshy ground at Miavaig, Mhorsgail, by Mr. H. H. Jones, and it appears to be local in Harris and Lewis. It was only twice observed near Stornoway by a very observant man, in 1879, and Mr. D. Mackenzie does not include it. Harvie-Brown has notes of it in North and South Uist.

Fulica atra, *L.* **Common Coot.**

Gaelic—*Lach-bhlar* = duck of the forehead spot.—A. C.

Earlier writers recognised the Coot as pretty common in suitable ground. By 1830 it is spoken of as abundant, betaking itself to the sea at Nibost in frosty weather. Commonest in Lewis and Harris and North Uist; rarer in Benbecnlay and southwards (Gray). Observed at Balranald in 1885 by Hugh G. Barclay. Observed in North Uist by Lieut.-Col. Feilden and Harvie-Brown, paired on a small pool of water in the machar, in May 1870, but not elsewhere, and Harvie-Brown has not, since that time, met with it anywhere abundantly, though there are many localities he has not visited where it may occur more numerously. It has been found "in considerable numbers" by Mr. J. MacDonald in the Sound of Berneray, and on Loch Bruist, a fresh-water loch in that island. This was during the last two winters, and Mr. MacDonald remarks this as unusual.

Family GRUIDÆ.

Grus communis, *Bechst.* **Common Crane.**

Grus virgo (*L.*). **Demoiselle Crane.**

Order 4. LIMICOLÆ.

Family OTIDÆ.

Otis tarda, *L.* **Great Bustard.**

Otis tetrax, *L.* **Little Bustard.**

Otis macqueeni, *J. E. Gray.* **Macqueen's Bustard.**

Family ŒDICNEMIDÆ.

Œdicnemus scolopax (*Gm.*). **Stone Curlew.**

Family GLAREOLIDÆ.

Glareola pratincola, *L.* **Common Pratincole.**

Family CHARADRIIDÆ.

Cursorius gallicus (*Gm.*). **Cream-coloured Courser.**

Charadrius pluvialis, *L.* **Golden Plover.**

Gaelic—*Feadag*=whistler: *Feadag Bhuidhe*=the yellow whistler.—A. C.

Very common, as recorded by all writers. Congregates, says John MacGillivray, in the end of July. Great numbers on the "Ebb of Berneray"[1]—a reef so called, of bare stones nearly one mile in length, covered at high tide.

The Golden Plover is not a regular migrant in the O. H., but may rather be looked upon as a permanent resident, or nearly so, merely shifting from the moors to the shore and back again, or migrating only in very unusually severe winters, when others take their place. We find we have noted it very many times in our journals from all parts of the islands, but in 1881 as especially abundant around the neighbourhood of Loch Hamanaway, south-west of Lewis, and noted as very common on the Mhorsgail ground, but all leaving in autumn. Common in the north of Lewis on the moors in the breeding season. In the very severe winter of 1880-81 (or 1878-79?), plover, after having been driven down to the tide edge, and getting wet with sea-spray, became so frozen as to be unable to fly, and were, in many instances, captured by hand. This happened, according to information, on the west coast of Lewis. They also become much emaciated in severe weather in winter.

Squatarola helvetica (*L.*). **Grey Plover.**

Gaelic—*Glas-Fheadag*=the grey whistler.—A. C.

W. MacGillivray speaks of the Grey Plover as very common in flocks from the middle of autumn to beginning of April, but Gray does

[1] Ebb is a name applied to all shore bare at low water, but the above reef is locally spoken of thus—"*The Ebb of Berneray.*"

not indorse this account; he says, "In the Outer Hebrides it occurs but sparingly," which is the more correct, and is what one would more naturally expect. We can record its occurrence at Skerryvore Lighthouse on one occasion with certainty. Gray adds that occasional specimens occur in Benbeculay. We have never ourselves met with it on the west coast, nor in the islands; and it is unknown to Mr. John MacDonald.

Ægialitis cantiana (*Lath.*). **Kentish Plover.**

Ægialitis curonica (*Gm.*). **Lesser Ringed Plover.**

Ægialitis hiaticula (*L.*). **Ringed Plover.**

Gaelic—*Bothag.*—A. C.

Abundant in every suitable place, especially all about the west coasts of the Uists and the machar lands and sandhills. A pair was seen in middle of June 1885 in North Ronay by Harvie-Brown, but the species was not observed there by either Swinburne in 1883 nor by Barrington during a three days' visit in the end of June 1886. The wind-stripped sandy and gravelly patch where Harvie-Brown saw the above pair was very limited in extent, and not far from the landing-place at Geodh-a-Stoth.

Eudromias morinellus (*L.*). **Dotterel.**

Gaelic—*Amadan mointich*=fool of the moor.—A. C.

[*Obs.*—The only discoverer, so far as we have been able to find, of this species in the O. H., is Professor Duns, but unfortunately he adds no particulars. We have utterly failed to obtain the slightest hint as to its presence from other sources. Possibly the Ringed Plover is intended under the misapplied name of "Dotterel."]

Vanellus vulgaris, *Bechst.* **Lapwing.**

Gaelic—*Curacag*=cappie: *Adharcag-luachrach*: *Adharcan-luachrach*=little horny of the rushes.—A. C.

While J. MacGillivray designates the Lapwing as very common in 1841, on some of the islands, as Ensay, Killegray, and Taransay, and also in the two Uists: and while Sir W. E. Milner reports it as very common in 1848, and it is also spoken of by Gray in the same terms in 1871; W. MacGillivray nevertheless considered it as very rare in the O. H. in 1852. We ourselves cannot look upon it as at all an abundant species, though resident, and present all the year. We have found it common, but not abundant as compared with more eastern localities. On Taransay only very few were observed in 1888.[1]

Strepsilas interpres (*L.*). Turnstone.

Although MacGillivray thought the Turnstone bred in the O. H., and shot specimens on Ensay in June, and even in the end of July and beginning of August on the reef of Berneray—elsewhere called the "Ebb of Berneray"—and it is undoubtedly a common bird in these islands; and although Gray prophesied the nesting of the species in Haskeir, St. Kilda, or Monach, such has not yet been realised. It is reported on all hands as a late spring and very early autumn migrant, and it may be resident. Nay, we will even go so far as to say it *may* yet be found on *rare* occasions to breed; but our own experience of this species in summer in the isles, which has been very considerable, has never given us the impression that they were breeding, nor that they had up to date extended their breeding range so far south. "When we bear in mind how little is known of the ornithology of the islands, or the wild west and north of Scotland," it may be interesting to have it recorded that our conviction is, that, though the birds are annually seen as late as the middle of June upon many of the islands, and portions of our western and northern coasts, it is exceedingly improbable that they have hitherto bred. At the same time, we are the last to oppose the opinion that they may yet some day be found breeding, because we have long since made up our minds that such may happen. Gray's prophecy may hold good enough, but Seebohm's statement "that it is difficult to believe that it has not bred in the Hebrides," must be the direct outcome of very superficial study of their habits there. Harvie-Brown found two in 1884—late in June—evidently a pair, and very tame, but their

[1] Taransay, not so very many years ago, however, was a well-populated island; and the "run-rig" system of cultivation has left the tell-tale evidence behind. The old "lazy-beds" are now for the most part covered with heather. What once may have been a suitable locality for Lapwings is scarcely a likely one now.

whole actions otherwise were directly opposed to all an ornithologist's experience of those of breeding birds, and in no case that Harvie-Brown has met with them, have these summer-staying birds' habits been otherwise.

Milner obtained Turnstones in full breeding plumage on the 1st of June, but in flocks of four or five. Barrington's party found a flock of five at the end of June on the rocks of North Ronay, which lie between high and low water mark, in which the summer plumage was not complete. This is usually the case in regard to Turnstones when seen in the O. H. and west of Scotland, as far as our experience teaches us, though we have on several occasions seen them in pairs *in full, or almost perfect summer plumage*. In no case did the actions of these birds lead to the belief that they were breeding.

In 1887 Harvie-Brown met with a large flock on Gaskeir, off the south-west coast of Lewis, as late as July 4th, as well as others at several localities amongst the islands and rocks of the archipelago.

The Turnstone is abundant in winter. W. MacGillivray found it on rocky shores from October to April, and Professor Duns also speaks of it as occurring at that season.

Hæmatopus ostralegus, *L.* Oystercatcher.

Gaelic—*Bridean* = St. Bridget's bird: *Gillebride* = St. Bridget's page: *Trilleachan traghad: An Trileachan*=the busy body or fussy body of the strand.—A. C.

"Very common everywhere" is the united verdict of Scottish ornithologists, who have or have not visited the O. H.; breeding, says J. MacGillivray, both on sea-rocks and on fresh-water lochs. We have not met with it on fresh water in the Hebrides, but we can readily believe it occasionally chooses such situations, as we know it does elsewhere in Scotland. Almost universally distributed, and breeding often at a considerable elevation above the sea, *i.e.* from almost sea-level up to the summits of rocks 80 to 100 feet above; Swinburne took a nest 200 feet above the sea on North Ronay in 1883.

The general abundance and widespread distribution of this species makes it almost invidious to particularise localities, but we may be permitted to make the observation that nowhere have we seen it nesting more abundantly than on North Ronay. Desiring fresh meat on board the yacht, Harvie-Brown was induced to add eight of these birds to the larder, which was accomplished in a few minutes. He could just as easily have doubled the number, but disliked the job. The flesh of the Oystercatcher is excellent food, if cooked in the proper way, like many others of our waders: plain roasted it tastes somewhat like Teal-duck. These birds do not mass so largely in most localities north of the Clyde as they do to the southwards, nor as upon the larger and oozier firths during the winter; and the ideas formed of vastness in the flocks are just according to the varying experiences of different observers in different localities. Oystercatchers are particularly fond of resting and feeding on what are technically called "mussel-scalps," or banks of mussels left bare in the estuaries of the larger rivers, and assemble, where these occur, in vast numbers. But in summer there is an abundance of other suitable food distributed over almost all their rockier breeding-haunts, as indeed seems to be a wise provision of Nature in regard to many species.[1] The enormous congregations seen upon the larger estuaries of the Solway and southwards, and upon the Eden and Tay and Forth, and especially *where mussel-scalps occur*, are greatly, of course, composed of the birds of the year, probably far the greater part of which are hatched out in our wilder Hebrides, and go to swell the flocks which delight the field-ornithologists of more southern and less barren lands.

Family SCOLOPACIDÆ.

Recurvirostra avocetta, *L.* Avocet.

Gaelic—*Gob Cearr: Cearra-ghob*=wrong-bill: *Mini-ghob*=awl-bill, from the resemblance, or supposed resemblance, to an awl.—A. C.

A specimen of this rare Scottish bird was shot at Stornoway on the 3d November 1879, and is now in the collection of Robert W. Chase, Esq., of Birmingham; it is a female, and weighed 14 oz. A grand male, shot by the same person, and at the same place, was obtained on the 3d April 1880, and is also in Mr. Chase's collection. Mr. Chase informs us that the man who shot these birds called them "*awl-birds*." Might not these birds have bred there,

[1] See under "Terna."

if unmolested: and should they not be deposited in the Stornoway Museum?

Himantopus candidus, *Bonnat.* **Black-winged Stilt.**

Phalaropus hyperboreus (*L.*). **Red-necked Phalarope.**

Gaelic—*Deargan-allt* = ruddy bird of the burn (Uist). NOTE.—*This name is elsewhere more aptly applied to the Kestrel.*—A. C.

Formerly bred numerously at several localities, but from all accounts has never been so plentiful since about the year 1869. Their numbers, however, undoubtedly vary year by year, as if they did not yet consider these localities sufficiently established nurseries. Some years they are very common: in others quite rare by comparison. Some years, too, they breed much earlier, and appear earlier at their breeding-grounds; and in other seasons unusually late. Sir W. E. Milner took eggs, and then only three pairs were seen. J. MacGillivray accords to Mr. D. Macrae the first discovery of the species as breeding at a locality in North Uist, where they have long since disappeared. Since then many raids have been made upon the several localities by collectors, even in defiance of the wishes of the proprietor, who desires to protect them, and by others also, who, having obtained leave, but were not limited, too fully availed themselves of their opportunities for wholesale spoliation. The station discovered by Captain Elwes in 1868, was deserted in 1870; it was pointed out to us by Mr. John MacDonald. On June 9, 1879, Harvie-Brown took a nest of three eggs, and saw about four pairs at one locality. It was believed at the time that "though some were laying, all had not yet returned," more especially as in that year all birds were later of nesting than usual. Our own first experience of the species was in 1870, though success in obtaining sets of eggs was limited. We quote from Harvie-Brown's journals: "Two pairs were swimming close to shore, not the least afraid of us. They are most lovely birds, especially the female, with the red neck, white marks under the eyes, and the back marked most richly with three broad stripes of buff.[1] They swam with great rapidity and ease against the somewhat strong ripple on the loch, and one of the two pairs we saw crept out upon a stone, and preened

[1] The female is the brighter-coloured of the two sexes.

the feathers under their wings, in total disregard of our close proximity." The sight was indeed lovely, never to be forgotten, and was often recalled some years after, when we met with these birds in much larger numbers on the tundras, "Kourias,"[1] and flooded lands of Siberia, in Europe. One nest of three eggs rewarded our search that year, but it was undoubtedly a season of scarcity. It is feared, alas! that other more direct agencies are at work. The once large colony at a certain place is now apparently permanently reduced in number to some three or four pairs, the boys of the district persecuting them, being bribed with large sums of money by collectors.

Phalaropus fulicarius, (*L.*). Grey Phalarope.

[Appears, so far as we are aware, to be absolutely unknown on the west coast of Scotland in summer plumage, and we are not possessed of any records whatever from the O. H., though MacGillivray says it "probably occurs there." Its line of migration is probably seaward of the O. H., but one line inside the islands is occasionally followed, as certain localities known to us are not once in a way only, but *repeatedly*, visited by the birds in autumn. We hope in a future volume to speak more fully of this somewhat peculiar migration-line of the Grey Phalarope.]

Scolopax rusticola, *L.* Woodcock.

Gaelic—*Creothar: Coilleach-coille: Fudagag: Udagag* (South Uist).—A. C.

Arrives in Lewis in October. Had never been known to breed up to the date of 1865 (Professor Duns).

In 1830 it is reported as "not abundant," but in 1871 is spoken of by Gray as common in North and South Uist and Benbeculay in winter, not remaining to breed. Its abundance now, in certain seasons, is everywhere spoken of, and in 1879, as we are informed by Mr. Greenwood, it bred—"supposed for the first time"—at Stornoway.

Harvie-Brown certainly has never seen a Woodcock in summer anywhere in the O. H., but it is just one of those species which may be expected to have prophecies made about it. It was shot on the Monach Isles in 1887.

[1] *Kouria* = a backwater from a flooded river, or any creek of the river-bank.

Gallinago major, (*Gm.*). Great Snipe.

Gallinago cœlestis, (*Frenzel.*) Common Snipe.

Gaelic—*Gobhar Athar* = air-goat: *Meannan-Athair* = air-kid: *Eun-ghobharag* = bird-goat: *Budagoc*: *Launrag* = also bird-goat: *Boc-saic* = buck: *Meagadan*: *Croman-loin* = marsh-hunchback: *Bog-an-loin*: *An Naosg.*—A. C.

All records agree as to its abundance "in all the bogs" on the west of North Uist, and it is said also to occur on St. Kilda, by Rev. Mr. Mackenzie. "Common in Benbeculay, South and North Uist, going in flocks on the ooze, like Dunlins. Breeds to the height of 1800 feet" (Gray). "Common at Mhorsgail" and at Carn. Up to the date of December 3, 1879, Mr. Greenwood shot in that season 527 Snipe. Even in MacGillivray's time (1852), it bears the early record "extremely abundant" and "seen in hundreds on the sands and fords at night in frosty weather, feeding in flocks like shore-sandpipers and *Limicolæ*."

Gallinago gallinula (*L.*). Jack Snipe.

This species was stated to have been seen in 1848 by Sir W. E. Milner, one rising on an island of Loch Scatavagh in June, but the nest was searched for in vain. If there is no error in the identification of the species, it is still the oft-repeated tale; no one has yet *taken* Jack Snipe's eggs in Britain, and our knowledge of its present distribution makes that achievement less and less likely, at least for a time.

It is common in winter, but not in flocks, scattered and occasional over the ground; identified in the autumn migration at Monach Island by Mr. Joseph Agnew in 1887.

Limicola platyrhyncha (*Temm.*). Broad-billed Sandpiper.

Tringa maculata, *Vieill.* Spotted Sandpiper.

Tringa fuscicollis, *Vieill.* Bonaparte's Sandpiper.

Tringa alpina, *L.* Dunlin.

Gaelic—*Pollaran*: *Graillcag*: *Graillig* (in summer): *Tarmachan Traghat* = the strand ptarmigan (in winter plumage).—A. C.

Common and resident. Leaves the shores for the heaths and machars in May. Returns to the shores with its young in the end of July. Dunlins are recorded as common by all authors. Thousands in autumn on the sand-fords (MacGillivray). As a breeding species it is common all over the islands, especially at certain particularly favoured localities in North and South Uist and Benbeculay. Not rare in St. Kilda, but less common in most of the remoter Hebrides. Found in North Ronay in 1887, but not ascertained to breed there.

Tringa minuta, *Leisl.* **Little Stint.**

Tringa temmincki, *Leisl.* **Temminck's Stint.**

Tringa minutilla, *Vieill.* **American Stint.**

Tringa subarquata (*Guld.*). **Pygmy Curlew.**

Gaelic—*Ruid-Ghuilbneach*: stunted curlew.—A. C.

[Seen by Gray on the ford of Benbeculay in September.]

Tringa striata, *L.* **Purple Sandpiper.**

Gaelic—*Cam glas*=the one-eyed grey (bird); from a belief in Uist that the bird is blind of an eye.—A. C.

The very late spring or early summer appearance of this species is noted by W. MacGillivray, though he does not seem to have met with it in summer in the O. H. in his later record.[1] Milner also saw it on the island of Scalpay, in "full breeding plumage," on 31st May 1848. John MacGillivray, however, notices it only as a winter visitant. Gray speaks of it as "common, and of late spring occurrence."

On the 27th May 1870, Lieut.-Col. Feilden and Harvie-Brown met with Purple Sandpipers on Barray and on Mingulay. On the first-named island Feilden shot two that were engaged in trimming one another's feathers, and had every appearance of being paired

[1] And the latest date at which he found it *anywhere* was on the 20th May, at the Bass Rock, Firth of Forth—a "side-light" which is of some interest for comparative purposes of east and west, and of *then* and *now*. No doubt, however, later records are forthcoming from the east coast of Scotland than the above, as the bird is abundant enough, especially about the rockier shores between North Berwick and Tantallon Castle.

birds. But on dissection they *both* proved to be females, having the ovaries somewhat distended, but with no appearance of eggs ready for extrusion. Shortly afterwards, and just before landing on Mingulay, Feilden shot two more, also females, though observed in pretty close proximity, and having the appearance of being paired birds. It appears possible that those birds had laid, and had come down to the shore to feed and wash, having left the duty of incubation to the males. If breeding anywhere on these islands, they are decidedly scarce, as these four birds were all that were observed on the two islands mentioned, and any others met with were on mere rocky islets. The breast of one specimen showed *decided* traces of incubation, which is perhaps the most important observation made at the time. Many times since, we have observed the Purple Sandpiper on the Western Isles of Scotland even later than above noted, and in almost, if not quite, perfect summer plumage, but in such cases invariably on similar ground to what they frequent in winter—viz., rocky shelving shores, and low reefs and barren rocks. In 1887, we again searched patiently, and kept a constant look-out, for Purple Sandpipers on Mingulay *in vain*, both alongshore and on the high summits. This was on the 16th of July, on which day we walked over a large part of this interesting island in company with our old friend Mr. Finlayson. Had Purple Sandpipers bred in Mingulay this season, surely the old birds would have been demonstrative enough to have shown themselves. They appear to be quite unknown to the inhabitants of both Barray Head and Mingulay.

It would be idle to enumerate all the skerries, rocky islets, and other localities, where, during the past twenty years and more, we have met with this species late in spring, and even in summer. The remarks already made under Turnstone (*antea*, p. 126) are almost equally applicable here. But at the same time the Purple Sandpiper is less common, and appears to linger less into the summer than the Turnstone does.

Tringa canutus, *L.* Knot.

It is certainly curious that there are so few occurrences of this species on record from the range of the Long Island, and that such a strange link in its migration appears almost wanting. Although we know of its occasional occurrence upon the shores of the Inner Hebrides, at certain points, we find it much scarcer on the whole west coast, suitable strands no doubt being somewhat scarce, and the superior attractions of the east-coast firths arresting their flights in autumn, while the still more direct return journey in spring is easily overlooked. As late as the 30th July 1870, Captain MacRae shot one in full summer dress in North Uist, and it was seen a few days afterwards by Mr. R. Gray; no doubt this belonged to the autumn migration. This bird is now in the collection of Sir John Campbell Orde, Bart. Although beyond this we have no other actual records, we believe it will be found not to be quite so scarce as it seems; meanwhile it is included as decidedly "not abundant." It has only been recognised once by Mr. John MacDonald, but he adds "it is very little known."

Machetes pugnax (*L.*). Ruff.

Calidris arenaria (*L.*). Sanderling.

Gaelic—*An Srillag.*

Earlier writers, presuming on the appearance of this bird as late as the middle of May, thought it possible it might be occasionally found remaining to breed. Our present knowledge of its general distribution, of course, make such statements rather antiquated, but it is curious to find authors, even of late years, harping upon such lines. If by any chance a pair did remain to breed, it could scarcely be put down at present to a natural extension of its range.

The Sanderling is abundant on the sands of Barray in the end of August. How is it that the Knot appears so rare or absent? Some difference, no doubt, in the areas of their breeding ranges, which are not yet worked out. But the statement by Duns, that Sanderlings arrive in September and depart in March is perhaps less incorrect than that they usually *remain* so late as the middle of May. Such as do remain till the middle of May are exceptionally late birds. It would be still more correct to say they arrive on passage in September, and pass again north on spring migration.

Tryngites rufescens (*Vieill.*). Buff-breasted Sandpiper.

Bartramia longicauda (*Bechst.*). Bartram's Sandpiper.

Totanus hypoleucos (*L.*). **Common Sandpiper.**

Gaelic—*Cam-glas*: *Boga-loin*=the dipper of the bogs: *An Camalubach*=the crooked bender (probably from its tortuous flight): *An Crithein*=the quiverer (from the quivering movements of the bird).—A. C.

Occurs commonly, breeding on the shores of lochs and streams. Not to be called very abundant, as it is on the mainland of Scotland, yet a fairly conspicuous species. Common in the west of Harris at Abhuinnsuidh, as observed by us, and reaching certain of the remote isles, though not all, at the date of 1887. Thus we found them on Shillay—the outermost of the Monach Isles—and on Shillay off the west end of the Sound of Harris, these sandy islands no doubt specially attracting them.

Totanus macularius (*L.*). **Spotted Sandpiper.**

Totanus ochropus (*L.*). **Green Sandpiper.**

Totanus glareola (*L.*). **Wood Sandpiper.**

Totanus flavipes (*Gm.*). **Yellowshank.**

Totanus calidris (*L.*). **Common Redshank.**

Gaelic—*Maor Cladaich*=shore officer.—A. C.

Recorded as occurring only in winter in 1830. Seen, however, in pairs at Taransay in June, and at Vallay in August, and shot on Berneray in May 1841, but no nests found. Professor Duns observed it near Stornoway in 1865. Occurred at Mhorsgail, and at or near Stornoway in 1879; and Mr. Greenwood found it breeding in the Hebrides, but not in great numbers. We cannot recall ever having met with the species in the O. H. in summer, not even during visits specially made to the localities mentioned by MacGillivray.

Totanus fuscus (*L.*). **Spotted Redshank.**

Totanus canescens (*Gm.*). **Greenshank.**

In 1830 it was considered a permanent resident. A pair of Greenshanks were seen at Loch Langabhat in 1818 by Sir W. E. Milner.

At present it is decidedly rare, and only found in Lewis and Harris with certainty, at the date of 1871. Mr. Greenwood only twice observed it near Stornoway. A pair were seen several times in May 1880 at a locality in the west of Lewis. We ourselves have never met with it in summer south of the Sound of Harris, and some mistake must surely have crept into MacGillivray's record when he speaks of their numbers in Uist, Harris, and Lewis as "astonishing," unless he refers to an autumn flight. This was in 1852, the date of his writing. Yet "astonishing" flights is not our own experience of the habits of the species in any locality we are acquainted with. J. MacGillivray in 1830 also speaks of it then as a "permanent resident." In 1887 we observed single pairs at two localities in the south of Lewis. They are considered to be rather more plentiful on the western side than on the eastern, and indeed only a pair or two are found over a very large extent of the Park of Lewis.

Totanus solitarius (*Wils.*). **Solitary Sandpiper.**

Macrorhamphus griseus (*Gm.*). **Red-breasted Snipe.**

Limosa lapponica (*L.*). **Bar-tailed Godwit.**

Gray records its great abundance on the fords of Benbeculay and South Uist in the end of July, and Dr. Dewar makes similar remarks as regards the Sound of Harris in 1858, when many of the birds appeared to be in full breeding plumage. Mr. Greenwood only once observed them at Stornoway, but it was probably not on account of their absence, but from want of opportunity that he did not see them oftener, though it is quite possible they may migrate northwards along the west side. We have one positive record only of its occurrence in the autumn.

Large numbers of Bar-tailed Godwits frequent the inner shore of Lingay Island, a seeming rendezvous of many other species, such as Curlews, Whimbrels, ducks, and geese. Mr. J. MacDonald says they occur regularly in the Sound of Harris, but he has not seen them elsewhere.

Limosa ægocephala (*L.*). **Black-tailed Godwit.**

Numenius borealis (*Forst.*). **Esquimaux Curlew.**

Numenius phæopus (*L.*). Whimbrel. May-fowl.

Gaelic—*Ian Bealtain* = Beltane bird.—A. C.

Accounted very common by all writers, arriving early in May (hence its name), and usually disappearing by the end of that month. About twenty or so are usually seen in each flock, though we have often seen more. MacGillivray never knew of any having been seen after the end of May; but W. MacGillivray speaks of it as arriving later—"middle of May and remaining five weeks,"—and he adds the note that it feeds almost entirely on *Helix ericetorum* and *Bulimus acutus*.[1] The fact appears to be that while nine-tenths of the birds arrive and depart within the month of May, large flocks are often seen well on in June. Lieut.-Col. Feilden saw a flock of fifty in the Sound of Wiay, on June 3d, 1870; they are also observed and noted at Stornoway, and they also are known to extend their flight westward to St. Kilda.

The instance of their evidently breeding on North Ronay seems of sufficient interest to make extended notice of it here desirable—*i.e.* if we accept our own experiences of extension of breeding range along the pathways of migration. We know that North Ronay lies either in the direct line of their spring migration, or very close to it; and that they should occasionally or regularly breed there can scarcely be wondered at, if we regard also their presently known breeding range.

The nearest point at which the Whimbrel is known to breed with certainty, to our knowledge, is on one of the southern isles of Orkney, and this they did certainly twenty or more years ago (in 1863). But they almost with equal certainty do not breed on the northern coasts of the mainland of Scotland, or only, if at all, most casually as yet. Faroe and Iceland seem to be their headquarters in the west of Europe. They pass most numerously north along the coasts of the O. H., are comparatively scarcer on the east coast-line than on the west, rarer on the west coast of the mainland and inner isles of the Hebrides, and still more rare on the east coast of Scotland. Yet certain passes are known to us by which they do cross over the backbone of Scotland, for instance, up the Spean and Pattack valleys, and down the Spey in spring. Distri-

[1] These appear to be favourite articles of diet with many different species, even of different genera; though in many cases they are most likely swallowed for purposes of trituration.

bution and migration are in great measure dependent upon one another—a simple fact, which is, we are glad to find, yearly more and more forcing its importance upon the notice of naturalists.

Barrington shot one male Whimbrel in North Ronay in 1886. The birds Harvie-Brown saw on the 16th June 1885 were *silent*, crouching and running, and did not get upon wing: all their actions were *distinctly* those of breeding birds. The birds Barrington saw, he *thinks, were not breeding*, and in 1887 Harvie-Brown found no birds there at all. He fears the link was broken. Why shoot a Whimbrel to identify it? There are few things more to be regretted in our personal experiences than having to leave Ronay without *any* attempt to find the nest or young in 1885.

Numenius arquata (*L.*). Common Curlew.

Gaelic—*Crotach-mhara*: *Crotach-mara*, ♂: *Crotag-mara*, ♀: *Guilbneach*: *Squilbneach*=the billed one.

W. MacGillivray says, in his earlier writing:—"Common, retiring to the heaths to breed in summer." It is included in our lists from Mhorsgail, and near Stornoway, but we find that MacGillivray makes no mention of it as a breeding species in 1841; indeed he says: "I did not find it breeding." Gray takes no notice whatever of it as occurring in the O. H. W. MacGillivray in his later and larger publication says it is "apparently common," but probably meaning in autumn, and says nothing as to its breeding. John Swinburne records having seen five or six pairs on North Ronay, but Harvie-Brown saw nothing of them in 1885; and Barrington records only one Curlew seen on North Ronay in 1886 (*fide* W. Williams, one of his party). The Curlews seen by Swinburne were stated to be "very noisy," and thus were quite different in their actions from the single pair of Whimbrels as related above. It is interesting to note from the schedules of the Migration Committee returned from Monach Island in 1887, that Curlews as well as several other species arrive at Shillay, the outermost of the Monach group, invariably from the north-east about the end of July or beginning of August, and that about this time they came daily and nightly, and that, after coming so far, *as invariably*, they turn inland; "and," adds Mr. Agnew, "I have no doubt but that the old parent birds are leading the young to previously well-known feeding-grounds.

Their direction on leaving Shillay is towards the south-east." This appears to us, taken in connection with other observations, to be one of considerable interest, though the natural trend of the land of the inner islands of the Monach group would perhaps offer the most likely line of advance from Shillay.

Order 5. GAVIÆ.

Family LARIDÆ.

Sub-family *STERNINÆ.*

Sterna macrura, *Naum.* **Arctic Tern.**

Gaelic—*Stearnal: Stearnain*=sea-swallow.—A. C.

Most writers agree as regards the superiority in point of numbers of this species over the next on the west coasts of Scotland and the Isles, and its generally more western distribution. Authors also speak of their both breeding together, but of this species being almost always the more abundant. Gray speaks of the Common Tern as breeding "in many of the islands of the Sound of Harris, and also on the rocky islands lying to the west of the principal group." And Captain Elwes in 1869 says, "The Common Tern, which is also common in the Outer Hebrides, usually lays two eggs." It is included in the Stornoway list by Mr. D. Mackenzie, to the exclusion of the Arctic Tern, which is not on his list; but Mr. H. H. Jones, speaking of the Arctic Tern in Loch Roag in 1881, says, "It is the only species of Tern I have seen there." Our own experience bears out the superabundance of the Arctic Tern, and inferior population, *if not utter absence, of the next species* from the list of Outer Hebridean species.

No tern is taken note of in North Ronay in 1883 by Swinburne, but Harvie-Brown found the Arctic Tern on the northern peninsula. Barrington found the nests and eggs of a colony of about thirty pairs in the end of June 1886, and one bird was "shot to make sure." In 1887 the species again observed on Ronay were undoubtedly Arctic, and they had abandoned the north peninsula and occupied the slope above the south-west promontory. Harvie-Brown found many nests amongst broken ground interspersed with bare rock, bits of sea-pink sward and tufts, and wind-swept breaks, but the birds seemed to prefer the sea-pink patches in the hollows, and to have the nests lined with small gravel or pebbles. The colony is a fair-sized one, but much scattered.[1] In a letter from Professor Newton, dated 31st July 1887, he states that during his cruise this year in Mr. Evans' yacht, the *Erne*, "we were not conscious of having seen a Common Tern at all north of Jura." In our cruise during June and July we may say we utterly failed to identify a single Common Tern anywhere to the north of the island of Coll, and we paid more careful attention to the comparative distribution of the species than usual, even going so far as to shoot specimens at most of the localities visited where Terns were breeding. Again in 1888 our experiences were similar. Altogether Harvie-Brown has visited a great many Hebridean terneries, and paid special attention to their inhabitants. We did not on this occasion visit Haskeir off Uist, where Elwes stated he found both species, nor Haskeir off Canna, where we found only the Arctic Tern in 1879. We are careful to mention such minutiæ, because, as we have always experienced, Terns are perhaps more "shifty" and uncertain in their breeding distribution in the isles of Scotland than most species which breed in colonies.

Terns usually arrive about the middle of May at or near their breeding stations. In 1888 they were much later, and however punctually they may make their first appearance, the actual time of their occupation of the nesting sites varies greatly with the weather, as also to some extent in different localities, especially as we have noticed in the Hebrides and Western districts of Scotland.

It is not until some time after the first appearance of the terns that the ova of many species of fish hatch out, nor do the fry at once approach shorewards and surfacewards. Many are hatched out on shoals and sand-banks, or even in deep water at a distance from land; and, in a late season especially, terns and other birds have to feed at a longer distance from their haunts on this account. Of course diving and swimming birds, such as Rock-birds, Shags, etc., are not influenced quite to the same extent as surface-feeding species. The time of the laying of the terns is coincident with the time of their food-supplies being most accessible to them. The natural history of birds

[1] The species observed on the Skerries (of the "Stack and Skerry group, west of Hoy Head) were, *as decidedly*, all Arctic Terns in 1887.

and fishes in similar respects is therefore closely connected with each other's existence.

Sterna fluviatilis, *Naum.* Common Tern.

Considerable confusion still exists amongst everyday observers and recorders as to the exact distribution of this bird as compared with the last, as will be evident from the quotations given under the allied form. As for ourselves, for years we have searched the Hebrides for a Common Tern, and never yet can we record a single occurrence we could positively identify. Yet we have often seen them together in other localities, and often, "to mak siccar," shot specimens of each.

Howard Saunders—our best authority on Terns and Gulls—found a wing of the Common Tern on Haskeir off Canna in 1886, where, in 1879, Harvie-Brown found nothing but Arctic Terns; but the following may explain many puzzling accounts. He writes: "The Arctic and Common Terns, where they are found breeding over the same area, not unfrequently change their ground, according to my experience. For instance, at the Farnes, in one year, a colony of Arctic Terns may be found, and on another occasion the same ground may be occupied by Common. I am inclined to think that the Common are gradually extending, but my data are limited. . . . On the wing, when near enough to see, the under parts of the Arctic Tern are slate-grey, and those of the Common are of a lighter and vinous grey; added to which, of course, there are the difference in colour of bill, length of wing and tail; but these points may be missed by the eye, as many birds are by the gun, because they won't come near enough, nor keep still long enough. The man who has been lying on the shingle watching two colonies for a quarter of an hour ought to have learnt to distinguish the species better than any book-maker can teach him." With all these remarks we perfectly agree, but still we have failed, utterly failed, to identify the Common Tern in the O. H., Elwes' and Saunders' remarks duly considered. Whereas Captain Elwes found a colony of Arctic Terns on the low portion of Haskeir on June 30th, 1868, of from 80 to 100 pairs, Harvie-Brown did not see a single bird of any species of Tern on 1st June 1881, and they had only just begun to lay nearer the mainland, as on the previous day one of his crew took a single egg of the Arctic

Tern on a rock in Newton Bay. Elwes got most of his eggs hard set. Howard Saunders never identified the Common Tern in the O. H., and only found the single wing on Haskeir off Canna, about which there can of course be no doubt.

In our Migration Reports are many instances of the irregularity of terns' behaviour at their nesting-places, whether of the Arctic or the Common species; and perhaps still more markedly in the Little Tern. This unsettled habit is worthy of remark. They often occupy and then abandon their nesting-places for apparently no particular reason, for it is not invariably because they suffer persecution, though they are more easily scared than most other sea-birds. In the Hebrides there are innumerable places where terns might breed, having, to all appearance, equal advantages with the selected spots; and possibly the very fact that they are naturally timid birds causes them to take advantage very frequently of a change of residence. In our Migration Report for 1886 we find, for instance, that a flock of terns arrived at Little Ross in the Solway Firth, remained a week, and then left. In this case, however, they were, of course, only resting and feeding, probably without any intention of breeding there; but in many other cases such movements take place suddenly, almost in mid-summer, or in the middle of their nesting season, the dates of which vary greatly at different stations.

For purposes of identification during his cruise in 1887, Harvie-Brown often shot some terns from a colony, *both adult and immature*, thereby bringing the whole tern population close about his ears. He scanned all carefully, then lifting the dead birds carried them to a distance, and, by throwing them up in the air, again brought the birds all around him. There need never be any difficulty in bringing terns thus close enough for identification. If a little cruel, it may, we think, in such a case be forgiven, from the naturalist's point of view.

In a subsequent communication from Captain H. J. Elwes (dated October 12th, 1887), in reply to our inquiries, he writes: "As to the Terns, it is so long since I attended to British Birds that I can tell you nothing worth knowing more than I have printed in the *Ibis*, but I distinctly remember noticing, *as far as I could judge without killing* [the italics are ours], the two Terns were found breeding in North Uist, and remarking that one usually laid two, and the other one three, eggs." No specimen of the Common

Tern occurs in the late Mr. Gray's collection from any locality in the Western Islands, "the only specimens being from Girvan, Burntisland, and Buddonness," as we are informed by Mr. Wm. Evans of Edinburgh, who made out a catalogue of the collections at Mrs. Gray's request. Under all these circumstances, and except for Howard Saunders' record of the wing on Haskeir, we would have no hesitation in rejecting it from our list for the present. If it is a spreading species it may before long obtain fuller admission.

Sterna dougalli, *Mont.* **Roseate Tern.**

Sterna minuta, *L.* **Little Tern.**

Sterna bergi, *Lichtenstein.* **Rüppell's Tern.**

Sterna caspia, *Pall.* **Caspian Tern.**

[*Obs.*—A naturalist told Mr. D. Mackenzie that he had seen a Caspian Tern on the west coast of Lewis "last season" (*i.e.* 1880), but we have no further evidence, nor do we know the name of the person who told Mr. Mackenzie this, nor at present are we able further to interrogate Mr. Mackenzie.]

Sterna anglica, *Mont.* **Gull-billed Tern.**

Sterna cantiaca, *Gm.* **Sandwich Tern.**

Sterna fuliginosa, *Gm.* **Sooty Tern.**

Sterna anæstheta, *Scop.* **Smaller Sooty Tern.**

Hydrochelidon hybrida (*Pall.*). **Whiskered Tern.**

Hydrochelidon leucoptera (*Schinz.*). **White-winged Black Tern.**

Hydrochelidon nigra (*L.*). **Black Tern.**

Anous stolidus (*L.*). **Noddy Tern.**

Sub-family *LARINÆ.*

Gaelic—*Faoileag : Faoilean : Glas-Fhaoileag*=grey gull.—A. C.

Xema sabinii (*Sabine*). **Sabine's Gull.**

Rhodostethia rosea (*MacGill.*). **Cuneate-tailed Gull.**

Pagophila eburnea (*Phipps*). **Ivory Gull.**

[*Obs.*—This is included without remark in Mr. D. Mackenzie's list (1881), Stornoway Museum.]

Larus atricilla, *L.* **Laughing Gull.**

Larus ridibundus, *L.* **Black-headed Gull.**

Gaelic—*Cra Fhaoileag : Faoilag-ceann-dubh*=gull of the black head. *Ceann-dubh : Ceann-dubhan.*—A. C.

In 1841 this species is quoted as abundant in all the marshes of the Uists and Benbeculay ; also near Rodel in Harris, and in Berneray, and as breeding in all these situations. And in 1830 a prior account by W. MacGillivray says : "Arrives in May and breeds on islets or lochs." Even in 1871 Gray speaks of it generally as common, and its "breeding haunts numerous," throughout the west of Scotland and the O. H. It is reported as breeding at Carn Cottage shootings near Stornoway in 1879 by Mr. Greenwood, who obtained eggs.

Near Newton, North Uist, on Loch-an-Dune, is a very fine colony of these birds, but this has been reported to Harvie-Brown, over and over again, as the *only* colony in that island. We visited it in company with Sir John Orde and Mr. John MacDonald on 10th June. A dozen pairs of Arctic Terns and Common Gulls were breeding there also. Black-headed Gulls were also seen in Stornoway harbour on 10th June 1885, probably belonging to the colony mentioned above by Mr. Greenwood. On the island of Loch-an-Dune just referred to, which is a tidal arm of Loch Maddy, and which runs almost across the island, there is a remarkably fine sample of a Pictish ruin, all over the ancient stonework of which are great masses of very long and pendent white lichens. On several occasions we found the eggs of the Black-headed Gull and of the Common Gull in the same nest—a sort of mutual-accommodation colony—and the same has been observed by us elsewhere. One nest was actually built *in* the water, in a small creek or bay in the peat, out of which it could not be floated by the tide, the latter not lifting it high enough to float it out of the little retort-shaped bay, and the nest itself being of greater diameter than the neck of the retort. On another island in the same loch Herring Gulls were breeding, and a pair of noisy Oyster-catchers. Of late years, owing to unusually high tides, a part of the peat island has been washed away, somewhat reducing the area of the Arctic Terns' and Black-headed Gulls' breeding-ground.

The only bird of this species ever seen on St. Kilda was on the 13th April 1887 by Mr. George Murray, who was teacher there, during the year June 1886 to June 1887.

Larus melanocephalus, *Natt.* Adriatic Gull.

Larus ichthyaëtus, *Pall.* Great Black-headed Gull.

Larus minutus, *Pall.* Little Gull.

A specimen of this rare British, and still rarer Scotch, bird was shot upon 1st November 1883 by Mr. John MacDonald, Sir John Orde's factor, at Newton, North Uist. It was seen on the sandy island of Berneray the day before—beside the lake which we have before spoken of as a great haunt of silvery eels—and when shot at Newton was found to have only one foot. The bird, which Harvie-Brown had an opportunity of examining in the flesh, is evidently a bird of the year. The specimen was exhibited, and a paper read upon it, by Harvie-Brown, at a meeting of the Royal Physical Society, Edinburgh.[1]

From the notes given in the paper referred to, it appears that this North Uist specimen gives the furthest westerly record in Europe of the species. The Sutherland specimen is one of the most northerly, the Wigtownshire one the most southerly, in the west of Scotland, East Lothian and Berwickshire specimens being nearest to their true main line of migration across the North Sea, which usually strikes the Northumberland coast near the Farne Islands.

[1] *Loc. cit.*, 19th December 1883—"On the Occurrence of the Little Gull (*Larus minutus*) in the island of North Uist, with remarks on the Objects of the International Ornithological Congress at Vienna, and on Uniformity of Method in recording rare species in future."

Larus philadelphia, *Ord.* **Bonaparte's Gull**

Larus canus, *L.* **Common Gull.**

Gaelic—*Crann Fhaoilag* or *Crion Fhaoileag* = small blue gull: *Faoileag Bheag an Sgadan* = small herring gull.

Appears, according to an early record, only to have bred inland occasionally, but was even then marked as very common. Eggs, said to have been obtained on Boreray, St. Kilda, appear to us to be of doubtful authenticity. But by 1871 Gray relates that its principal breeding-haunts in the Hebrides were in the interior, "as there were no suitable cliffs to occupy." Mr. H. H. Jones found it commonly breeding "on the coast near Mhorsgail, on the west of Lewis, but I have never found them nesting inland here." We have found that it accommodates itself to nesting-places on islands off the coast, or inland lochs at different places, but we have never found it breeding anywhere *on cliffs*: notwithstanding all the statements of other observers, we still hold this habit must be exceedingly exceptional.

In 1870 we visited perhaps one of the most strikingly picturesque colonies of this bird in the Hebrides. Lieut.-Col. Feilden writes, and we quote his description *verbatim*: "A very large colony of *Larus canus* breeds on this island, and the nests are thickly scattered over the whole of it. All the eggs were fresh (this was on the 13th May 1870). Perhaps one-half of the nests had their full complement of three eggs, many had two, and a few only one. We gathered 300 eggs of *Larus canus* on this island, and nearer to the water's edge we took a dozen nests or so of Lesser Black-backed Gulls." The factor, Mr. John MacDonald, assisted us; indeed, the eggs were collected by orders of the proprietor, and were, for the most part, preserved in lime-water to be sent to him. From the whole we selected what now constitutes our series of eggs of the species. It is upon this island that the finest deer-grazing is found in North Uist.

Larus argentatus, *Gm.* **Herring Gull.**

Gaelic—*Faoileag Mhor an Sgadan* = the large herring gull; large gull of the herring.

In the times of the earlier records, the Herring Gull appears not to have been found breeding at inland localities, at least so far as

J. MacGillivray's experiences went, though he adds significantly, "as the Common Gull occasionally was." "Recorded at St. Kilda and Hasgeir Rocks," and "betake themselves in autumn to the west side of the Isle of Harris, North Uist, Benbeculay, and South Uist." Very abundant on the west coast of Lewis and the Mhorsgail deer-forest, and also on the fresh-water lochs. But a most curious note is that by Mr. Greenwood, who says it is "never seen by him at Carn." This seems very strange, if true.

The following is a description of the nesting-place of a colony of Herring Gulls in North Uist, which was quite unmixed with the allied Lesser Black-backed Gull. It is by Lieut.-Col. H. W. Feilden, 1870: "This islet being uneven, and covered with *Luzula sylvatica*, the nests of the Herring Gulls were more diversified than is usually found on flat grassy islands; nests were perched here and there on ups and downs, and some were composed of heather, some of bracken, and others of *Luzula*; in all, however, there being at the bottom a thick substratum of *Luzula*. One nest attracted my special attention; placed under a large shelving stone, covered with rich green moss for a background, whilst a plant of bright green bracken waved over it on either side. Three eggs laid on a bed of fresh *Luzula* completed the picture."

Fairly common on the Flannan Isles. In North Ronay Swinburne saw several pairs, Harvie-Brown found a few, and Barrington, who spent several days on the island, found "several breeding, but not so common as the Lesser Black-backed Gull."

On Eilean Mhûire, of the Shiant group, we found it common, but nowhere really abundant, simply, we believe, because every available vantage was occupied by Puffins. Very different is this state of their colonisation from what is the case, for instance, on the coasts of the Dunnet Head peninsula of Caithness and elsewhere on our coasts, where they are assertively abundant.

Larus fuscus, *L.* Lesser Black-backed Gull.

Not so common as others of the genus, but there are large colonies found on the fresh-water lochs, and it is always recorded as common, except by J. MacGillivray, who, writing in 1830, says, "it is of rare occurrence."

In Mhorsgail deer-forest a few pairs occur among the Herring Gulls; it also occurs on the shooting of Carn, near Stornoway. It is, however, rarer on the west coast of the Hebrides, and the

Herring Gull would appear to take its place, and be of more oceanic, or at least maritime, habits than this species. Yet we find this species recorded at St. Kilda also. On Haskeir, off Uist, this and the previous species appear to be in about equal proportions.

They are comparatively scarce on the Flannan Isles, though there are considerable colonies, and a few closely-packed slopes of, say, a few hundred pairs in all. On a gun being fired, probably about one-half of the birds flew off, giving a rough idea of their numbers, leaving the sitting birds, or those tending their young, on the rocks. On North Ronay, whilst Swinburne "did not notice any Lesser Black-backed Gulls," though "he had no doubt they were there!" and though Harvie-Brown also saw nothing of them during his hurried visit in 1885, Barrington marks them as plentiful, "breeding on the south side," and again in 1887 Harvie-Brown saw a few on the slope of the south-west promontory.

On Eilean Mhŭire a small colony—about forty pairs or so—occupy the eastern extremity beyond the col mentioned in our descriptive chapter. Here the slope becomes abrupt near the extremity, towards the bare rocky point.

Larus marinus, *L.* Great Black-backed Gull.

Gaelic—*Farspag: Farspag: Farspach: Farspreig Sgluirach* (Argyll) = in first year's plumage.

Abundant. Found on Shillay, at Towhead, and in a considerable colony on an inland loch in North Uist. Has been stated to lay usually only two eggs, whereas other gulls lay three, but this does not hold good. Mr. H. H. Jones writes that a good many breed on certain lochs in Lewis. Usually this bird breeds in single pairs, slightly apart from other species, but in the same gullery. We know, however, of many other colonies of this species in Scotland. They are also found breeding more or less around the coasts, and not far from Stornoway. On Haskeir, and many other such islands, perhaps only one pair or so have taken up the most commanding positions. On the last-named island a pair were breeding in 1881, near to the western end of the main island, amongst a crowd of Herring and Lesser Black-backed Gulls, at a point above the brackish puddle of water, which is frequented by the seals during their breeding season.[1] Only a few pairs were observed on the Flannan group of islands, on Eilean Gobha and Bron-na-cleit. They are abundant in a colony in North Ronay.

The Great Black-backed Gull is abundant in Sulisgeir, and we saw a few of the Herring, but none of the Lesser Black-backed Gulls, nor were Kittiwakes very abundant; and on Eilean Mhūire, and elsewhere in the Shiant group of islands, there are a good many scattered pairs, but no colonies, so far as we could observe.

Our greatest personal experience of a colony of these birds was on one of the Outer Hebrides. On the 14th May 1870, on arriving at the inland loch, we found "about twenty pairs" (*sic*) occupying three low green islands, but the day was too stormy for us to attempt to reach the colony in our india-rubber boat, so we tried again two days later, and this time successfully. We counted the birds several times, and came to the conclusion that twenty-five pairs constituted the colony. Our assistant, who went out in the boat, brought back seventy-four eggs, varying wonderfully in colour, about half of which now represent our series of their eggs in our collection. Our collector reported that only four nests contained two eggs, and all the rest had three. (Allowances are made for slight discrepancies in our estimate, or his report.)

Larus glaucus, *Fabr.* Glaucous Gull.

In 1830 W. MacGillivray mentions the Glaucous Gull as "not at all common on the west coast." In 1862-63 considerable numbers were seen in North Uist, chiefly on the west side and in the Sound of Harris; and Gray says they also visit the shores of Benbeculay and South Uist. Mr. Bertram B. Hagen records one in his possession, obtained in November 1885,[2] and he met with another there in the following January. "A Glaucous Gull visited the islands of St. Kilda about the middle of November 1886, and remained for a week. I failed to kill it" (George Murray, schoolmaster, *fide* Rev. H. A. MacPherson, *in lit.*).

Larus leucopterus, *Faber.* Iceland Gull.

Whilst earlier writers are silent as to its occurrence, and all accounts are negative up to 1871, Mr. D. Mackenzie reports that several

[1] See under Great Grey Seal, *antea*, p. 23.

[2] Mr. B. B. Hagen, present (1887) lessee of Gress shootings, Stornoway.

are seen every winter at Stornoway (1881). Mr. B. B. Hagen observed one Iceland Gull at Gress, and as he knows the Glaucous Gull (*q.v.*), this may be considered as of value in accentuating its occurrence in these parts.

Rissa tridactyla (*L.*). Kittiwake.

Gaelic—*Tarroch: Scolhag: Ruideag.* In Barray, *Sgaireag: Seigire.* Mr. Finlayson gives *Seigir*, from Mingulay.—A. C.

According to J. MacGillivray, only breeds in Haskeir and St. Kilda, in the latter place in large colonies; and W. MacGillivray partly confirms this by saying that, in 1830, it was only found in some of the great breeding-places of the birds, where it arrives in May.

It is very abundant in the North Minch after the breeding ledges are abandoned. Its winter residence did not appear to have been located, at the time Gray wrote, to any portion of the Scottish coasts, except as an unusual occurrence. *Now* it may be looked upon as a regular winter visitant in greater or less abundance. Elwes found a colony on Haskeir off Uist, in 1868, but in 1881 Harvie-Brown saw nothing of them there; but Elwes was there in the latter end of June, and Harvie-Brown at the beginning of the month. A considerable colony was noticed in 1881 at Flannan Isles, on a "clett" or "stack" close to Eilean-Mòr. Kittiwakes breed on the north side of Sulisgeir, and also on the north-west and north-east sides of North Ronay (Swinburne, Harvie-Brown, and Barrington). There are large colonies on the west horn of Ronay, occupying as usual the lower ledges, especially the sides of a great cave to the eastward of the granite vein of MacCulloch, and there are very fine colonies on Eilean Mhùire, especially that on the south side of the col.

Speaking of this species, Mr. J. Finlayson, schoolmaster at Mingulay (whose knowledge of the native bird-life is accurate, and is the result of continued residence amongst their haunts), says: "The Kittiwake begins early in summer to build a nest of mud and weeds. It builds it in ugly sloping places, so that heavy rains sometimes carry away the nests. The loss of their nests, I believe, compels them often to be the last of the rock-birds to leave with their young. The time required for hatching is about four weeks. They leave as soon as ever the young ones fly off the rocks. Every bird returns to the exactly same spot of the cliff every year, and this remark applies also to the other species of rock-birds."

Sub-family *STERCORARIINÆ.*

Stercorarius catarrhactes (*L.*). Great Skua.

Stercorarius pomatorhinus (*Temm.*). Pomatorhine Skua.

Though believed to breed, or to have bred, in the O. H., there has been no corroborative evidence since Gray wrote; but there cannot be any doubt as to its frequent, if not regular, summer visits to the coasts of these islands, and the seas to the west of Lewis, most of such birds proving to be old birds in most perfect plumage. It frequents the harbour of Carloway on the west of Lewis, and the seas over the great cod-banks between that and the Flannan Islands, as also still further to the westward, where we have shot them from the deck of our yacht. It is reported also as occurring some years in autumn by Mr. H. H. Jones, who has shot specimens on the west coast in October 1879, and has seen several others. We have seen it flying eastward during a gorgeous sunset and rainbow effect in July 1887, across Loch Eport in North Uist, and we have also met with straggling birds as far east as the Sound of Rum amongst the Inner Hebrides. We have also inspected an immature specimen, alas! in a high state of decomposition, forwarded by Mr. John MacDonald of North Uist, and shot November 1887 on the west coast of that island.

Stercorarius crepidatus (*Banks*). Richardson's Skua.

Gaelic—probably generic also—*An Fasgadair*=the squeezer.—A. C.

Breeds on a few spots in the interior of the islands, and is seen chasing terns on the coast. "Unusually abundant in the Sound of Harris in 1848," as witnessed by Sir W. E. Milner, "pursuing gannets." At Loch Eoin—the bird loch—several Richardson's Skuas were seen and one egg obtained in the end of May 1871. They had just begun to lay, three weeks after the other gulls. Local observers state that a pair of skuas usually consists of a black- and a white-breasted bird, but a pair which the Rev. H. A. MacPherson watched in 1886 were both white-breasted. Though we happen to be aware of where these birds breed regularly in large numbers in Scotland, we nevertheless consider them rare enough to deserve consideration by proprietors and shooting tenants, who ought to prevent them from being slaughtered indiscriminately in the breeding season at their nesting sites.

Stercorarius parasiticus (*L.*). Buffon's Skua.

Messrs. Gray and Dewar record it, or speak of it, as breeding on Wiay. Lieut.-Col. Feilden spent a long, hot, hopeless day in search thereof, but found no trace of it. No doubt, however, in 1863, a pair was shot there by Dr. Colin MacRury, now of the Indian Army, and was examined by Gray. Wiay is not a small island, however, as Lieut.-Col. Feilden found out to his cost, after he had traversed it in all directions, although it is designated as such by Gray.

Order 6. TUBINARES.

Family PROCELLARIIDÆ.

Procellaria pelagica, *L.* Storm-Petrel.

Gaelic—*Amhlag-mhara* = sea-swallow: *Luaireag: Luaireagan: Famhlag: Aisileag* (St. Kilda): *Loireag.*—A. C.

The complete distribution of this species is not yet accurately known, but there can be little doubt it is much more common, and more widely distributed, than is at present generally supposed, though decidedly pelagic in its habits, and in its choice of breeding sites. We find that it breeds on Shillay, as recorded in 1830, and subsequently that it was found on Boreray, near St. Kilda, where, however, it had not begun to lay by the 13th June. We consider it a much later breeder than the next species; and we are personally acquainted with many of its nesting haunts. It breeds commonly on North Ronay, as discovered by Barrington in 1886, and even, in isolated cases, among the galleries of the allied species as observed by H. G. Barclay and Harvie-Brown. Colonies may be quite truly described as difficult to find, but are very generally common. We had very little time or opportunity to find it on the Flannan Isles, but it is reported to us as occurring there by Mr. H. H. Jones.

At Soay, St. Kilda, a party on 13th June got one Storm-

Petrel's egg and three old birds, and this was considered unusually early, and also, on the same date, but in another year, one Fork-tailed Petrel only.

In 1886 a regular stream of migration of Petrels seems to have taken place with the "great rush" of other species on the 5th-6th October, as they were reported from several stations, as striking the lighthouses for several nights in succession about that time. A winter rendezvous or station of these birds was also discovered in South Uist, in the days of the MacGillivrays, as noted by Alexander Carmichael. Little has been recorded of the winter haunts of the Stormy Petrel by ornithologists, so the note which I extract from Carmichael's MS. in 1886 may prove of interest. After giving the various Gaelic names by which it is known, Carmichael adds the note: "They remained on Harstamal, Ru-nan-ordaig, in South Uist, during the months of November, December, and January 1822-23. John MacDonald, usually called John the Pilot, from his being the trusted pilot of Captain Otter during the whole Admiralty Survey of the Western Isles, used to hear them 'crooning' under him as he lay there watching the wreck."[1] Similar accounts have reached us from other localities of their winter habitations and colonies, about which we may have more to say at some future time.

Information to date of 1886, is that of Mr. John Finlayson, a gentleman whose abilities as a field-observer, according to his opportunities and location, have already been taken notice of; and he informs us in a letter, dated 9th April 1887, that "petrels' nests used, of old, to be found in the rents of peat-banks (*i.e.* in Mingulay), but none now." And in 1887 we made the following note when visiting the same island: "The old peat-banks in which they used to breed are nearly exhausted, and, though new ones have been opened out, these have not as yet been taken possession of." There can scarcely be any doubt that there are innumerable colonies of Petrels throughout the isles, as yet unknown to us. Their nocturnal habits, and general silence in their holes till dusk begins to creep down, render their discovery not always easy, and they are later in breeding than their allies,

[1] The wreck of a slate-laden (?) ship in which he had a pecuniary interest. He is now captain of Harvie-Brown's yacht, and volunteered the above information to us on more than one occasion.

the Fork-tailed Petrels. The Storm-Petrel is occasionally found dead on the shore of North Uist.

The following note cannot fail to prove of interest in the life-history of a very interesting species. In a letter from C. Darwin to F. D. Hooker, March 2, 1859, occurs the passage: "I think it would be hardly possible to name a bird which apparently could have less to do with distribution than a petrel. Sir W. E. Milner, at St. Kilda, cut open some young nestling Petrels, and he found large curious nuts in their crops; I suspect picked up by parent birds from the Gulf Stream, etc. . . . Petrels at St. Kilda apparently being fed by seeds raised in the West Indies." [1]

Procellaria leucorrhoa, *Vieill.* Leach's Petrel. Fork-tailed Petrel.[2]

Gaelic—*An Gobhlan Mara* = the fork-tailed of the sea.—A. C.

Discovered on St. Kilda by Bullock in 1818;[2] but not taken notice of in 1841 as resident or as occurring there. Sir W. E. Milner, however, in 1848, found a colony of the Fork-tailed Petrels on the Dun (or Dune) of St. Kilda. In one hole only did Sir W. E. Milner find a male and female together. He considered this species quite three weeks earlier in laying than the Stormy Petrel, and where the former were found, not a single specimen of the latter was discovered (but see "North Ronay"). There is a specimen of the bird in Sir W. E. Milner's collection in the Leeds Museum from St. Kilda, obtained in 1847. But, since Sir W. E. Milner's visit, this species has been found most abundantly on Boreray, another of the St. Kilda group, and it is recorded from several other localities, such as Mingulay and Barray—"a few pairs in holes and cracks in the peat" (Elwes). Gray also includes Benbeculay, Rum, Barray, Pabbay, Skye, Eigg, and Cannay, as localities whence he has obtained specimens. Of course their distribution is very general over the west coast of Scotland in winter, but more restricted in summer, and not yet thoroughly worked out; and we consider it to be more western—or more pelagic—than the last in its choice of breeding haunts.

[1] *Life and Letters of Charles Darwin*, vol. ii. pp. 147-148, second edition.

[2] See Fleming's *British Animals*, p. 136.

[illegible]

[illegible] Petrel is [illegible]

[illegible] of [illegible]

[illegible] C. [illegible]

[illegible] 2. [illegible] the passage: [illegible]

[illegible] Sir W. [illegible]

[illegible] Petrel [illegible]

[illegible] St. [illegible]

[illegible] in [illegible]

From [illegible] Petrel [illegible] Fork-tailed [illegible]

[illegible] A. C.

Bullock in [illegible] Sir W. E. M. [illegible] the Fork-tailed [illegible] the [illegible] W. [illegible] Sir W. [illegible] M [illegible] tion [illegible] St. [illegible] W. E. M. [illegible] few [illegible] in [illegible] ation [illegible] in [illegible] and [illegible]

[illegible] p. [illegible]

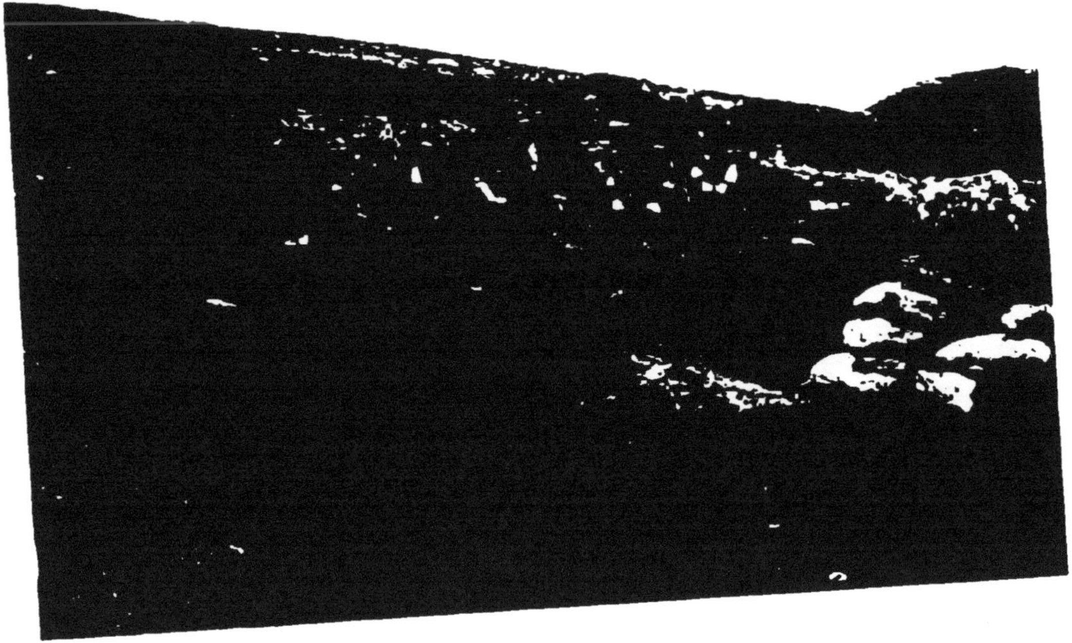

OLD CHAPEL IN RONAY. BREEDING HAUNT [illegible] TAILED PETREL

We have already given most of our own experiences of the species (*antea*, p. li.).

Puffinus anglorum (*Temm.*). Manx Shearwater.

Gaelic—*Sgrab* (Barray): *Sgrabail Sgrabaire* (St. Kilda): *Fachach* = (the young).

In 1841, recorded as seen occasionally at sea, but on land only at St. Kilda; these remarks now, however, give way to a great extent to those of later historians. In the spring they are found commonly off Gallon Head in Lewis (*auct.* H. H. Jones), and even, but less commonly, between the Butt of Lewis and North Ronay. Gray reports Pabbay, Barray, and St. Kilda as breeding stations, and that Barray Head or Berneray has been deserted by them since the lighthouse was built, none nesting there since 1843 (*fide* Elwes); and in some localities, which are known to us, they are often driven out, in the struggle for existence, by the Puffin. They are said by Mr. D. Mackenzie to breed in the Flannan Isles, but they are certainly more numerous amongst the Inner than the Outer Hebrides. At St. Kilda, Soay is the main breeding-place, and there are none on the principal island, nor on Boreray, and but very few, if any, now on the Dune.

Shearwaters are said to have been scarcer for some years past off Coll and Tiree and off Colonsay. We only saw *one* in June 1886, whereas usually in passing at the same time of year, we have found that portion of the sea covered with them, during their ceaseless flight. This observation has also been made by others who are in the habit of passing these points twice every week during the year—officers on board the S.S. *Dunara Castle*. Whether the decrease is only local, caused by change of food-supplies, or is a permanent decrease in their numbers at their breeding-places throughout the Outer and Inner Hebrides, is of course, difficult to determine. We shall have more to say of this species on another occasion.

They are also much scarcer on Mingulay and Barray since Elwes' visit. In Eigg and Cannay the natives take some for food, but this cannot sufficiently explain such an apparently patent decrease. Swinburne saw the species within a few miles of North Ronay, but searched there in vain for the nest; and Barrington met with it near the Butt of Lewis; but, during repeated attempts to reach North Ronay in 1885, Harvie-Brown saw nothing of it in that direction. In 1887, however, numbers were seen *west* of the Lewis, not, however, in flocks or large parties, but always single birds, or at most in pairs, pursuing their ceaseless flight, and never, as seen among the Inner Hebrides, sitting in vast flocks on the surface of the water. The "Fachachs," known by this name both in the Outer and Inner Hebrides, were at one time very abundant on the Stack of Lianamul among the cliffs of Mingulay, but they have long since been driven out by Puffins. Upon a certain slope of detritus in an adjoining island, however, there are still known to be a few pairs.

Puffinus griseus (*Gm.*). **Sooty Shearwater.**

Puffinus major, *Faber*. **Great Shearwater.**

[*Obs.*—That a large grey Petrel, twice the size of body of the Manx Shearwater, with wings shorter in proportion to its body, without any white parts visible, darker brown on the back, lighter on the lower parts, and flying often higher, and with the skimming phase of flight of shorter duration between the wing-strokes, has several times been seen (as in 1887 by our yacht's company), cannot be doubted, but the species has not yet been clearly identified.]

Fulmarus glacialis (*L.*). **Fulmar.**

Gaelic—*Fulmair : Fulmaire.*—A. C.

A speciality of St. Kilda, and seldom seen near the Long Island unless in gloomy weather, or in early morning or evening, or when following the s.s. *Dunara Castle*, on her return voyage from St. Kilda, as they almost invariably do, leaving, however, regularly, off the western entrance of the Sound of Harris: this bird is invariably seen and noted by all naturalists who have *done* the St. Kilda trip.

Mr. Henry Evans informs us he has a perfectly white Fulmar from St. Kilda, and had some of the dark specimens, which latter he gave to Howard Saunders (*in lit.* 25th November 1886). A favourite but cruel amusement of tourists is fishing for Fulmars whilst making the voyage returning from St. Kilda to the Sound of Harris. As the Fulmar has been known to extend its range of late years, within the memory of Herr Müller, the veteran ornithologist of Faröe, and has also taken up its quarters at Foula in Shetland, it is not unreasonable to expect its increase and extension to much nearer islands of the Hebrides. It regularly visits the west end of the Sound of Harris, and is a wide ranging bird over the ocean, but is rarer on the east side of the Long Island. When visiting the Flannan Isles (or Seven Hunters) in a yacht in 1881, we saw a few individuals close to these islands; and there was a report current that a branch-colony had arrived there and taken possession; but we failed at the time to make a note of the particulars. This was stated by several persons quite independently of each other. Corroborative evidence and proof are still wanting.

In North Ronay, Barrington saw *one* specimen of the Fulmar circling round the highest cliff of the island, and also met with a few between the Butt of Lewis and Ronay.

Harvie-Brown did not meet with a single specimen in 1885 during all the time he essayed to land on North Ronay, nor whilst on the island; but a more careful survey in 1887 revealed quite a number flying along the cliffs of the western horn, and alighting upon, and resting in, nest-like depressions in the sea-pink of the slopes as related at greater length in our descriptive chapter upon Ronay (*antea*, p. xlviii). It was also seen, even abundantly, *resting* on, and flying around, Sulisgeir.

Œstrelata hæsitata (*Kuhl*). **Capped Petrel.**

Bulweria columbina (*Moq.-Tand.*). **Bulwer's Petrel.**

Order 7. ALCÆ.

Family ALCIDÆ.

Alca torda, *L.* **Razorbill.**

Gaelic—*Duibheineach* = blackbird (in Mingulay, *auct.* J. Finlayson): *Am Falc: Lamhaidh* (St. Kilda): *Ian Dubh an Sgadan* = the black bird of the herring.—A. C.

Common on St. Kilda; scarcer on Haskeir. Common on the Flannan Isles. Not so abundant perhaps in O. H. as in many other parts of Scotland, but very numerous on Barray Head and Mingulay. Breed in very easily accessible ground on many of the smaller and less accessible islands, as, for instance, under loose boulders on the surface, or within arm's-length of the cliff tops. Common in North Ronay. Common all round Eilean Mhuire of the Shiant group, perhaps most prominent in numbers at the eastern horn or col, but mingling throughout all the puffin cliffs, as well as among the talus slopes. Possibly more abundant in Mingulay than in most of the O. H., or indeed than in most parts of Scotland, a continuous stream often passing overhead from the higher crevices of the göe—the waters of which lave the base of Aoinaig—and also at the north-west cliffs of the island.

Alca impennis, *L.* Garefowl.

Gaelic—*Gearr bhul: An Gearra-bhal: An Gearrabhal*=the strong stout bird with the spot: *An Gearrabhul.*—A. C.

[*Obs.*—To repeat all the history of the Garefowl or Great Auk to some may not seem necessary, after the exhaustive accounts by Professor Newton, Gray, and the later publication of Mr. Symington Grieve, but the following abstract of, or extracts from, their works may prove interesting, and preserve the continuity of our volume, as belonging entirely to St. Kildian and O. Hebridean records. The account here given has been repeatedly told to tourist visitors and ornithologists visiting St. Kilda. We ourselves, when we visited St. Kilda in 1879, devoted more of our attention to a survey of the island, spending our short four hours on shore in endeavouring to see as much of it as possible, and ascending Connachar; and in this way we did not interview the old St. Kildian personally on that occasion; but the tale told by him to Mr. Boyd and other listeners varies little from the accounts given at later dates to other people by the same individual. We are obliged to Professor Newton for condensing the accounts, and the following is very nearly the exact contents of a letter we have received from him upon the subject.

"The bird, frightened by men on the cliff, jumped into a boat in which was a boy of fourteen years of age, named Donald MacQueen, whose son of the same name—now a man of from fifty to fifty-five years of age—gave Mr. [Henry] Evans these particulars, and heard his father say he caught the bird thus. It was on the main island, *i.e.* St. Kilda itself.

"But it also seems, from Mr. Evans' information, that another bird was caught on Stack-an-Armine, in or about 1840, by some five men who were stopping there for a few days. Three of them were Lauchlan M'Kinnon, about thirty years of age—and now, or till recently, alive—his father-in-law, and the elder Donald MacQueen before mentioned—both now dead. M'Kinnon told Mr. Evans that they found the bird on a ledge of rock, that they caught it asleep, tied its legs together, took it up to their bothy, kept it alive for three days, and then killed it with a stick, *thinking it might be a witch.* They threw the body behind the bothy and left it there. M'Kinnon described the bird to Mr. Evans, so that the latter has no doubt about its having been a Garefowl.

"It was Malcolm M'Donald who actually laid hold of the bird, and held it by the neck with his two hands, till others came up and tied its legs. It used to make a great noise, like that made by a gannet, but much louder, when shutting its mouth. It opened its mouth when any one came near it. It nearly cut the rope with its bill. A storm arose, and that, together with the size of the bird and the noise it made, *caused them to think it was a witch.* It was killed on the third day after it was caught, and M'Kinnon declares they were beating it for an hour with two large stones before it was dead: he was the most frightened of all the men, and advised the killing of it. The capture took place in July. The bird was about halfway up the Stack. ('This last statement,' says Professor Newton, 'is, to me, conclusive that the bird could not have been a Great Northern or other Diver, as some have suggested.') That side of the Stack slopes up, so that a man can fairly easily walk up. There is grass upon it, and a little soil up to the point where they found the bird. Mr. Evans says that he knows there is a good ledge of rock at the sea-level, from which a bird might start to climb to the place. Mr. Evans tried in vain to fix the *exact* year in which this event happened, but could only get 1840 as an approximate estimate."[1]

Professor Newton adds: "I have an extract from the *Glasgow*

[1] We believe there are few people in Britain who have had longer or more intimate acquaintance with every cave and stack and intricacy of shore in St. Kilda than Mr. Henry Evans.

Herald of 14th June 1886 (sent me by Robert Gray) containing, among other things, a version of the 1821 story, which differs in some slight particulars from Mr. Evans. This I sent to him, and he (Mr. Evans) prefers his own."][1]

Lomvia troile (*L.*). Common Guillemot.

Gaelic—*Eun an't a Sgadan*=the bird of the herring: *Lamhaidh* (St. Kilda): *Langaidh* (Barray): *Langach* (Barray); or, *Langidh* (Finlayson).

Haunts the Little Minch at times in vast numbers along with Razorbills (see "The Shiant Isles and their Bird Life"). This is the most abundant of the true Alcidæ of the O. H. and west of Scotland. The bridled variety (*Lomvia lacrymans*) is also abundant, and Harvie-Brown and other writers reckoned their percentage at one in ten or twelve.[2] The following are notes taken at Barray Head, by Feilden and Harvie-Brown in 1871 :—

"*Ledge* 1, which we examined, contained in all twenty-eight Guillemots, of which five were bridled birds. Two bridled birds were evidently paired, as they appeared to be much taken up with themselves, to the disdainful exclusion of others on the ledge. One bridled and one of the typical form were also evidently paired, and the *typical male* was distinctly seen to tread the *bridled* female.

"*Ledge* 2 examined, contained twenty-nine Guillemots, of which five again were bridled.

"*Ledge* 3.—On this ledge two bridled birds rose off their eggs, and, as we observed on previous occasions, these were blotched and marked like other Guillemots' eggs. They were dark green in colour.

[1] Fleming, in his *History of British Animals*, says: "In winter the brownish-black of the throat and fore-neck is replaced by white; and I had an opportunity of observing this in a living bird brought from St. Kilda in 1822 (*op. cit.* p. 130). See also Rev. Dr. Fleming in his *Gleanings of Natural History* (*Edin. Phil. Journal*, vol. x. p. 97). This bird was captured by Mr. MacLellan, the tacksman of Glass, "some time previous" to 1822.

[2] By a slip this appears in Gray's *Birds of the West of Scotland* as *one in one hundred*. As will be seen, Harvie-Brown has since then ascertained their numbers to be even greater than the first imperfect record, or otherwise (?) that it is an increasing variety.

"*Ledge* 4	contained 12	birds, of which		3 were bridled.
5	" 16	" "		4 " "
6	" 10	" "	only	1 was "
and 7	" 21	" "		4 were "

On this last ledge a *bridled male* paired with a *typical female*, and was *seen to tread* after feeding her with a fish. Thus is shown that both males and females may be bridled birds."

From these and other statistics collected for many years, we now arrive at the conclusion that one bridled bird to *five* of the type is the average in the western islands, certainly in the O. H.

This bridled variety was also found at Soay by Sir W. E. Milner in 1848, and an egg taken. The supposed difference in the eggs of the two will not now be much reckoned upon by naturalists; indeed, it is scarcely worth while to discuss this point, at least that is our opinion, after repeated observations. At North Ronay, in 1886, R. N. Barrington places the average as seventeen out of sixty-four on one ledge. Guillemots are abundant—in keeping with extent of area—on all the suitable ledges round the main island of Haskeir off Uist; the eggs are, as a rule, more safely deposited than those of the Razorbill, and more difficult to reach, as is, we think, usually the case.

On the occasion of Harvie-Brown's first visit to Haskeir, he unfortunately forgot to carry his binoculars from the yacht, so he lost the opportunity of scanning all the further-off ledges for bridled birds. He only succeeded in making out one ledge as containing one bridled bird in six, close to the stone arch (mentioned in our description), but he has little doubt the average here will prove to be as great as at other western localities, such as Handa in Sutherland, or at Barray Head in the Hebrides. On Eilean Mhüire of the Shiant Isles, Guillemots were most abundant on the north side, where, if any difference really was appreciable, Puffins were scarcer than elsewhere throughout the circumference, and Guillemots were also abundant on the south side, at, and near the col. Harvie-Brown counted six ledges, and found the proportion of bridled birds at eleven of the bridled to 108 of the total. By similar observations in North Ronay he found the average to be about one bridled in every nine or ten, counting six ledges.[1]

[1] These notes must be taken along with many others, if it be desired to arrive at actual average over all the islands and west of Scotland. The bridled variety certainly seems to increase in numbers towards the west.

Speaking of this species, along with the Razorbill and Puffin, Mr. John Finlayson, schoolmaster at Mingulay, says: "These birds arrive on our coasts early in spring and occasionally visit the rock, making a short stay upon it each time, till they permanently settle upon it about the middle of May, when laying commences. The rock is now taken possession of by a population of birds representing two birds (male and female) for each single egg. The one bird is of course, and of necessity, absent, providing food for the other, and when the young are hatched out, for them also. On special days—perhaps states of the weather—the numbers of birds "in the rocks" are more than trebled by the visitation of millions of fellows, who act no part in the responsibilities of either father or mother. Of these additions to the inhabitants of the rocks the Razorbills predominate. The Guillemots are the first to appear, and they come in February. The Guillemots sit on their eggs quite closely packed one against another, and choosing the broadest shelves. In some instances," continues Mr. Finlayson, "they lay on sloping shelves, and in such cases it seems that the birds are destined never to stir off the eggs, for as soon as they do so the eggs invariably roll over the edge and perish. The male attends to the wants of the female, and in consequence is lean and light, while the latter is sottish and fat to an extreme. The female often allows itself to be caught by the hand, but that only at the time when the young one is nearly being hatched. The Guillemots and also the Razorbills are so blinded by their affection for their young that, during the week before and the week after the little ones are hatched, they allow themselves to be captured in hundreds. The way of capturing them practised in Mingulay and Barray is by a lasso of horsehair stuck to the top of fishing-rods. The Guillemot can carry only one fish at a time with the tail projecting, but the Razorbill and Puffin carry as many as twelve or more small herrings every trip. All the species lay several times again if their eggs are lost. I have seen the same shelf robbed three times of its eggs. The time allowed between each lifting may be fifteen or sixteen days.

"Razorbills hatch more isolated from one another. They are too vicious to suffer contact with a neighbour. Puffins burrow; never hatch in exposed situations. They are the last to have their young ripe for migration. The young ones are quite as large as their parents before they leave. Guillemots' and Razor-

[illegible]

[illegible] many [illegible]
[illegible] years [illegible]
[illegible] when [illegible]
[illegible] by a [illegible]
[illegible] families [illegible]
[illegible] more [illegible]
[illegible] looked out for the [illegible]
[illegible] of the [illegible] and [illegible]
[illegible] are more than [illegible] by the [illegible] of
[illegible] in the responsibilities of either
[illegible] as to the [illegible] of the
[illegible] The Girl [illegible] the best to
[illegible] The [illegible] that
[illegible] against [illegible] and [illegible]
[illegible] instances," continued Mr. [illegible]
[illegible] it [illegible]
[illegible] as they are
[illegible] people [illegible]
[illegible] and [illegible] is [illegible] and
[illegible] for an [illegible]
[illegible] the band but that [illegible]
[illegible] one [illegible] being [illegible]
[illegible] [illegible]
[illegible] would [illegible] the [illegible] after
[illegible] they allow them [illegible] capture.
[illegible] the [illegible] Mingulay
and [illegible] stuck to the [illegible]
[illegible] only [illegible] fish [illegible]
[illegible] people [illegible] tree [illegible] and [illegible]
[illegible]
several [illegible] seen [illegible]
[illegible]
[illegible] days.

"[illegible]
[illegible]
[illegible]
[illegible]
[illegible]

bills' young are about the size of blackbirds when they leave. The Puffin clears out his nesting hole before leaving. He can scarcely ever be captured by the lasso.

"Another method followed at Mingulay," continues Mr. Finlayson, of capturing the birds, is also "by means of a heavy pole. The natives sit on the verge of the cliff, and the birds come hovering above and within blow-distance. No blow on the body appears to disable the birds, but the least knock or blow on the heads or necks finishes them, though no blow, however hard, kills them outright. They are apparently dead when they fall down, but if the necks be not broken they will soon recover. The moment the little ones open their eyes on the world around them, they are as sensible of danger as a man of sixty. They invariably keep away from the danger side of the rock." Finally, Mr. J. Finlayson tells us: "Guillemots and Razorbills all leave Mingulay by the 15th of August, leaving a majority of the Puffins behind, and all the Kittiwakes."

The Guillemot is certainly more densely packed on the ledges of Lianamull and Mingulay than in most rock-bird stations, the great breadth and regularity of the horizontal layers of rock, and the subsequent widening by air and water of the crevices of junction, affording great foot-room for the birds, as many as 200 or 300 jostling together on a single ledge. On parts of the cliff in the middle of July 1887, many young were hatched out and were sitting head inwards to the cliff, but in other parts but few young were to be seen.

As many as 2000 (*sic*) are noosed off during a day's raid by the men of Pabbay, but this year, having been hurried in their operations on the Stack, only some 800 Guillemots were taken. The figures can be depended upon, thanks to the accuracy of Mr. Finlayson. Few of the Mingulay people now do any cliff-work systematically, so the men of Pabbay avail themselves of the spoil.

Lomvia bruennichi (*Sabine*). Brünnich's Guillemot.

[*Obs.*—We can find no reference made to the occurrence of this arctic species except that by Sir W. E. Milner, who said it was found at Soay, St. Kilda, but we have never been able to verify this, nor have we ever heard of any other records.]

Uria grylle (*L.*). **Black Guillemot.**

Gaelic—*Callag: Calltag: An gearra-Breac; Ian Dubh a Chrulainn: Gearra-gleas Gearr-Gals an Sgadan: Craigeach* (in Eigg): *Cala-Bheag an Sgadain: Cala-Bheag nan Cudigean.*—A. C.

At some future time we shall have a better opportunity of describing what Harvie-Brown believes to be the largest colony of this species in Britain. Meantime it is enough to say it is everywhere common where suitable ground for it occurs along the shores of the O. H.; but perhaps not so numerous as among some of the Inner Hebrides. Nor is it so plentiful over all the west coast islands as on the mainland of the Long Island, nor so abundant on the west side as on the east. But it is found from north to south, generally distributed, and in no locality can it be called scarce. On the west coast it is perhaps commonest at Loch Roag. Naturally it is rarer on the sandy coasts of the west of the Uists, and on similar coast-lines. In summer it is seen commonly in the Sound of Harris, mostly single birds "off duty," for, as we have shown elsewhere,[1] the males take part in incubation.

A tolerably sufficient plan of arriving at an estimate of the numbers in any one colony is simply to count the birds seen sitting on the rocks or in the water any day in June, and, by doubling that number, one can obtain a very fair estimate of the population.

In North Ronay Swinburne found a good many about the west end, and Barrington found them breeding in the walls of the old dwellings 100 yards from the sea, and adds the note, "During the day they sat sunning themselves on the grassy roofs of old houses. I never found this species breeding before except in clefts of rocks and cliffs not far from high-water mark." Harvie-Brown saw nothing of this at Ronay in 1885, but on Shillay, off the Sound of Harris, and other of the western islands, he has often taken the eggs from under large rocks and loose boulders 80 to 100 feet above the sea and on the level. Nor in June 1887 did he see any Black Guillemots amongst the ruins of the old houses of Ronay.

Mergulus alle (*L.*). **Little Auk.**

Of very rare occurrence in the O. H., and, up to the time Gray wrote, only some four or five were recorded. Two are reported

[1] *Fauna of Sutherland, Caithness, and West Cromarty*, p. 241.

from North Uist in the winter of 1868-69. The line of migration seems to be *down the east coast of the main chain of* the O. H., and the species is much rarer on the west side. Two in the Museum at Stornoway Castle were taken on the moor at Mhoragail on 7th January 1878 (*auct.* D. M'Kenzie).

Fratercula arctica (*L.*). Puffin.

Gaelic—*Coltrachan* : *Contrachan* : *Comhdachan* : *Coltair-cheannach* : *Seumas Ruadh* = Red James : *Peta Ruadh* = red pet (*auct.* Mr. John Finlayson, Mingulay) : *Bugaire* or *Buigire* (St. Kilda) : *Fachach* (more correctly, to the Manx Shearwater) : *Colcair* : *Colgaire* (Harris) : *Boganach* (the young).—A. C.

Myriads and myriads on St. Kilda, where MacGillivray procured a pure white one.[1]

Myriads and myriads on the Shiant Isles. Abundant on the Flannans, North Ronay, and other suitable localities—the most abundant of all rock-birds in Scotland generally. The Stack of Lianamull, at the back of Mingulay, and Barray Head, also swarm with them, making it a somewhat difficult matter to state with any accuracy where the most populous colony exists. We would name four of the largest we have seen and visited to date of 1887,—St. Kilda, The Shiant Isles, the coast of Sutherland between Cearvaig Bay and opposite Garbh Island (the highest cliffs on the mainland of Scotland, populous throughout their height, and their whole range of three miles), and Ailsa Craig. On St. Kilda their name is also truly legion ; but perhaps on the Shiants their density is greater, if their area is more restricted. But we do find the record of the earlier authors, viz. :—"Not generally diffused," correct enough in a comparative sense. At Haskeir apparently the most abundant species at both ends of the main island, tunnelling deep into sea-pink hummocks, or under the loose stones : with a spade or pick hundreds of eggs could, we believe, easily be gathered if wanted. We extract the following short note from our journals of a visit to the Shiant Isles :—

"In countless thousands. The sea, the sky, and the land seemed populated by equal proportions, each vast in itself—a

[1] It is rather rare to find albinos among rock and sea birds, but we have a series of Puffins, about a dozen in all, showing almost every phase of variation.

constantly moving, whirling, eddying, seething throng of life, drifting, and swooping, and swinging in the wind, or pitching and heaving on the water, or crowding and jostling on the ledges and rocks, arising from, and alighting on, the boulder-strewn slope, or perched like small white specks far up in the cliff-face amongst the giant basalt columns." We fully indorse the remarks of other Scottish writers upon birds, that the Puffin is the most abundant species of rock-bird on the west coast of Scotland. In 1887 we found the whole circumference of Eilean Mhŭire densely colonised—some three miles of cliff and cliff-top, of *débris* and talus, several hundred feet in altitude. In Mingulay they have self-consequently displaced the true *Fachach* on the Stack of Lianamull, but the natives, in revenge, have extirpated them on the larger Stack of Arnamull, in order to preserve the grazing for about a score of sheep. The Puffin has complete hold over the whole upper crust of Lianamull; and over their heads waves a dense crop of red-seeding sorrel in summer. Later in the year, as Mr. Finlayson informed us, the whole surface is one sticky compound of mud and dung, feathers, bad eggs, and defunct young Puffins, ankle-deep or deeper—waiting perhaps to be scraped away some day from the rocky floor on which it rests, and be spread far and near over the worn-out pastures by future generations of farmers—truly a filthy, if a fruitful compost.[1]

Order 8. PYGOPODES.

Family COLYMBIDÆ.

Colymbus glacialis, *L.* Great Northern Diver.

Gaelic—*Bunabhuachaille : Am Murbhuachaille* = the shepherd; the sea shepherd.—A. C.

"Plentiful until the beginning of June, then all disappeared." So reported in 1841; but there is no doubt they are often seen much

[1] The sorrel has its uses in such localities, more particularly perhaps as an article of diet for the Manx Shearwaters which formerly occupied this stack, if we entirely credit the statements of how the papas and mammas feed their young "fachachs" on sorrel leaves to reduce their "sottish fatness," and enable them to creep forth at last from their holes, as is invariably related of the species in the Inner Hebrides.

later in the year. In 1848 one was seen, 2d June, by Sir W. E. Milner in the Sound of Harris. Another account says, "abundant all the year, except July; collecting in June, the most leaving before the 15th of that month" (Gray). Assembles in flocks. Seen in Loch Roag in October. We have frequently met with them in the O. H. in summer, but invariably on salt water.

Colymbus arcticus, *L.* Black-throated Diver.

Gaelic—*Learg: Learg-Uisge: Learga Fairge:* also *Brollach bothan: Learga Mhor: An Learga Dubh.*—A. C.

Curiously, MacGillivray does not mention this species, but speaks of the last as just noted. The present species betakes itself to the fresh-water lochs early in the season, and remains all summer to breed. Generally distributed, and not uncommon. Mr. Heywood H. Jones, however, only saw it twice in Mhorsgail—once on Loch Langabhat, and once on a loch in Sobhal, where he obtained the eggs. In this district they seem to be less regular in appearance, but there is no doubt they breed occasionally, if not annually. Over most of the main chain they are generally distributed, if not very abundant. They vary in their numbers in different breeding seasons. In 1879 only about three pairs were known to be breeding in North Uist, whilst in other years we have known them to be more abundant.

Colymbus septentrionalis, *L.* Red-throated Diver.

Gaelic—*An Learga: An Learga Chaol*=slender learg: *An Learga Ruadh*=the red learg.—A. C.

Common, and even very abundant in some districts. Some seasons they are much more numerous than in others, and, at times, are even scarce. In 1881 they were very abundant in the west of Lewis, and in that year bred not far from Stornoway.

Family PODICIPITIDÆ.

Podiceps cristatus (*L.*). **Crested Grebe.**

Podiceps griseigena (*Bodd.*). **Red-necked Grebe**

Podiceps auritus (*L.*). **Sclavonian Grebe.**

MacGillivray received a Sclavonian Grebe from a gentleman in North Uist, but appears at that time to have considered them rare. Since then Gray has received it also from North Uist from Alex. Carmichael, who shot it at Loch Maddy in March. Mr. H. H. Jones considers it as not rare in Loch Roag in the west of Lewis in October. Harvie-Brown found two specimens stuffed in a glass case at Loch Maddy in 1886, and Sheriff Webster spoke of them as far from uncommon, knowing the Little Grebe quite well. It does not seem certain that they breed anywhere in the O. H., so far as we know, but there is some probability that they do, as there can scarcely be any doubt that they breed occasionally, if not regularly, on the mainland of Scotland. Gray's collection contains specimens from the O. H., as we are informed by Mr. W. Evans.

Podiceps nigricollis (*C. L. Brehm*). **Eared Grebe.**

Not the least likely to breed, and almost equally unlikely to occur, if we consider its known distribution, though we find a record by MacGillivray, who had one sent him from Ormaclate in South Uist (*op. cit.* p. 409).

Podiceps fluviatilis (*Tunstall*). **Little Grebe.**

Gaelic—*Gobhachan*=the little smith.

Said only to be seen on Loch Roag and in Mhorsgail, in the autumn, but its annual local migration from fresh-water lochs to the sea in winter may account for this. In hard winters, however, it frequently returns to the interior, and to running streams on the mainland; and when shot in such places is often found to contain remains of small fresh-water fish—sticklebacks, and the like—and no doubt this habit holds good in the Isles.

Common all over the O. H. in summer, in all suitable localities, and generally quoted as such by previous authors.

Class 3. REPTILIA.

Sub-class *SQUAMATA*.

Order **OPHIDIA**.

Family **COLUBRIDÆ**.

Tropidonotus natrix (*L.*). **Common Snake.**

Gaelic—*Nathair.*

Coronella lævis *Lacép* **Smooth Snake.**

Family **VIPERIDÆ**.

Vipera berus *L.* **Viper. Adder.**

Gaelic—*Beithir Nathair-nimhe.*

Order **LACERTILIA**.

Family **LACERTIDÆ**.

Lacerta vivipara *Jacq.* **Common Lizard.**

Gaelic—*Dearc-luachrach.*

Lacerta agilis *L.* **Sand Lizard.**

Lacerta viridis *L.* **Green Lizard.**

Family **SCINCIDÆ**.

Anguis fragilis *L.* **Slow-worm. Blind-worm.**

[The chapter upon Reptiles in the O. H. is almost of necessity as short as the famous one relating to Iceland. The Frog and the Toad are unknown in the O. H., and we have no positive records regarding the Lizards.

Not rare in some localities, but local in its distribution.

Sub-class *CATAPHRACTA.*

Order **CHELONIA.**

Family **CHELONIDÆ.**

Dermatochelys coriacea (*L.*). **Leathery Turtle.**

Chelone imbricata (*Schweigg.*). **Hawk's Bill Turtle.**

Class 4. AMPHIBIA.

Order **BATRACHIA URODELA.**

Family **SALAMANDRIDÆ.**

Triton cristatus *Laur.* **Great Crested Newt.**

Triton tæniatus (*Schneid.*). **Smooth Newt.**

Triton palmipes (*Latr.*). **Palmated Newt.**

Order **BATRACHIA ANURA.**

Family **BUFONIDÆ.**

Bufo vulgaris *Laur.* **Common Toad.**

Gaelic—*Losgann* : *Magan* : *Craigean* : *Gille-craigean.*

Bufo calamita *Laur.* **Natterjack.**

Family **RANIDÆ.**

Rana temporaria *L.* **Common Frog.**

Gaelic—*Magag* : *Mach-ruhag* : *Muileag* : *Much-ruhag* : *Leumachan.*

Rana esculenta *L.* **Edible Frog.**

FISHES.

The Fishes of the West of Scotland have unfortunately been little studied outside the Mull of Kintyre, even from an economic point of view, and still less by Naturalists. Desultory efforts at odd seasons are really of little value, as I have steadily examined the same stretch of water year after year, and then suddenly found a species new to me to be plentiful. Yet I had not met it before! The immediate entrance to the whole Atlantic, as well as direct entrance of the district both to southern and northern areas of sea, ought to lead us to expect a wide variety of species and a great wealth of forms as visitors. At the same time the groups of islets cut off portions of sea, and isolate them sufficiently to possibly create interesting varieties, if not new species. The lochs are so numerous, and so varying in conditions, that a comparison of their several faunas must necessarily be interesting, as we know that each loch supplies a herring capable of being distinguished from that of any other loch, by a trained eye, without having yet developed any specific distinctions. The seas over a great portion of this western area are remarkable for their little variation in temperature throughout the year, compared with the more Arctic German Ocean; and some portions are so protected as to ensure a fauna such as could not be expected off ruder and more exposed coasts. On the other hand, the wild shores of the Outer Hebrides, or South-West Mull, are ocean-buttresses against which the full fury of the Atlantic is constantly expended. Many of the lochs, such as Hourn, from their depth and wild surroundings, introduce conditions widely differing from those waters open to the warm Gulf Stream, and tenderly treated by the lolling land around. Over a coast-line roughly estimated at 1500 miles, the varying conditions must be great, and the fauna proportionately novel. When to this is added a foreshore of great extent, through the great rise and fall of tides, of great richness of vegetation and invertebrate life, and in all positions of exposure, the possibilities are great. But we must look especially to enormous bodies of "insectivorous" fish, such as Herring or Mackerel; seeing that the warm waters, driven in by the prevailing south-west winds, bring inexhaustible supplies of minute crustacea, and the general floating life of a richly-dowered element. Following and preying upon these shoals come countless white fish; and the fauna of the West may thus be expected to yield to earnest workers a valuable harvest.

At present the means of putting together a fish fauna are very meagre. Although I examined all fishes taken off the port of Carloway, Lewis, for two years, and carefully hunted the shore, the number of species was very limited. Those I have secured in Loch Creran over a series of years are far more numerous, and probably fairly represent the fauna of protected lochs. The south of Mull yielded but few species in winter and spring, but these were plentiful, and gave promise of more on their return from the depths, as the pelagic ova were numerous. Barray has as yet been inadequately explored. During my special investigations I did not obtain so many species as I anticipated; but this is quite in accordance with the unsteady nature of the West Coast fishes, which seem more migratory than those of the German Ocean, probably because the food-supply is more pelagic than local, the fish moving off- and in- shore in pursuit. From the warm waters of Islay to the cold lochs of Skye, or the Minch, there is a wide field for observation, and only a large body of accurate observers can cover it properly. Few stationary naturalists have worked upon it, and flying observers can do but little. Captain Swinburne, Eilan Shona, has been one of the most long-continued and reliable observers in his own district, supplemented by Mr. William Blackwood; while Captain Campbell of Inverneil has intelligently worked the Knapdale coast; but all I can gather can only be considered a point of departure, a sort of skeleton to be filled up as our observers and recorders increase. We have not sufficient data to theorise upon, as I have recently met fishes in numbers that I had previously failed to find a trace of.

There does not appear to be any reason to believe in important movements of the land or sea-bottom to cause sudden isolation, or other fluctuations in the fauna of a particular bit of water. Portions of The Lews have subsided apparently, and the theory of the great rise of portions of Scotland in historic times seems to be without adequate foundation, so that meanwhile we are not disturbed in our examination of these regions by cataclysms apart from the rest of Great Britain. The hundred-fathom line is outside the region we cover, and the few isolated depressions that reach or exceed this depth, although interesting to the dredger and naturalist, did not add apparently to the fish problem, until Mr. Murray's trawlings in deep water, referred to elsewhere. At the same time, the depths within easy reach of the shore are far in excess of anything the

East Coast, or England, can supply, and introduce conditions that will no doubt materially affect the food-supply, and consequently the fish fauna.

The absence of a mackerel-fishery does not mean the absence of mackerel schools, as these fish are taken in the herring-nets all over the region, and the persistent neglect by the resident population of the wealth of the sea, curtails our opportunities of obtaining special knowledge. The natives have become so accustomed to heavy hauls of Herring, enabling them to tide over a year with a short season's labour, that the steady pursuit of the fisheries has never been attempted. Hence all subsidiary fisheries have been neglected, and when times of herring scarcity came, the general impoverishment was as complete as if they had again relied alone upon the potato. Macleod of Harris, and other proprietors, early in the century, sought to establish fishing colonies, but without success. The people have not shown special adaptation to a sea-faring life, and consequently we find the main marine wealth of those seas reaped intermittently by East Coast crews, in place of being steadily worked by their proper owners.

The great fisheries of the West of Scotland at this time are the herring, and the cod and ling fisheries. Like oats and potatoes, they are great staples, demanding no great skill in their pursuit in a perfunctory way; while those fisheries that may be compared more to special husbandry, and "market-gardening," are disregarded by a people ignorant of the first elements of thrift. The absolute necessity for the complete organisation of this great industry, if the chronic poverty of these regions is to be removed, is self-evident. We should have liked, if our space permitted, to have introduced a full *résumé* of the herring question, so as to show the advantage of securing its local regularity in the various lochs, by attending to its artificial reproduction in quantity. The Herrings of great reaches of water have disappeared suddenly, and apparently unaccountably; but it may have been a question of food-supply. The best observers attributed the absence of the usual supply of Sea-trout on the West Coast in 1886 to the long spell of east wind in the spring, that had driven the food-supply out to sea or into deep waters.[1] It is either by food, or necessity for reproduction, that their movements are regulated; and the ripeness required for the latter must be dependent partly on food and partly on temperature. The propagation of an entomostracan food-supply must again be largely a matter of temperature, so that to this one main element in the problem we must specially look for guidance. The Herring of our coasts has been steadily driven into deeper and deeper water to spawn, and so become

[1] Vide *Scottish Geographical Magazine*, article by Dr. Mill, Oct. 1888.—(Ed.)

more and more pelagic, and less of a local loch fish. This should be remedied, and the remedy is readily within reach. The reckless disposal of fish-refuse also, at our curing stations, has gathered after a time such a school of ruthless enemies of the Herring, that the shoals have been driven away, not to return for a generation. The introduction of great establishments for the utilisation of refuse shows how to remedy this result of careless and wasteful working. The working out of the thorough distribution of our commercial fishes will possibly show us that, in our comparatively mild western waters, the ordinary fishes cover most of the ground, as the variation in the outer waters is not important. In the *New Statistical Account* a general statement is made, that the Northern Tusk is taken "west of Kintyre," but we are wanting in information as to its capture south of Ardnamurchan. The capture in numbers of a West Indian fish somewhere near Stornoway (*Holacanthus tricolor*) points to possibilities of rare visitants following the Gulf Stream to unusual latitudes, just as the beans of plants from the Gulf of Mexico are thrown on the Lewis shore. I have seen more specimens of *Notidanus griseus* caught on the coast of Lewis than are recorded in Couch as British specimens, so that we must of necessity have visitors direct from the pure Atlantic fauna, in a way that the English and East Coasts can scarcely be expected to have. It will probably be found that the peculiarly unstable character of the haddock fishery over most of the West Coast is largely dependent upon the herring shoals, upon the spawn of which they feed greedily. If the Herrings have been driven to spawn on distant banks, the Haddocks have probably followed them to the banquet they affect. I am on the whole, therefore, disposed to look upon the fishery problem (in its economic aspect) as one more capable of being handled than we are commonly willing to admit. The Herring is not only our food-supply, but that of our greater fish-fauna generally in the West; and once the movements of the Herring are tabulated, and, if possible, regulated to a degree, we shall have a more thorough knowledge of the movements of those fishes that mainly subsist upon them. At the same time, there are subsidiary dishes that have to be investigated; as when we find the Cod of a district living almost wholly for a period of six or eight weeks upon the Norwegian lobster; and others with nothing to show over a lengthened period but the crustacea of the laminarian zone.

Many minor questions of special interest to the naturalist will suggest themselves, as our inquiry proceeds and investigators increase. Our ignorance of certain forms is frequently caused by ignorance of their habits. More than once I have taken a fish recorded for the first time in the region under examination; but no sooner was the one specimen captured than many more were forthcoming from the proper habitat. The fish taken in the seine-net differ from those taken in the beam-trawl, or on the long lines; and the class of fishing pursued in a district will regulate the species recorded. It is consequently important that all means of capture be exhausted, over the whole season, before a negative statement be accepted as of value. Premising, therefore, that the district from the Mull of Kintyre to Cape Wrath has only been tapped here and there intermittently by naturalists, I present my inadequate list as a rude preliminary effort to stimulate and aid investigation.

In this connection I must call especial attention to a new direction in which inquiries may be made, and to the important results already obtained by Mr. John Murray of the *Challenger*, by means of the trawl in deep water. As the dredge does not usually capture fish in any quantity, it may safely be concluded that important results yet remain to be obtained by a fuller system of investigation. But meantime Mr. Murray has made a new departure that has already led to most interesting, and may lead to most valuable, discoveries. The paper that is the advance-courier of these researches was read at the Royal Society of Edinburgh this year, and is explained in the accompanying editorial note:—

1888. GÜNTHER, A.—*Report on the Fishes obtained by Mr. J. Murray in Deep Water on the North-West Coast of Scotland, between April* 1887 *and February* 1888. By Dr. A. GÜNTHER, F.R.S. (Royal Soc. Ed. 1888).

We are indebted to Dr. Gunther and Mr. J. Murray for a sight of the above paper whilst still in MS., and Mr. W. Anderson Smith has incorporated the notes referring to species obtained at localities outside of the Clyde area, *i.e.* in other words, north of the Mull of Kintyre. Future reports of a similar nature are alluded to in the introductory portion, as being 'in course of preparation.' By far the larger number of these fish obtained by Mr. Murray, and the dredgings of the *Medusa*, come to be included more correctly in the Clyde area. The number of species obtained was thirty-one, of which three have been found for the first time in British waters, or at least close to the mainland.

There are many interesting finds also, besides those we refer to here relating to the Clyde area, as regards the young of many common species, taken at recorded depths. Such bathymetrical records cannot fail to be of great interest in such carefully conducted reports as Mr. Murray's. In conducting any further investigations in the West of Scotland, it may be well to study this paper apart, and look forward to the rapid progress which may now be expected since Mr. Murray has given the impetus, and to the future reports also alluded to.

In more senses than one, it is almost impossible to disassociate the Clyde area's fauna from those outside Kintyre, but by instituting comparative faunas, some good service may be done in drawing out from other sources more complete comparative lists. This must be our reason for restricting our present area.—(Ed.)

WEST OF SCOTLAND AND HEBRIDEAN FISHES, NORTH OF, AND OUTSIDE, MULL OF KINTYRE.

Class 5. PISCES.

Sub-class 1. *PALÆICHTHYES.*

Order 1. **CHONDROPTERYGII.**

Sub-order *PLAGIOSTOMATA.*

Division SELACHOIDEI.

Family **CARCHARIIDÆ.**

Carcharias glaucus (*L.*). Blue Shark.

Numerous off Lewis, but seldom captured larger than from 4 to 6 feet. These show plenty of fight, and are very troublesome. Mostly taken in winter. One of large size taken in Loch Linnhe. West of Barray, and off Muck (T. A. S.).[1] Off Moidart (W. B.). Taken in Lochfyne.

Forty or fifty years ago many "Cearban" came into Mingulay Bay, and the natives killed large numbers for the sake of the oil found in their livers. The natives found this lucrative, so much so, that they procured harpoons on purpose by steamers from Glasgow. These sharks lingered, in decreasing numbers, on this coast till about twenty years ago; but since ten years ago, none have been seen until last autumn (1887), when *one* was brought into Castle

[1] For List of Abbreviations, see p. 226.

Bay, Barray. None were known to have frequented the coast before this great invasion, nor since (H-B.). This probably refers to *Selache maxima* (Gunner).

Galeus canis *Bonap.* **Common Tope.**

Plentiful off Canna Island (*nobis*). Sound of Jura (C. J.).

Zygæna malleus (*Risso.*). **Hammer-headed Shark.**

Mustelus vulgaris *Müll. and Henle.* **Smooth Hound.**

"It is not rare among the Hebrides, where it is used as food and esteemed a very delicate fish, its difference in this respect from the other sharks being no doubt occasioned by the different nature of its food" (N. L.). Minch (H-B.). Very numerous off Carsaig, Sound of Mull, more so as you proceed west. Comes in shoals separate from the picked dog (M'L. C.). Off Moidart (W. B.).

Family **LAMNIDÆ.**

Lamna cornubica (*Gm.*). **Porbeagle.**

A specimen in Dunrobin Museum is labelled "Loch Inver, 1875," and Harvie-Brown caught another on long line at Loch Inver in 1881, 56 lb. weight. "Minch," in Hunterian Museum. Off Moidart (W. B.) one 8 feet 10 inches long in 1883.

Alopecias vulpes (*Gm.*). **Thrasher, or Fox-Shark.**

One taken south of Mull, 9 feet 6 inches without tail fluke (M'L. C.). "One morning," as stated by Captain Crow, in a work lately published,[1] "during a calm when near the Hebrides, all hands were called up at three A.M. to witness a battle between several fish, called Thrashers or Fox Sharks, and some Swordfish on the one side, and an enormous Whale on the other. It was in the middle of summer; and the weather being clear, and the fish close to the vessel, all had a fine opportunity of witnessing the contest. As soon as the Whale's back appeared above the surface, the Thrashers springing several yards into the air, descended with great violence upon the object of their rancour, and inflicted upon

[1] *Memoirs of the late Captain Hugh Crow of Liverpool, compiled chiefly from his own MSS.* London, 1830. 8vo.

him the most severe slaps with their tails, the sounds of which resembled the reports of muskets fired at a distance. The Sword-fish, in their turn, attacked the distressed Whale, stabbing from beneath,—and thus beset on all sides, and wounded whenever the poor creature appeared, the water round him was dyed with blood" (N. L., 1843). One taken off Coll, June 1888, 14 feet long (Colonel Stewart, Coll). This is quite an exceptional size.

Selache maxima (*Gunner*). Basking Shark.

Gaelic—*Cearban.*

"It is said by Dr. Fleming to be common on the west coast of Scotland, particularly during the prevalence of a west wind" (N. L.). West of Barray (T. A. S.). Appears in Hebrides in June in small droves of seven or eight, but oftener in pairs, and continues in those seas till the latter end of July, when they disappear (P.). "Sunfish" appear off Small Isles, and remain till July (O. S. A.). "There is a kind of fish which was formerly pretty often seen on this coast (Tiree and Coll), but seems for the last thirty or forty years to have almost entirely disappeared." "It is frequently known by the name of the Sunfish or Basking Shark, from its practice of floating on the surface of the water during warm weather or sunshine. These were caught with harpoons and lines in somewhat the same style as the Greenland Whale, and were valuable from the quantity of oil extracted from their liver. I recollect, when a boy, seeing one of them taken, not reckoned a large one, the liver of which filled eight barrels, and might have been estimated at £25. Since that period I have seen only one of them (about four or five years ago), which was amusing itself during the greater part of a day in the bay opposite to the manse" (N. S. A., 1840). In Barray formerly they were very successful in spearing *sail-fish*, Cearbans; but they have no tackling now, and although *hundreds* of these fish appeared last season (1842?) on the coast, no one could take advantage of their appearance" (N. S. A.)

Family NOTIDANIDÆ.

Notidanus griseus (*Gm.*). Grey Notidanus. Six-gilled Shark.

Four large specimens from 9 to 12 feet long were taken in Lewis on the deep-sea lines during 1871 and 1872. The fishermen were quite familiar with them, as they took them frequently. Small specimen 2 feet 10 inches in length was taken off mouth of Lochbuie in Mull, also on long lines, in May 1887. It is a sluggish ground-shark, with magnificent iridescent eyes, saw-like teeth, and six gills. It is an Atlantic fish.

Dr. Günther now considers this species as a regular inhabitant ("regular occurrence") in British waters. "Well known to the fishermen who frequent the banks between the Orkneys and Shetland. From this locality Mr. Wm. Cowan obtained two adult specimens last summer, 1887, one of which is now in British Museum." (H-B.).

Family SCYLLIIDÆ.

Scyllium canicula (*L.*). Small-spotted Dogfish.

Gaelic—Old, *Biraich*: juv., *Biagish*: large, *Biast gorm*.

Common all over the west coast, where the eggs are found attached to sea-ware at low-water of springs. Breeds freely in Loch Creran at all seasons of the year; also in Lochbuie, Mull; and indeed throughout the west. Off Lewis took one at end of September with two egg cases ready for expulsion, two large ova entering the tubes, and a large quantity of ova in all stages of advancement (*nobis*). Minch (H-B.). Occurs commonly north of Mull (J. N. F.). Shiant Isles (H-B.). General (called King-fish) (T. A. S.). Killisport, with young Gurnet, crustacea, and operculæ in stomach; Sound of Jura, Portree, Canna, Barray, etc. (*nobis*). Off Moidart (W. B.).[1]

Scyllium stellare (*L.*). Large-spotted Dogfish.

Also widely diffused throughout the region, but not so numerous as above.

Pristiurus melanostomus *Bonap.* Black-mouthed Dogfish.

This fish was very numerous about Portree in August 1886, when it was throwing eggs; and it seemed also plentiful off Canna. The fishermen say it is not uncommon. The eggs more resemble those of Skate, being devoid of the gelatinous cords at the corners for binding them to seaware, and are probably deposited like Skates'

[1] Sound of Sanda, Kintyre, 20 fathoms: *juv.* (Murray).

in somewhat deeper water, and glued to the fronds of seaware by a gelatinous slime (*nobis*). "This species is not uncommon on the west coast of Scotland" (N. L.). West coast of Scotland (Malcolm), one of two examples from there having been sent to Yarrell (F. D.).

Family SPINACIDÆ.

Acanthias vulgaris *Risso*. **Picked Dogfish.**

Gaelic—*Gobag*.

Plentiful over all the region at different seasons of the year. Most destructive to the liners in the Cod and Ling fisheries. "Particularly abundant on the west coast of Sutherland in July 1882" (H-B. & B.). It is ovo-viviparous, and the female throws young every month in the year. The young are from 6 to 9 inches long when thrown. "Countless at times south of Mull. Has passed through such shoals that those in the boat seized some by the tail, smashed their heads on the gunwale, and tossed them back to be at once devoured by the shoal" (M'L. C.). These shoals are generally on the outside of a shoal of herring. The people of Ness, Lewis, used at one time to pay their rents from dogfish oil. Off Moidart (W. B.).

Læmargus borealis (*Scoresby*). **Greenland Shark.**

One 13 feet long was, in 1824, found dead at Barray Firth, Uist.

Echinorhinus spinosus (*Gm.*). **Spinous Shark.**

"We were informed of the capture of a Spinous Shark in Stornoway Bay some years ago" (Lewsiana).

Family RHINIDÆ.

Rhina squatina (*L.*). **Angel-Fish. Monk-Fish.**

Sound of Jura (C. J.). Two in Hunterian Museum from west coast.

Division BATOIDEI.

Family TORPEDINIDÆ.

Torpedo hebetans *Bonap.* **Cramp-Ray. Torpedo.**

Family RAJIDÆ.

Raja clavata *L.* **Thornback Ray.**

Gaelic—*Sonan.*

Very plentiful all over the west from St. Kilda to Loch Linnhe. Throws its eggs in pairs in shallow water—May, June, July more especially. They take six months to incubate. "We cut an egg from *R. clavata* at end of March, the capsule fully formed" (Lewsiana). Throwing eggs in August at Portree (*nobis*). These give a wider period of spawning over the whole coast than we are acquainted with in Loch Creran. General (T. A. S.). Sound of Jura (C. J.).

Raja maculata *L.* **Homelyn Ray.**

Plentiful in shallow water at Barray. Several taken near 30 fathoms in Loch Alsh (*nobis*). Loch Moidart (Hunterian Museum). One juv. taken in Loch Leven on 27th August 1887 (J. M.). One taken off Storr Head, Sutherland (H-B.). Loch Leven, 25 fathoms (Murray).

Raja radiata *Donovan.* **Starry Ray.**

Raja circularis *Couch.* **Sandy Ray.**

West of Lewis not uncommon (Lewsiana). Sound of Sanda, Mull of Kintyre, 20 and 49 fathoms, feeding on prawns and sand-eels (Murray).

Raja batis *L.* **Skate.**

Gaelic—*Sgàt.*

Abundant, and reach a great size off O. H. "We have seen livers frequently 10 to 12 lb. weight, and in one instance weighed one of 17 lb. full of the richest oil" (Lewsiana). St. Kilda, Sound of Mull, Loch Linnhe, Loch Alsh, and Portree, very large, Canna, etc. (*nobis*). Loch Carron (T. E. B.). Sound of Jura (C. J.).

Mr. Alexander Allan of Aros informed Harvie-Brown that he once killed a Skate in the Sound of Mull which measured 7 feet 6 inches by 5 feet 6 inches, which is, he thought, unusual. The seas around Rum are said to be famous on account of the large Skate there commonly obtained.

Raja marginata *Lacép.* **Bordered Ray.**

Raja lintea *Fries.* **Sharp-nosed Ray.**

Raja fullonica *L.* **Shagreen Ray.**

West of Lewis, not uncommon, Lochbuie, Mull, Loch Hourn (*nobis*). One of 70 lb. taken in Loch Leven (Nether Lochaber). Has he not mistaken this for another larger species (?).[1]

Raja vomer *Fries.* **Long-nosed Skate.**

Off Lewis (*nobis*). Sound of Jura (C. J.).[2]

Family **TRYGONIDÆ.**

Trygon pastinaca (*L.*). **Sting Ray.**

Family **MYLIOBATIDÆ.**

Myliobatis aquila (*L.*). **Eagle Ray.**

Dicerobatis giornæ (*Lacép.*). **Horned Ray. Ox Ray.**

Sub-order *HOLOCEPHALA.*

Family **CHIMÆRIDÆ.**

Chimæra monstrosa *L.* **Northern Chimæra.**

Gaelic—*Buachaille-an-Sgadain.*

One taken in Sallachan Bay, Loch Linnhe, full of herring fry. It follows the Herring, and is called in Gaelic *Buachaille-an-Sgadain* = the herring-herd, or herdsman (Nether Lochaber).

Order 2. GANOIDEI.

Sub-order *CHONDROSTEI.*

Family **ACIPENSERIDÆ.**

Acipenser sturio *L.* **Sturgeon.**

Gaelic—*Stirean.*—A. C.

"At Barvas a Sturgeon was taken some years ago" (Lewsiana). "A Sturgeon was found on the beach of North Uist in 1887, as I am

[1] In Clyde area (Murray).

[2] One 9 feet across (!) taken off entrance to Firth of Clyde, April 1883, by steam-trawler "Wallace."

informed by the late Mr. John Macdonald of Newton. I have some of the scales he sent me by which it was identified" (H-B.). Sound of Jura (C. J.).

Sub-class *TELEOSTEI.*

Order 1. ACANTHOPTERYGII.

Division ACANTHOPTERYGII PERCIFORMES.

Family PERCIDÆ.

Perca fluviatilis *L.* **Perch.**

Gaelic—*Creagag.*

"Mr. Colquhoun, as we understand him, has caught Perch of 3 lb. weight in Loch Awe;" Lochbroom, Ross-shire. No Perch in Kintyre in 1843 (N. S. A.).

Labrax lupus (*Lacép.*). **Basse.**

Not uncommon, Loch Carron (T. E. B.). I believe them to be the species most commonly designated "Sea Perch" in the west of Scotland, but demanding more definite identification.

Acerina cernua (*L.*). **Ruffe. Pope.**

Serranus cabrilla (*L.*). **Smooth Serranus.**

Serranus gigas (*Brünnich*). **Dusky Serranus.**

Polyprion cernium, *Val.* **Stone Basse.**[1]

Dentex vulgaris *C. and V.* **Sparus. Dentex.**

Family SQUAMIPINNES.

First Group CHÆTODONTINA.

Holacanthus tricolor.

["The specimen of this fish now in the Elgin Museum was brought many years ago in a *fresh* state from Stornoway (or its neighbourhood) by a Brandenburgh (Lossiemouth) fisherman. I was told

[1] Taken in Clyde, off Little Cumbrae, 1870, by Dr. J. Young.

that more than one was brought, and that they had been seen 'jumping out of the sea.' Stupidly mistaking the specimen sent to the Museum for the European File Fish (*Balistes capriscus*), it lay a long while there under this name. At length I began to suspect that it was some other species—a fish not to be seen in Yarrell,—and sent a coloured sketch of it to Dr. Günther, who immediately gave me the correct name, and would have noticed its occurrence as British, in a work then about to issue from the press, had I been able to supply some fuller information."—Rev. George Gordon, Birnie, Elgin, *in lit.* to H-B., 25th February 1888.

"The preopercular spine is grooved, and reaches to the vertical from the posterior margins of the operculum: the soft dorsal and anal and the upper caudal ray moderately produced. Head, anterior part of the trunk, caudal, and the margins of the soft caudal and anal fins, yellow; the remainder brownish-black." "Atlantic coasts of tropical America." Those in British Museum from West Indies mainly (Günther).]

Family **MULLIDÆ**.

Mullus barbatus *L.* Surmullet. Red Mullet.

Gaelic—*Jasg Driomnom.*

"Heard of this species being taken on coast of Scotland" (P.). Off Moidart (W. B.) two specimens in 1882.

In Carsaig Bay, "I myself saw a shoal of fish in on the sand, which the lobster-man said were Red Mullet. I was within a few yards of them, and they had a red appearance on the sides and reddish-brown on the back, but I am not sufficiently acquainted with the habits of Red Mullet to know if they come into the shallow water in this way. I caught one in Lochfyne; I think it is described as a Striped Surmullet (variety), in a trammel net, in 10 fathoms, and though I tried in many places for them, I never caught another" (C. J.).

Family **SPARIDÆ**.

Group CANTHARINA.

Cantharus lineatus (*Mart.*). Black Sea-Bream.

Box vulgaris *C. and V.* Bogue.

Group PAGRINA.

Pagrus vulgaris *C. and V.* **Braize. Becker.**

Pagrus auratus (*L.*). **Gilthead.**

Pagellus centrodontus (*De la Roche*). **Common Sea-Bream.**

Gaelic—*Carbhanach.*

North of Ireland called *Muranoe*, *Barwin*, and *Gunner.*

Locally numerous throughout the west of Scotland, Loch Linnhe, Ardmucknish Bay, Loch Don, in Mull, etc. (*nobis*). Loch Carron (T. E. B.). Eilan Shona (T. A. S.). Plentiful south of Mull (M'L. C.). Sound of Jura (C. J.). Kilmuir, Skye (N. S. A.). "The spawn is shed in the beginning of winter, in deep water, and the young are called *chads*" (N. L.). "Very abundant in the sea-lochs of the west of Sutherland, and known by the local name of 'Barbarian Haddies.' We have caught many of these fish at the head of Glen Coul, and also at Loch Inver" (H-B. & B.).

Pagellus bagaraveo (*Brünn.*). **Spanish Bream.**

Pagellus owenii *Günth.* **Axillary Bream.**

Pagellus acarne (*Cuv.*).

Pagellus erythrinus (*L.*).

Family **SCORPÆNIDÆ.**

Sebastes norvegicus (*Müll.*). **Bergylt. Norway Haddock.**

Division ACANTHOPTERYGII SCIÆNIFORMES.

Family **SCIÆNIDÆ.**

Sciæna aquila (*Lacép.*). **Maigre.**

Division ACANTHOPTERYGII XIPHIIFORMES.

Family **XIPHIIDÆ.**

Xiphias gladius *L.* **Swordfish.**

Gaelic—*Iasg e claidheamh.*

"Nether Lochaber" mentions having seen four specimens in 20 years in his district, Loch Linnhe. Off Kilmelfort (N. S. A.). Hebrides, see "Thrasher Shark."

Division ACANTHOPTERYGII TRICHIURIFORMES.

Family TRICHIURIDÆ.

Lepidopus caudatus (*Euphr.*). **Scabbard Fish.**

Trichiurus lepturus *L.* **Hair-tail.**

Division ACANTHOPTERYGII COTTO-SCOMBRIFORMES.

Family CARANGIDÆ.

Caranx trachurus (*L.*). **Horse Mackerel. Scad.**

Gaelic—*Crea-Rionnach* : *Cneamh-Rionnach.*

Generally diffused on west coast; supposed to follow the herring shoals. Have taken them in Loch Hourn; and Loch Slapin, Skye (*nobis*). Kilmuir, Skye (N. S. A.). Found all round the coasts of Great Britain (F. D.). Off Moidart (W. B.).

Naucrates ductor (*L.*). **Pilot Fish.**

Lichia glauca (*L.*). **Derbio.**

Capros aper (*L.*). **Boar Fish.**

Family CYTTIDÆ.

Zeus faber, *L.* **Doree.**

More occasional than rare. "We were informed of the capture of a great number of John Dorys of late years, but never saw any, nor were they made use of by the people" (Lewsiana). "The John Dory is occasionally caught" in Killisport, Loch Linnhe, Strath, Skye (N. S. A.). Loch Carron (T. E. B.). Eilan Shona (T. A. S.).[1]

Family STROMATEIDÆ.

Centrolophus britannicus *Günth.*

Centrolophus pompilus (*L.*). **Black Fish.**

[1] Has been taken off Rothesay.

Family CORYPHÆNIDÆ.

Brama raii *Bl.* **Ray's Sea-Bream.**

Gaelic—*Cearc-maradh* (?).

It is met with at Belfast, where it is called the *Hen-fish;* and it is frequently found on the west coast of Scotland (N. L.). Common all down the west coast. Seen in fishermen's boats at Stornoway by Dr. Heddle in June 1887, and caught out in herring-nets in Broad Bay (H-B.). Scouler considered it to be not uncommon on the west coast of Scotland (F. D.).[1]

Lampris luna (*Gm.*). **Opah. King Fish.**

One in Hunterian Museum from near Campbeltown, but inside Mull of Kintyre. Another from Andersonian Museum: probably therefore a west-coast fish.

Luvarus imperialis *Rafin.*

Family SCOMBRIDÆ.

Scomber scomber *L.* **Mackerel.**

Gaelic—*Rionnach.*

Common throughout the west-coast waters. Not generally taken so large as the Irish, but a good fish when fresh. Not such a favourite in Scotland as in the south. Large shoals occasionally in Loch Linnhe, Sound of Mull, and around that island, Kintyre, and West Loch Tarbert, Sunart, and Barray (*nobis*). Large Mackerel in Loch Craignish and Kilmuir, Skye (N. S. A.). It visits the Western Isles of Scotland, but not in great abundance (N. L.). Loch Carron (T. E. B.). General (T. A. S.). Off Moidart (W. B.).

Scomber colias *L.* **Spanish Mackerel.**

Orcynus thynnus (*L.*). **Tunny.**

Known as *Mackrelsture.* They are not uncommon in the lochs of the west of Scotland, where they come in pursuit of Herrings (P.). Vicinity of the islands on the north and west coast of Scotland. (F. D.).[2]

[1] Specimen in Hunterian Museum taken near Ayr.

[2] One taken at Garelochhead, Clyde.

Orcynus germo (*Lacép.*). **Germon.**

Thynnus pelamys *C. and V.* **Bonito.**[1]

Pelamys sarda (*Bl.*). **Pelamid. Belted Bonito.**

Auxis rochei (*Risso*). **Plain Bonito.**

Echeneis remora *L.* **Remora.**

Family TRACHINIDÆ.

Trachinus draco *L.* **Greater Weever.**

I took a young specimen in the dredge from 7 or 8 fathoms at the mouth of Loch Creran, in August 1882, which is the only recorded instance of its capture in the district. "We understand the larger Weever is also found, but have never met it" (Lewsiana).

Trachinus vipera *C. and V.* **Lesser or Viper Weever.**

Gaelic—*Tarbh Shiolag* : *Tarbh nibhe* or *nimhe*.

In Lewis, "The little Weevers are most dangerous frequenters of the pools, the wound they inflict being exceedingly severe" (Lewsiana). Dr. Macleod, Hawick, a native of North Uist, says the people call it the male of the Sand Eels, and dread it much, because they sometimes lose a finger from its contact (H-B.). Very numerous and troublesome off Carsaig (M'L. C.).

Family PEDICULATI.

Lophius piscatorius *L.* **The Angler or Toad-fish.**

Mulrein and Merlin, Edinburgh; and in Northern Isles, *Wide-gape*. Plentiful all over the west coast. Frequently driven ashore, probably when chasing their prey, as they have then often Skate in their stomachs almost as large as themselves. Have taken them from Lewis to Loch Creran. General (T. A. S.). Loch Carron (T. E. B.). Sound of Jura (C. J.).

[1] Taken in Clyde area.

Family COTTIDÆ.

Cottus gobio *L.* **River Bullhead.**

Miller's Thumb. Eilan Shona (T. A. S.).

Cottus scorpius *L.* **Short-spined Sea-Bullhead.**

Loch Creran, Mull, Barray (*nobis*). North Uist (Dr. M'In.). Loch Carron (T. E. B.). Sound of Jura (C. J.).

Cottus bubalis *Euphrasen.* **Long-spined Sea-Bullhead. Father Lasher.**

Loch Creran, Mull, Barray (*nobis*). Sound of Jura (C. J.). Probably quite general. "A very young specimen; in the Mull of Kintyre, from 60 fathoms; February" (Murray—Günther).[1]

Cottus quadricornis *L.* **Four-horned Bullhead.**

Loch Carron (T. E. B.). Sound of Jura (C. J.). Threw spawn in confinement in Lochbuie, April 4th, 1887. Mull, Barray, Loch Creran, of very large size, and rich colouring, var. *Greenlandicus* (*nobis*).

Cottus Lilljeborgii *Günther* MS. **Norway Bullhead.**

New to the British Fauna.—Günther—Murray (*per* H-B.). "This species is allied to *C. bubalis*, but distinguished by the lesser development of the armature of the head." "This is a new addition to the fish-fauna of Great Britain. Previously this species had been found on various parts of the coast of Norway, and near the Faroe Islands." Mull of Kintyre.[2]

Trigla cuculus *L.* **Red Gurnard.**

"Crooner" (T. E. B.). Sound of Jura (C. J.). Loch Carron (T. E. B.). Common, Eilan Shona (T. A. S.). Kilmuir, Skye; Strath, Skye. Sometimes taken in Glenshiel, Ross, etc. (N. S. A.).

Trigla lineata *Gm.* **Streaked Gurnard.**

Gaelic—*Gromard*: *Cnodan.*

Sound of Jura (MacCulloch, per H-B.), and C. J.

[1] An immature specimen; Sound of Sanda, 20 fathoms; March (Murray—Günther).

[2] Sound of Sanda, 20 fathoms.

Trigla hirundo *L.* **Sapphirine Gurnard. Tubfish.**

Loch Linnhe, Ardmucknish Bay (*nobis*). Loch Leven (Nether Lochaber). "We recollect observing the sports of shoals of this species when on an excursion to the Western Isles, during a week of beautiful and too calm weather; for it was before steamboats plied. They were often discovered by their noise, a dull croak or croon, or by the ripple or plough of their nose on the surface of the calm sea; thus they would swim for a few yards, and then languidly sink for a foot or 18 inches, display or stretch their lovely fins, and again rise to the top. Boats were out with hand-lines; almost all were half-full, the men having little to do but bait the hooks and pull up. We resorted to our guns, and killed sufficient for dinner from the deck of the vessel" (N. L.). I believe all the species are called "Croonan" in the Highlands from this crooning sound in fine weather. Off Moidart (W. B.).

Trigla gurnardus *L.* **Grey Gurnard.**

Gaelic—*Crodan: Crunan.—Nobis.*

Common around the west coast, and pretty generally distributed. Loch Linnhe; Lochbuie, Mull, with ripe ova 14th April 1887. West of Mull; Killisport; Knapdale, Lewis (*nobis*). North coast of Mull (J. N. F.). Sound of Jura (MacCulloch and C. J.). Loch Carron (T. E. B.). Numerous off Carsaig, Sound of Mull (M'L. C.). Common (T. A. S.). Numerous off Moidart (W. B.).[1]

Trigla lyra *L.* **Piper.**

Trigla obscura *L.* **Shining Gurnard.**

Triglops Murrayi. *Sp. n.* D. 10/19. A. 19. P. 17-18, v. 3. c. 17.

"Although this species is closely allied to *T. pingellii*, it may be readily distinguished, not merely by the less number of fin rays, but also by the different form of the head, size of the eye, and more compressed tail, which in *T. pingellii* is singularly depressed. Several specimens, from 2½ to 4 in. long, were obtained at the Mull of Kintyre, at a depth of 64 fathoms, in the months of February and March" (Murray—Günther).

[1] Adults ready to spawn in March, Firth of Clyde (Murray).

[2] And also 4 miles south-east of the island of Sanda, in 35 fathoms, in the middle of March (Murray—Günther).

Family CATAPHRACTI.

Agonus cataphractus (*L.*). Pogge.

In Scotland it is said to be common on the west sands after storms (F. D.). Specimens from the Mull of Kintyre, 49 to 50 fathoms, *Medusa* (H-B.).

Peristethus cataphractus (*L.*). Mailed Gurnard.

Division ACANTHOPTERYGII GOBIIFORMES.

Family DISCOBOLI.

Cyclopterus lumpus *L.* Lumpsucker. Cock Paidle. Paidle Fish.

Common all along the west coast. The large cast in Frank Buckland's museum from one taken in Lochbuie. Taken in myriads in Lobster creels in the spring, of less than an inch in length. Took similar size in Tobermory Bay Aug. 3d, and in Loch Alsh in townet Aug. 7th; plentiful in Loch Creran still later in the season. Have taken them also from Lewis to Barray in the Hebrides (*nobis*). The large ones exceedingly numerous off Carsaig in salmon-nets, one season the shore literally covered with their spawn (M'L. C.). Sound of Jura (C. J.). Loch Carron (T. E. B.). Off Moidart (W. B.). "They resort in multitudes to the coast of Sutherland, near Ord of Caithness. The Seals which swarm beneath prey greatly on them, leaving the skins, numbers of which thus emptied float at that season ashore" (Pennant). They are sluggish fish, and supposed mainly to live as sea-scavengers. The male is brilliantly coloured in the season, and the rich-coloured ova in masses on the foreshore cannot be mistaken.

Liparis vulgaris *Flem.* Sea-Snail. Unctuous Sucker.

"Of infrequent occurrence" on west as on east, but I have obtained a specimen from Ardmucknish Bay, Loch Linnhe. North Uist (Dr. M'In.). Probably common in suitable localities.

Liparis montagui (*Donovan*). Montagu's Sucker.

Numerous of all the varieties—including so-called *Network Sucker* and var. *picta*—on the shores of Black Island, in Loch Linnhe, and around the shores of Lochbuie, Mull. It is a comparatively active fish, and will not endure confinement well.

Liparis liparis *Günther* MS. **Sucker.**

Many specimens from 49 and 64 fathoms at the Mull of Cantyre (Murray—Günther).[1]

Family GOBIIDÆ.[2]

Gobius niger (*L.*). **Black Goby. Rock Goby.**

Gaelic—*Buidhleis* or *Buillis.*—A. C.

Lochs Creran and Etive; plentiful in places; seems to prefer brackish water, or fresh flowing into the sea (*nobis*). Islay (Pennant).[3]

Gobius rhodopterus *Günth.* **Speckled Goby.**

Loch Creran; plentiful in brackish water on a shore with entering stream.

Gobius paganellus *L.*

Loch Creran; low tide; rocky ground.

Gobius auratus (*Risso*).

Loch Creran; deeper water (*nobis*). The distinctive beautiful dorsal fin of this species marks it out from the others.

Gobius minutus *Gm.* **Spotted Goby. Little Goby.**

Loch Creran; sandy bay with much fresh water; it differs little in colouring from the sandy bottom, and is not readily observed when still. A very quick species in its movements (*nobis*). North Uist (Dr. M'In.). West of Scotland (N. L.).[4]

[1] Three others inside the Clyde area, at 30 to 40 fathoms (Murray—Günther).

[2] The Gobies will require special treatment for West of Scotland, as this nomenclature is not satisfactory (W. A. S.).

[3] Taken in Clyde area.

[4] Plentiful in deep water. Clyde area (Murray).

Gobius ruthensparri *Euphr.* **Two-spotted Goby.**

Loch Creran, Barray, Canna (*nobis*). North Uist (Dr. M'In.). Extends round the coast of Great Britain, being very common in suitable places (F. D.).

"Took a male guarding a mussel-shell, which was lined with the eggs, in which tail and eyes of embryo, under microscope showed distinctly. A small portion of the lips of both valves of the empty shell was broken away, thus giving ingress to the little fish.

"The fish was preserved in glycerine, sea-water, and methylated spirit in equal proportions, in a test-tube. The spawn was carefully taken off the shell and similarly preserved.

"Locality Loch Skiport, South Uist, 25th May 1888. A magnificent bed of mussels exists at Loch Skiport Head" (H-B.).

Latrunculus albus (*Parnell*). **White Goby.**

Crystallogobius nillsoni *Day.*

Callionymus lyra *L.* **Dusky Sculpin. Dragonet. Gowdie.**

Not uncommon on west of Lewis; Loch Killisport, Knapdale, ♂; Lochbuie, Mull, *juv.* ♂; from stomach of cod, Loch Linnhe, ♀; widely diffused and frequently taken from stomach of cod. Captured commonly with lug-worms (*nobis*). One taken in North Uist, sent me by the late Mr. J. Macdonald[1] (H-B.). Off Carsaig, Mull (M'L. C.). Sound of Jura, male and female (C. J.). Islay (Pennant).[2]

Callionymus maculatus (*Bonap.*). **Spotted Dragonet.**

Added to the British Fauna by Dr. Günther in 1867 (*Ann. and Mag. Nat. Hist.*, vol. xx. p. 289, where also an adult ♂ is figured in Pl. V. fig. A). He writes: "The three specimens then known to me, and placed by me in the British Museum, came from the neighbourhood of the Hebrides, from a depth of 80 to 90 fathoms."[3]

[1] Since this went to press, it is our melancholy duty to record the death of our friend Mr. John Macdonald, on 21st August 1888 (H-B.).

[2] Clyde area.

[3] "Mr. Murray has now rediscovered this beautiful species in Kilbrannan Sound (Firth of Clyde), where it seems to be rather abundant at a depth of 26 fathoms. The largest male measures 4½ in." (H-B.). Other specimens were obtained in the Sound of Sanda, 24 to 28 fathoms (Murray).

Division ACANTHOPTERYGII BLENNIIFORMES.

Family CEPOLIDÆ.

Cepola rubescens *L.* **Red-band Fish.**[1]

Family BLENNIIDÆ.

Anarrhichas lupus *L.* **Wolf-fish or Cat-fish.**

Gaelic—*Cat maradh* (nobis). *Corran-greusaiche* (Fraser).

Common off Lewis (*nobis*). One captured off Oban (Nether Lochaber). Off Muck (T. A. S.). The female deposits her ova on marine plants in the months of May and June in Iceland (N. L.).

Blennius gattorugine *Bl.* **Gattoruginous Blenny.**

"Dr. Fleming, under the name of Crested Blenny, seems also to have described this species from a specimen which he found in Loch Broom" (N. L.). As it is common off Portrush, it is probably in the western waters also.

Blennius ocellaris *L.* **Ocellated Blenny.**

"Spotted Blenny," Glenshiel, Ross-shire (N. S. A.).

Blennius galerita *L.* **Montagu's Blenny.**

Blennius pholis *L.* **Shanny.**

Numerous in Canna, Lochbuie, Mull, Barray, in rock-pools. Loch Linnhe, scarce. Loch Creran, a single specimen. One in Canna threw up a 15-spined stickleback when captured (*nobis*). "Smooth Blenny" (?), Glenshiel, Ross-shire (N. S. A.).[2]

[1] Two have been taken on Ayrshire coast, just inside Clyde area.

[2] In Clyde area.

Carelophus ascanii (*Walbaum*). **Yarrell's Blenny. Crested Blenny.**

Numerous in certain seasons in Loch Creran at low water of springs, at other times unobtainable. Several taken in dredge from 8 to 10 fathoms (*nobis*). Loch Broom, Ross-shire (Fleming, *per* F. D.). North Uist (Dr. M'In.).

Stichæus lampetriformis. *Günther* MS.[1]

Centronotus gunellus *L.* **Butterfish.**

Numerous all over the West Coast, from Cape Wrath to the Clyde. Under every stone in sound between Scalpa and Skye at low water (*nobis*). Eilean Shona (T. A. S.). Sound of Jura (C. J.).

Zoarces viviparus (*L.*). **Viviparous Blenny.**

I have taken two specimens in Loch Creran, but it seems rare on the West Coast (*nobis*). Sound of Jura (C. J.). Glenshiel, Ross-shire (N. S. A.).

Division ACANTHOPTERYGII MUGILIFORMES.

Family **ATHERINIDÆ.**

Atherina presbyter *Cuv.* **Sand-Smelt. Atherine.**

I have captured it from shoals at mouth of stream in Loch Creran. A specimen from Argyllshire in Edinburgh Museum. Specimen recorded from Loch Linnhe by "Nether Lochaber." Rare on the Scottish shores (N. L.). Plentiful west of Mull; calls it "smelt" (?) (M'L. C.).

Atherina boyeri *Risso.* **Boyer's Atherine.**

[[1] Discovered by Mr. Sim, off Aberdeen. This species proved to occur also on the west coast of Scotland; three adult specimens having been found between Cumbrae and Skelmorlie Light in 20 fathoms in April; and at Cumbrae Lighthouse in 60 athoms in February" (Murray—Günther). (H-B. and B.).]

Family MUGILIDÆ.

Mugil octo-radiatus *Günther*. **Eight-rayed Mullet.**

Mugil capito *Cuv.* **Grey Mullet.**

Gaelic—*Iasg drimionn*.—N. S. A.

Loch Etive (*nobis*). North coast of Mull (J. N. F.). Eilan Shona (T. A. S.). Carsaig, Mull (M'L. C.). Islay (Pennant). Sound of Jura; Loch Bee in North Uist, abounds in fine Mullet (N. S. A.). "The females shed their spawn about midsummer" (N. L.).

Mugil auratus *Risso*. **Long-finned Grey Mullet.**

Mugil septentrionalis *Günther* (*Cuv.*). **Lesser Grey Mullet or Thick-lipped Grey Mullet.**

Not uncommon in Lewis Bays; found locally all over area (*nobis*). Loch Carron (T. E. B.). Sound of Jura (C. J.). "I have found it common on West Coast of Scotland" (Dr. Parnell—N. L.). Mullet, Strath, Skye (N. S. A.). The above species is probably confounded with this under the general name Grey Mullet.

Division ACANTHOPTERYGII GASTROSTEIFORMES.

Family GASTROSTEIDÆ.

Gasterosteus aculeatus *L.* **Three-spined Stickleback.**

Common in salt, fresh, and brackish water all around the coast from Lewis to Loch Creran, Canna, Mull, Barray, etc. (*nobis*).

Gasterosteus brachycentrus *C. and V.* **Short-spined Stickleback.**

In North Uist (Dr. M'In. and H-B.).

Gasterosteus spinulosus *Ten.* **Four-spined Stickleback.**

Gasterosteus pungitius *L.* **Ten-spined Stickleback.**

Gasterosteus spinachia (*L.*). **Fifteen-spined Stickleback.**

Gaelic—*Carran* (*Creaige ?*).

Common from Lewis to Barray in Hebrides, and all over the inner waters; Canna, Mull, etc. Very plentiful on the verge of the laminarian zone, where the more luxuriant sea-weeds abound.

Division ACANTHOPTERYGII CENTRISCIFORMES.

Family CENTRISCIDÆ.

Centriscus scolopax *L.* **Trumpet-fish. Sea-Snipe.**

Division ACANTHOPTERYGII GOBIESOCIFORMES.

Family GOBIESOCIDÆ.

Lepadogaster gouanii (*Lacép.*). **Cornish Sucker.**

In Sound of Jura (Pennant). He says in Sound of Jura and Cornwall ?[1]

Lepadogaster candollii *Risso.* **Connemara Sucker.**

Very numerous in the spring in Loch Creran at low tide in the breeding season. Took a very brilliant spectacle-marked specimen in Canna Island, evidently the Cornish Sucker of Pennant, and Connemara Sucker of Day.

Lepadogaster bimaculatus (*Penn.*). **Doubly-spotted Sucker. Network Sucker. Bimaculated Sucker.**

Plentiful in Loch Creran, Mull, Barray. I have taken them at low water, but they are mostly taken in the dredge from scallop ground in 8 to 12 fathoms. In the empty valves of this shell it is fond of spreading its ova in a fine layer, and remaining to watch over them.

North Uist, abundant amongst tangle forests and seaware (Dr. M'In.). This he calls *network sucker*, which is commonly

[1] Taken at Millport, Clyde.

applied to a variety of *Liparis montagui*; but this variable little fish may take similar markings.[1]

Division ACANTHOPTERYGII TÆNIIFORMES.

Family TRACHYPTERIDÆ.

Trachypterus arcticus (*Brünn*). **Deal-fish.**

Regalecus banksii (*C. and V.*). **Bank's Oar-fish.**

Regalecus grillii (*Lindr.*).

Order 2.

ACANTHOPTERYGII PHARYNGOGNATHI.

Family LABRIDÆ.

Labrus maculatus *Bl.* **Ballan Wrasse.**

Gaelic—*Muc-creaige.*

Not uncommon in Lewis, Mull, Loch Creran (*nobis*). "Common on all the rocky coasts from Cape Wrath to Mull of Galloway. We caught them at the Shiant Isles in the trammel-net" (H-B.). Loch Carron (T. E. B.). M'Intosh records it from North Uist, Sound of Jura (C. J.). Glenshiel, Ross-shire (N. S. A.). Common off Moidart (W. B.). The various Wrasses (three species undefined by captor) are so numerous off Moidart that Mr. W. Blackburn took 470 in the trammel-net during 1887, and proportionate numbers other years.[2]

Labrus mixtus *L.* **Cook Wrasse. Blue-striped Wrasse.**

North coast of Mull (J. N. F.). Loch Carron (T. E. B.). Sound of Jura (C. J.). At North Uist it is found in shoals at the margins of the rocks, or lurking under sea-weeds in rock-pools (Dr. M'In. *fide* F. D.).[3]

[1] Taken in Clyde area. See "Notes on Liparis and Lepadogaster," Royal Phys. Soc. Edinburgh, 16th December 1885.

[2] In Clyde area (*fide* Dr. J. Young).

[3] *Ibid.*

Crenilabrus melops (*L.*). **Goldsinny. Corkwing Baillon's Wrasse.**

Not uncommon in Loch Creran (*nobis*). Spawns in April (N. L.). Dr. M'In. includes *Ctenolabrus pusillus* (?) in North Uist list (H-B.). This has no resemblance to the "Baillon's Wrasse" of Couch. *C. melops* (Corkwing).[1]

Crenilabrus rupestris (*L.*). **Jago's Goldsinny.**

In swarms round the rocks of North Uist, Dr. M'In. (H-B.).[2]

Acantholabrus palloni (*Risso*).

Centrolabrus exoletus (*L.*). **Small-mouthed Wrasse. Rock Cook.**

Several specimens taken by me in Loch Creran, largest 7½ inches. Loch Carron (T. E. B.). Sound of Jura (C. J.).

Coris julis (*L.*). **Rainbow Wrasse.**

"Bimaculated Wrasse" (?) sometimes taken in Glenshiel, Ross-shire (N. S. A.).[3]

Coris giofredi (*Risso*).

Order 3. ANACANTHINI.

Division 1. ANACANTHINI GADOIDEI.

Family LYCODIDÆ.

Gymnalis imberbis (*L.*). **Beardless Ophidium.**

Family GADIDÆ (*Cuv.*).

Gadus morrhua *L.* **Common Cod.**

Gaelic—*Trosg-bodlach.* Young or rock cod—*Boduch ruadh.*

A plentiful western fish. Finest banks, Barray; west of Lewis; about St. Kilda; around Canna; off Lochbuie, 60 fathoms; numerous off Moidart; Sound of Mull; west of Mull. The red

[1] Clyde area (Murray—Günther).
[2] In Clyde area (Murray—Günther; Dr. J. Young).
[3] In Clyde area (Dr. J. Young).

inshore cod, called Rock Cod. These remain distinct in character to a large size about the Barray shores (*nobis*). Red Rock Cod of the largest size off Kintyre; good cod banks two or three miles from each end of the island of Gigha. Cod-fishing carried on to a considerable extent by people of Colonsay, and its banks are frequented by the boats of Islay, Gigha, and Kintyre; banks of Islay. (N. S. A.). During this century our finest fishing-grounds have been from forty to sixty miles west of Lewis, but the weather in the winter is commonly so stormy that the utmost advantage cannot be taken of them. The cod-fishing of Rockall in the Atlantic was quite phenomenal for a time, but it was said to be fished out by the fleet of vessels that frequented it. This, however, is not credited, but the rendezvous of the crews in the Hebrides did not please the fishermen. These Cod were very large and coarse. It could still be possible to prosecute a considerable fishing in the vicinity of this rock in the Atlantic. The neighbourhood of Dhuheartach is also good ground.

Gadus æglefinus *L.* Haddock.

A local and uncertain fish in west of Scotland. The best and most regular fishing off Stornoway. I have taken them in quantity and of fine quality off west of Lewis, but very unreliable there. Bank of note for large Haddock in Loch Duich; off Isle Ornsay; Sound of Jura; Kyleakin in Skye; occasionally fine off Oban and in Lochs Linnhe and Sunart (*nobis*). Loch Killisport; Sound of Mull; west of Kintyre; large Haddock caught on the banks off Gigha; Loch Carron; Broadford, Skye. In Kilmuir, Skye, Haddocks are comparatively rare, but were about half a century ago the most numerous of all fish (N. S. A., 1845.). Throw spawn February to April.[1]

Gadus minutus *L.* Power Cod.

This brilliant little fish was plentiful about Skye in 1886; brilliant from the tangle, Loch Creran (*nobis*). "Sound of Mull, in 70 fathoms" (J. M.). Sound of Jura (C. J.). Clyde area, "The specimens obtained on March 10 and 17 were ready to spawn, and had fed on *Nyctiphanes*, sand-eels, and *Aphrodyte*" (Murray—Günther.)

[1] Young, 2¼ in. to 3¼ in. long obtained off Ardrossan in 10 to 15 fathoms in April. Off Cumbrae in 90 fathoms, August; 3 half-grown in 26 fathoms, December. in Kilbrannan Sound. One young 4 in. specimen between Cumbrae and Wemyss Point, in 30 and 40 fathoms, in February (Murray).

Gadus merlangus *L.* **Whiting.**

Gaelic—*Cuiteag.*

Abundant in Loch Linnhe and the lochs opening on to it; taken by me in Killisport, Sound of Jura, Loch Creran, Loch Hourn, Loch Alsh, Broadford, Skye; small Whiting ate all the bait off the hooks in Loch Slapin, Skye (1886, September). Bloody Bay in Mull, and indeed over all the region from Lewis (Loch Roag) south. The young were exceedingly numerous under *Medusæ* in August and September 1886 over a very considerable range (*nobis*). West of Kintyre (N. S. A.) Upper Loch Nevis, 50 fathoms (J. M.). Loch Carron (T. E. B.). Two, each 3 lbs. weight, in Loch Swein, Knapdale (C. J.). Mr. Blackburn mentions he has never caught Whiting off Roshven, Moidart!

Gadus virens *L.* **Coal-fish.**

Gaelic—*Steinloch* (nobis). *Suian* (A. C.). Half-grown, Sillock = Gaelic *Smalak*. *Cudainn* or *Cudhic*—hence cuddy for the young.

Numberless all over the west. I am not aware of any district they do not frequent. But this and the following species are most plentiful and of finest quality in the Outer Hebrides. There a "Saithe" or half-grown fish that is tasteless and watery amongst the inner lochs is firm and well tasted. I have taken them from Lewis to the Clyde along the whole water way (*nobis*). "In some of the lakes of North Uist with which the sea communicates they are found of large size and fine quality, and partaking in some degree of the flavour of fresh-water trout" (N. S. A.). Sound of Mull, 70 fathoms, Upper Loch Nevis, 50 fathoms, 2 *juv.* on 3d September. Loch Sunart, 45 to 50 fathoms (J. M.). In Kilmuir, Skye, the livers of "80 to 100 Sythes will yield an imperial gallon of oil" (N. S. A.). Very numerous off Moidart (W. B.).

Gadus pollachius *L.* **Pollack or Lythe.**

Gaelic—*Luith.*

Almost equally numerous with the Coal-fish wherever rocks rise from the sea. Those of Hebrides specially fine. Much superior in its adult state to the Coal-fish, and I am half inclined to indorse the statement that in Lewis "the Laithe far surpasses the Whiting in delicacy and sweetness" (N. S. A.). They are taken in vast quantities with the seine-net along with the former, their livers giving a rich supply of oil, formerly of great value in the Highlands. Plentiful, but not so common as the preceding, off Moidart (W. B.).

June 14, 1879.—"I caught four Lythe at Rodel in the evening, none bigger than about 5 or 6 lbs., but on a good evening they are caught more abundantly and weigh up to 13 and even 18 lbs. (?!) and as many as fourteen fish taken in an hour or so. The red-coloured smaller Lythe are the best for the table and are excellent fish, but should not be cooked in steaks but *whole*, with the skin on, and they should lie in salt for a day or longer before use" (H-B.).

Gadus luscus *L.* Bib-Pout. Whiting-Pout.

Glenshiel, Ross-shire (N. S. A.). Not uncommon off Moidart (W. B.).

Gadus poutassou *Risso.* Couch's Whiting.

It has been taken off North Uist in the Hebrides (F. D.).

Gadus esmarkii (*Nilss.*). Norway Pout.

New to British Fauna.—Sound of Mull, etc. Clyde area "along with a host of other *Gadi*, especially *G. minutus*, with which it may be readily confounded. The species does not appear to be unfrequent in Kilbrannan Sound. The specimens measure from 3½ to 7 inches." The Norway Pout has been recognised for many years on the coasts of Scandinavia, where it occurs locally in deep water during the winter months. Dr. Lutken records its occurrence near the Faroe Islands. The characteristics by which it can be distinguished from its British congeners are:—The lower jaw, which projects beyond the upper; the dentition, the teeth of the outer series in the upper jaw being a little larger than the inner ones; the length of the snout, which is almost equal to the length of the diameter of the eye; the large size of the eye, which is a little less than one-third of the length of the head; the slender barbel, which is about half as long as the eye; and, finally, the fin formula, —it being D. 15-16, 23-25, 22-25; A. 27-29, 23-25. Dr. Günther then goes on to draw varietal distinctions between Scandinavian specimens and these British ones, which perhaps are only necessary to notice here, with the reference to his paper direct. Young specimens of this fish were also found in tolerable abundance in

Lower Loch Fyne,	80	fathoms,	January.
Sound of Mull,	70	„	September.
Upper Loch Nevis,	150	„	„
Loch Sunart,	45-50	„	„
Lochaber,	70-80	„	„

(Murray—Günther.)

Merluccius vulgaris *Flem.* **Hake.**

Gaelic—*Calamor.*

Plentiful off Stornoway and very fine. Called Herring-Hake from following the Herring. Taken by me off Portree and in Loch Alsh. They are plentiful off Skye at times. They were formerly abundant in Loch Slapin, but Dr. Mackinnon informed me in 1886 that not one had been taken for many years previously (*nobis*). Loch Carron (T. E. B.). Caught by us in Loch Inver on the long lines (H-B. and B.). Eilan Shona and general (T. A. S.). Sound of Jura (C. J.).[1]

Phycis blennioides (*Brünn.*). **Greater Forked Beard.**

I obtained one in August 1886 off Portree, Skye. Others I have been assured have been taken by the fishermen, who say they obtain one or two every season. One was captured in Loch Hourn.

Lota vulgaris *Cuv.* **Burbot. Eel-pout.**

Molva vulgaris *Flem.* **Ling.**

Gaelic—*Langa*, juv. *Donnag.*

General; approach nearer shore in winter, when they are in best condition. I have taken them off Skye, off Canna, off Lewis, magnificent fish upwards of five feet long. They are plentiful off Barray, but generally much smaller than those of Lewis. One 52 lbs. weight was however taken, February 1888, when boats were catching 10 to 20 cwt. each. "Common at Loch Inver, and caught by using the long sea lines" (H-B. & B.). Killisport, Sound of Mull, west of Mull, west of Kintyre; banks two to three miles from the two ends of Island of Gigha; banks off Islay (N. S. A.). Loch Carron

[1] Clyde area (Murray—Gunther).

(T. E. B.). Eilan Shona (T. A. S.). Sound of Jura, good winter fishing (C. J.). Of recent years it is more frequently seen in the market in a fresh state, although its eel-like form is objectionable to some.

Motella mustela (*L.*). Five-bearded Rockling.

Rare in Loch Creran; plentiful at low-tide in west of Lewis; north and south of Skye; Barra; Loch Linnhe; Mull, north and south (*nobis*). North Uist (Dr. M'In. *fide* H-B.). Carsaig, Mull (M'L. C.). Sound of Jura (C. J.). Glenshiel, Ross-shire (N. S. A.).

Motella cimbria (*L.*). Four-bearded Rockling.

Very common, and generally distributed in Clyde area at depths varying from 6-18 to 56 fathoms in April; to 70-90 in July and August; to 100 fathoms in November at Upper Loch Fyne and Kilbrannan Sound; at 37 to 46 fathoms in December, etc. (J. M.). (*Onus cimbrius*) Günther.

Motella maculata (*Risso*). Spotted Rockling.

Motella tricirrata (*Bl.*). Three-bearded Rockling.

I have taken them over most of the region. Flannan Isles, west of Lewis, Loch Creran, Tobermory, Lochbuie, Mull; took one in foreshore, 6th April 1887; very dull in colour. Young taken from under *Medusæ* in July. Sound of Jura, and off Rum in September. Very common off rocky coast of Carsaig, Mull. A heavy, rich fish to eat (M'L. C.). Sound of Jura (C. J.). "It seems to be rare in every part of Scotland" (N. L.). I am disposed to look upon this as a mistake, arising from the nature of the rocky ground frequented. Off Moidart every season (W. B.). Günther makes two species; the larger he calls *Onus maculatus*, but we have always considered the rich colour and spotting the result of maturity.

Motella macrophthalma *Günth.*

Raniceps trifurcus (*Walb.*). Lesser Forkbeard. Tadpole Hake.

The west of Scotland (N. L.).

Brosmius brosme (*Mull*). **Torsk. Tusk.**

Gaelic—*Traille-Manach.*—A. C.

I have taken them off Lewis, where they are occasionally thrown ashore by the waves. It seldom exceeds two feet in length. More especially a northern fish, and I have not known it taken south of Ardnamurchan, although N. S. A. says "West of Kintyre" (*nobis*). West of Muck, off Barray, St. Kilda, etc. (T. A. S.). Minch (H-B.). Being a deep-sea fish, it rarely comes close to land, but has been caught by fishermen in the Minch from Loch Inver, and we ourselves caught four in one day in July 1886, on the Stoir Cod banks, in 40 fathoms of water (H-B. & B.). It is also common in the Atlantic, west of the Hebrides, where we have ourselves met with them, and considerable numbers are sent to Castle Bay, Barray, to the curers, which are caught principally to the westward (H-B.).

Family OPHIDIIDÆ.

Ophidium broussonettii *Mull.* **Bearded Ophidium.**

Fierasfer dentatus *Cuv.* **Drummond's Fierasfer.**

Ammodytes lanceolatus *Lesauvage.* **Larger Launce. Greater Sand-Launce.**

Gaelic—*Siolag, Saundag, Sachasan.*—A. C.

Numerous in Lewis sands, Loch Linnhe, Easdale (*nobis*). North Uist (Dr. M'In.). Loch Carron (T. E. B.).[1]

Ammodytes tobianus (*L.*). **Lesser Launce.**

Gaelic—*Easgann* (Fraser) *Siolag.*

Numerous in sandy bays throughout the region. Loch Creran, Mull, Lewis (*nobis*). Seen abundantly in Newton Bay, or Lingay Sound, North Uist; also common on all sandy shores Outer and Inner Hebrides, Eigg, etc. (H-B.). Loch Carron (T. E. B.). Eilan Shona (T. A. S.). Strath, Skye (N. S. A.).

Ammodytes cicerellus (*Rafin*)

[1] Sound of Sanda, Kintyre, young numerous end of March (Murray—Günther).

Division ANACANTHINI PLEURONECTOIDEI.

Family PLEURONECTIDÆ.

Hippoglossus vulgaris (*Flem.*). **Holibut.**

Gaelic—*Bradan Leathann.*—A. C.

Plentiful and of great size off St. Kilda, where an attempted Cod-fishing was partially frustrated by the immense takes of this fish that they could not dispose of. Off Barray; Lewis; I have taken young a few days old in the tow-net off Lochbuie in March; west of Muck (T. A. S.). Sound of Jura (C. J.). Common in Atlantic west of the Hebrides, where we saw immense specimens on board fishermen's boats; also Minch (H-B.). They have been captured upwards of 2 cwt. in weight.

Rhombus maximus (*L.*). **Turbot.**

Gaelic—*Liabac Brathain* (nobis). *Bradan Brathain.*—A. C. Also Gaelic—*Bacach-Bacach Ceaw-Pac.*—A. C.

Not uncommon all around Lewis (*nobis*). Sleat, Skye; west of Mull, off Ulva; off Islay, on banks; Loch Melfort; Loch Killisport; rare off Tiree and Coll (N. S. A.). Off north of Mull (J. N. F.). West of Muck (T. A. S.). Scattered all over the region, but not so numerous in localities as in east; off Moidart (W. B.).[1]

Rhombus maximus (norvegicus). Norway Topknot.[2]

[1] One captured in Clyde area 25th August 1888; formerly numerous there.

[2] "Known to Scandinavian ichthyologists for last fifty years as a rare fish on the coast of Sweden and Norway. Günther first pointed it out as distinct from *Pleuronectes cardinia* of Cuvier. More recently found most abundantly on west coasts of Norway. (Previous British localities—Bristol Channel, Couch, 1863, 5 in. long; Shetland, Günther, 1868, 2 in. long, at about 90 fathoms depth. The third is now from Lamlash Bay (Clyde), from 6 to 18 fathoms, 3¼ in. long, and in excellent condition. Dr. Gunther describes this last-mentioned specimen at length. See paper.) —Murray—Günther."

Phrynorhombus unimaculatus (*Blochs*). Topknot.[1]

Hippoglossoides limandoides (*Bl.*). Rough Dab.

Off Lismore in 50 fathoms (*nobis*). Sound of Mull in 70 fathoms (J. M.). Upper Loch Nevis; Loch Sunart, 45 to 50 fathoms; Loch Duich, in 60 fathoms, *juv.* 31st August 1887; Loch Hourn, 70 fathoms, 29th August 1887 (Murray—Günther).[2]

Rhombus lævis (*L.*). Brill.

Off Lewis; off Mull, rare (*nobis*). Loch Carron (T. E. B.). Kintra Bay, Ardnamurchan (T. A. S.).

Zeugopterus unimaculatus (*Risso*). Eckstrom's Topknot.

Sound of Jura (C. J.). Is this *R. Norvegicus* (Günther), now *R. maximus*?

Zeugopterus punctatus *Bl.* Müller's Topknot.

Off Lewis "was only represented to our eyes by one specimen, about 5 in. long, taken on the long lines. It seemed unknown to the fishermen, and must be exceedingly rare in these seas" (Lewsiana). Sound of Jura (C. J.). Loch Carron, Topknot (?) (T. E. B.). Var. *papillosus* (Brook) plentiful in Creran and Linnhe. I submitted coloured drawing identical with that in Fishery Board Report to Dr. Day in 1882, who pronounced it *punctatus*. Off Moidart (W. B.).[3]

[1] One specimen in 10 fathoms off Ardrossan, Clyde area, in April (Murray—Gunther).

[2] "Most common flatfish on west coast, more at least entered the dredge than any other species of this family. The majority were, however, young, from 2 in. to 9 in. long. Many adults and young (2 in. in length) in 26 and 46 fathoms in Kilbrannan Sound in December. Many *ad. et juv.* between Cumbrae and Wemyss Point, 30 to 40 fathoms, in February. In these specimens the ovaries showed conspicuous signs of enlargement, whilst the testicles were in a collapsed condition. In all the food was too much digested for discrimination" (Murray—Gunther).

[3] Clyde area, 60 fathoms (Murray).

Arnoglossus megastoma (*Donovan*). **Sail-Fluke. Whiff.**

Loch Linnhe; off Oban; south of Mull; not common. I obtained one full of roe in Lochbuie, 14th April 1887.[1]

Arnoglossus laterna (*Walb.*). **Scaldfish.**

Sound of Jura (C. J.).[1]

Pleuronectes platessa *L.* **Plaice.**

Abundant over the greater part of the region Lewis to Killisport; Sound of Mull, west of Mull, Loch Alsh, Barray, numerous but small at Broadford, etc. Spawns in February and March, have taken little slips in April (*nobis*). Banks off Islay (N. S. A.). Loch Carron (T. E. B.). Islay (P.). Sound of Jura (C. J.). The commonest of our flat fish, largely used as bait in long-line fishing. Goes under the generic name of "flounder" in the west. Common off Moidart (W. B.).

Pleuronectes microcephalus *Donovan.* **Smear Dab. Smooth Dab, or Lemon Dab of Yarrell.**

Very fine in Lewis—one 16 in., well filled with roe, in March (Lewsiana). Caught in Loch Eport in trammel-net, near entrance to Loch Obisary, July 1887 (H-B.). Islay (Pennant). Sound of Jura, Lemon Dab! (C. J.).[2]

Pleuronectes cynoglossus *L.* **Craig-Fluke. Pole.**

Loch Creran, 15 fathoms; also off Oban (*nobis*). Loch Hourn, in 70 fathoms, 29th August 1887. Loch Carron, 2d September 1887, in 60 fathoms. Lochfyne, 100 fathoms, November 6th. Lower Lochfyne, 80 fathoms, January (J. M.). Sound of Jura (C. J.). Those taken, March 22, had finished spawning (Murray —Günther).

This species, Dr. Günther tells us, is known to descend to a depth of 200 fathoms on the Norwegian coast, and is reported from the north-west Atlantic to descend to a greater depth than any other flat fish: viz., to more than 700 fathoms. Those obtained by J. M., as above, also came from great depths (Murray —Günther).

[1] Clyde area (Murray—Günther).

[2] Clyde area, feeding on *Solens* and *Annelids* (Murray—Günther).

The stomach contained pieces of a green *Medusa*, of *Crangon Allmanni*, and *Nephrops*, and a large number of *Nereis*.

Pleuronectes limanda *L.* **Common Dab.**

Lewis—common. Mull, Loch Creran (*nobis*). North Uist (Dr. M'In.). Minch—common in most, if not all, of the sea-lochs of the Hebrides—Lochs Bracadale, Eynort, etc. (H.-B.). Loch Carron (T. E. B.). Sound of Jura (C. J.). This species more affects brackish water. Very numerous off Moidart (W. B.).[1]

Pleuronectes flesus *L.* **Flounder.**

Gaelic—*Liabac.*

Numerous in suitable situations all over the region. I have taken them from Lewis to Islay. Sound of Mull; "the Broad Bay Flounder, Stornoway, the finest in the world"; in Loch Bu, South Uist, a brackish water loch (N. S. A.). North Uist (Dr. M'In.). Minch (H-B.). Loch Carron (T. E. B.). Eilan Shona (T. A. S.) Sound of Jura (C. J.). Off Moidart (W. B.).

Solea vulgaris *Quensel.* **Sole.**

A number of boxes of fine soles were trawled on the banks off Colonsay. Has been trawled off Oban. The fishery cruiser trawled them in Stornoway Bay (*nobis*). It is taken all round the Sutherland and Caithness coasts (H-B. and B.). West of Mull; off Ulva; Loch Creran; banks off Islay; Killisport; Strath, Skye (N. S. A.). Kintra Bay, Ardnamurchan (T. A. S.). Off Moidart (W. B.). This fish is much wider spread in the west than is commonly supposed.[1]

Solea aurantiaca *Günth.* **Lemon Sole.**

"Lemon Sole," Kintra Bay, Ardnamurchan (T. A. S.).

Solea variegata (*Donovan*). **Variegated Sole.**

Mull of Kintyre, 65 fathoms (Murray—Günther).[1]

Solea lutea *Bonap.* **Solenette.**

[1] Clyde area (Murray—Gunther). [2] [Clyde area.]
[(Off Rothesay and the Isle of Bute in Scotland. Scouler, *fide* F. D.).]

Order 4. PHYSOSTOMI.

Family CYPRINIDÆ.

Sub-family *CYPRININA.*

Cyprinus carpio *L.* **Common Carp.**

"Carp will not thrive in Scotland until some means be discovered for meliorating the climate and giving a soft quality to the water" (N. L.).

Carassius vulgaris *Nilss.* **Crucian Carp.**

Carassius auratus (*L.*). **Gold Carp.**

Common in domestication.

Barbus vulgaris *Flem.* **Barbel.**

Gobio fluviatilis *Flem.* **Gudgeon.**

Sub-family *LEUCISCINA.*

Leuciscus rutilus (*L.*). **Roach.**

Leuciscus cephalus (*L.*). **Chub.**

Leuciscus vulgaris *Flem.* **Dace.**

Leuciscus erythrophthalmus (*L.*). **Rudd.**

Tinca vulgaris *Cuv.* **Tench.**

Leuciscus phoxinus (*L.*). **Minnow.**

I have never met with the Minnow over the region in question, but in N. S. A. I find it noted as in Ardchattan district, and Lochbroom, Ross-shire.

Sub-family *ABRAMIDINA.*

Abramis brama (*L.*). **Bream.**

Abramis blicca (*Bl.*). **White Bream.**

Alburnus lucidus *Heckel and Kner.* **Bleak.**

Sub-family *COBITIDINA*.

Nemachilus barbatulus (*L.*). **Loach.**

Cobitis tænia (*L.*). **Spinous Loach.**

Family **SCOMBRESOCIDÆ**.

Belone vulgaris *Flem.* **Garfish.**

Taken off Oban in 1887 (*nobis*). Only one seen in Loch Linnhe in twenty years by "Nether Lochaber." Loch Carron (T. E. B.). Occasionally numerous (M'L. C.). As they are too thin to be caught in ordinary nets they probably escape notice in the west.[1]

Scombresox saurus (*Walb.*). **Saury. Skipper.**

Probably confused with the above (as they have been taken in quantities in Orkney; to be looked for in the west).[2]

Exocœtus evolans *L.* **Flying Fish.**

Exocœtus volitans (*L.*). **Greater Flying Fish.**

Family **ESOCIDÆ**.

Esox lucius *L.* **Pike.**

Gaelic—*Gead-iasg.*—Fraser.

This fish was introduced into a small loch in Benderloch by Mr. Cameron of Barcaldine (Lochan Dhu), but seems to have died out. They are only of recent introduction into Loch Awe, says N. S. A., 1840. Colonel Thornton, of sporting celebrity, caught one by trolling in Loch Awe, after a struggle of one hour and a quarter, which weighed 50 pounds; it measured exactly 4 feet 4 in. from eye to fork, and jaws and tail included could scarcely be less than 5 feet. "So dreadful a forest of teeth or tusks," exclaims the Colonel, "I think I never beheld." Compare with this Loch Ken, 61 lbs.; Loch Lomond, 79 lbs. (N. L.). Five lakes in Kilninian and Kilmore (Mull), all abounding with excellent Pike. Pike

[1] Taken in Clyde area, occasionally, in shoals.

[2] Taken in Clyde area.

abound in some of the lochs of Kildalton parish, Islay; Lochbroom, Ross-shire (N. S. A.). None in Kintyre, 1843 (N. S. A.). Gedd Lochs, Attadale (T. E. B.).

Family STERNOPTYCHIDÆ.

Maurolicus borealis (*Nilss.*). **Argentine.**

Argyropelecus hemigymnus (*Cuv.*).

Family SALMONIDÆ.

Salmo salar *L.* **Salmon.**

Gaelic—*Bradan.* Parr = *Gille ruadh.*

The Salmon rivers of the West Coast have been so recently and fully investigated by Mr. Archibald Young that we must refer to his report (included in those of the Fishery Board) for information. The West Coast Salmon, as a rule, are not so large as those of the East, although fish weighing over 40 lbs. have been taken in the river Awe, the principal river. Some of the Lewis rivers, such as the Grimersta, the Blackwaters, the Creed, and the Laxay, are noted for the great takes of fish obtainable in the fresh water; while off Barvas river the capture of fish by the sweep-net has occasionally been almost incredible.[1] The fish frequenting the smaller rivers of the Islands are usually diminutive, but finely shaped, and the feeding of the waters they frequent must be very rich, when we consider the numbers of fish in spite of the persistent poaching by yachtsmen and native scringers. At the Obb of Harris is a noted salmon-cast in brackish water, another being on the coast of Sutherland. Salmon have been frequently captured elsewhere with bait in salt and brackish water, but these are the most noted places where they are known to rise to the angler. Sir John Morris, in Stornoway Bay, took five salmon with a salmon-fly, in August 1888, from pure salt water. The bay was full of fish at the time. The Salmon spawns early—the latter end of Sep-

[1] The Salmon of The Lews, of which Mr. A. Williamson in one season killed 90—getting in one day 19—averaged only 6¾ lbs.—(H-B.) One day, this season—August 1888—one angler took 54 salmon to his own rod in the Grimersta.

tember—in many of these rivers, and their spawn might be usefully employed to introduce earlier migrations into later rivers on the mainland.[1] From Loch-in-daal in Islay to Loch Roag in Lewis, Salmon are plentiful, and generally healthy, in the Island waters, although I have seen disease prevalent in Lewis many years ago. The fine inland fresh-water lochs, when of sufficient extent, seem to give vigour to the fish, if they do not linger too long. The fish of the Grimersta are specially fine; yet in Skye those that spawned in the small river which runs into the Bay of Altivaig—Kilmuir—were considered the finest in Skye. In Portree parish they do not run until late in July.

With regard to the Salmonidæ of the Uists, as to which there has been some question, I cannot do better than quote Sheriff Webster, Lochmaddy—a skilled and enthusiastic angler:—"As to the *Salmo salar* in North Uist, there are some, but not a great many, and chiefly, at any rate, in the lochs draining into Loch-na-Kiste, at Lochmaddy, into which the sea flows at spring-tides. Personally, I have never killed a Salmon in any other range of lochs. I have no doubt a few find their way into Loch Obisary, though I have never killed one in it; but this, as its name imports, is a sea-loch, into which you can sail at spring-tides. I have never seen a *Salmo salar* in any of the lochs communicating with the west or Atlantic side of North Uist, and I do not believe there are any, though the Eriox and Sea Trout run to a large size. As to South Uist, from the information I have, I have no doubt that the *Salmo salar* enters Loch Barp from the Minch at Lochboisdale, but into this loch also the tide flows at spring-tides. I have never myself caught one in it, but I have rarely fished it. I however caught two undoubted Salmon in the Howmore (so-called) *river* in South Uist last October. I killed ten Eriox, weighing 50 lbs., and two Salmon of 5 and 6 lbs. I had a long conversation with John Lamont, the fisherman who has been all his life on the water, about this very matter, as *I* had never previously killed what I called a *Salmo salar* on it, and he informed me that in the late John Gordon's time the same question arose, and that three typical fish were sent, I think, to Frank Buckland, who pronounced them to be *Salmo salar*, *eriox*, and *trutta*. If this is

[1] This has been attempted unsuccessfully on the Inver, and on the Kirking, by, in the latter case, putting in fry above a fall of 60 feet. There have been no appreciable results.—H-B.

correct it settles the matter, but I have no doubt that the fish I got were *Salmo salar*. One of the *erior* I got that day was 8½ lbs."

Salmo trutta (*Flem.*). Sea-Trout. Salmon-Trout.

Gaelic—*Gealag.*—D. C.

This fine fish is very plentiful throughout the Hebridean waters, and, but for the incessant "scringing" by all and sundry, would be a very important element of wealth. They feed freely on Cuttle-fish, Sand-eels, Herring-sile, and the smaller crustacea along the coast, and ought to be encouraged and cultivated in every stream that has communication with the sea. These fish vary greatly with the locality, but those of the same locality have frequently a remarkable resemblance to each other. Thus we have taken at one spot well on to a hundred fish of the same average size, half being what are locally called Salmon-trout and half Sea-trout, hardly to be distinguished from each other except from the variable characteristic of numerical prevalence of spots. On the other hand, the so-called Sea-trout of our seas may have many varieties, including *estuarinus*, and the common *Salmo fario*, gone down for a change of diet. They differ widely in the several streams of the same river, and the beautiful Sea-trout of Glenure water, that flows into Glen Creran, are in form and quality incomparably superior to any others in this sea-loch. The species partakes of all degrees of delicacy, and frequently attains a large size. Mr. Allan caught a Sea-trout 9¾ lbs. weight at the Obb of Harris, a long and ugly fish compared to the finely shaped 4½ lb. Salmon caught at the same time. Sea-trout have been taken up to 23 and 24 lbs. in the Tweed.

This fish has been frequently isolated amongst the fresh-water lochs of the region. In Loch Fiart, in Lismore, the fish are very fine, and considered to be isolated Sea-trout, with what justice I am not aware. In a fresh-water loch over Loch Slapin, in Skye, Dr. Mackinnon, the late minister of Strath, informed me that Sea-trout had been isolated for over a hundred years, and that at a certain season of the year they still took on a silver coating before spawning. They have also been successfully isolated in a small loch near Carsaig, Mull, attaining the weight of 5 lbs. in four years. They have thriven in a loch in Rum, and afford good sport. The keeper explained their condition by the loch being rich in caddis worms, on which the trout fatten greatly.

"A form of Estuary Trout is apparently peculiar to Loch Inver and Kirkaig—locally known as 'Fossack.' The Trout of the tidal Loch Obisary, in North Uist, at the base of Ben Gabhal, I fancy are like those of Loch Bee. So also the Trout of several sandy lochs in west side of Uist" (H-B.).

Salmo cambricus *Donovan.* **Sewin.**

Salmo fario *L.* **Common Trout.**

Gaelic—*Breac*=brown trout: *Bricean.*—Fraser and D. C.
Ceabhanach—A. C.

There is scarcely a fresh-water stream or fresh-water loch in the West in which Trout are not found. They vary so greatly, even in the same stretch of water, and especially in lochs isolated from each other, that species might very well be multiplied indefinitely. Perhaps the most remarkable variety in the region is the so-called Tailless Trout of Islay, which has rendered permanent a strange malformation. Loch Finlagan, in which it is found, is of no great extent, and yet this interesting deformity has managed to perpetuate itself within its precincts. A specially fine variety of Trout is found in Loch St. Clair, Barray, which, from the peculiar formation of the slight barrier between it and the sea, might well be also an isolated Sea-trout! The Trout of Loch Bee, in South Uist, which becomes almost brackish water, may be looked upon as living under similar conditions to the Sea-trout. They are of noted superior quality. It was remarked in the New Statistical Account that "the lochs in the moor in South Uist abound in black Trout of a watery, insipid taste; but in most of the lakes on the west side, surrounded by arable land, and having a sandy or muddy bottom, the Trout are of a delicious flavour, weighing between 1 and 3 lbs." I go further, and from my experience declare that both classes of fish frequent the same loch in many of the various islands, those black and worthless being found at the mossy portion, but where there is a sandy or gravelly beach, as frequently happens, the fish are brighter and richer in colour of flesh. This probably arises, not solely from the fish taking the colour of the ground, but from the marked absence of food in peat or peaty waters. The great variety of Common Trout taken in the islands of South and North Uist can be best seen in the angling season, when they are displayed on the green in front of the hotels, in the evening, by the returned anglers. They might well be looked upon as separate species, so marked is the colour and form variation. I have seldom seen a finer show of *Salmo fario* than can be displayed by either of these watery islands. Loch Maree Trout have sometimes weighed as much as 30 lbs. The so-called *Salmo ferox* of Loch Awe reaches a great size; but the labours of Dr. Francis Day and others have inclined us to reduce the species of Salmonidæ, and it is probable that not only this but other acknowledged species may be relegated to mere varieties. I do not myself incline to treat *Salmo trutta* as other than *Salmo fario* under changed conditions, and all stages of progress between may be found.

Sheriff Webster says of brackish Loch Obisary, in North Uist: "There are plenty of fresh-water trout in it, running from ½ to ¾ lb., those at the end nearest the sea in very good condition, those at the other end the reverse. I have killed lythe, saithe, and codling in it with a fly!"

Harvie-Brown's notes on the Trout of North Uist are incorporated in Day's *Salmonidæ*, q.v.

Salmo fontinalis *Mitchell.* American Brook Trout.

Introduced into many of the Mull lochs. "Lord Dunmore some years ago put some *S. fontinalis* into a loch (a very small one). We netted it last year, but could not find any trace of them, but, so far as I could see, it was quite unsuited for them" (Sheriff Webster, North Uist).

Salmo alpinus *L.* The Alpine Charr.

Gaelic—*Tarragheal.*

In the district of Loch Carron (T. E. B.). Loch Maree (P.). Introduced into the various lochs on the Lochbuie estate, Mull. "In one lake Charr is to be met with," Craignish parish; Loch Awe, now rather a scarce fish, 1843; plentiful in Loch Ederline, Argyllshire. None found in Kintyre, 1843.—(N. S. A.). Mr. Webster says, "I have only caught two or three, but at any rate they are in two different lochs in North Uist."

Charr occur in many of the lochs of the O. H., of which several can be named in North Uist, but they are not known to frequent any of those lochs which at the present time communicate with the sea. They must be abundant, but are, as elsewhere, seldom caught by rod and line. It may be of scientific interest to catalogue all the lochs *known* to contain Charr, against such as have never yielded any to anglers. Such a list might become useful in the future.[1]

Osmerus eperlanus (*L.*). **Smelt.**

[I am not satisfied that this fish has been noted, as it seems to be confounded with the Sand Smelt or Atherine.][2]

Coregonus clupeoides *Lacép.* **Gwyniad. Powan.**

Coregonus vandesius *Rich.* **Vendace.**[3]

Coregonus pollan *Thomp.* **Pollan.**[4]

Thymallus vulgaris *Nilss.* **Grayling.**[5]

Argentina sphyræna *L.* **Hebridal Smelt of Couch.**

Loch Alsh, in October 1879, as figured in Day.

This deep-sea fish does not seem to be at all uncommon on the North-West Coast. By sinking a small-meshed trammel net to a depth of 30 and more fathoms, Mr. Murray succeeded in getting a number of specimens, which probably would have been still larger but for the attacks on the captured specimens by crustaceans and star-fishes. The specimens obtained in February had not yet spawned; there were three obtained, in 32 fathoms, between Little

[1] Sheriff Webster had promised to forward in spirits the first Charr he got, and H.-B. has since received one from Loch Fad, about ½ lb. weight, from him.

[2] [One taken with mussel bait at Brodick, Arran, October 1886. Not uncommon.]

[3] [Hitherto found only in the neighbourhood of Loch Maben.]

[4] (Only found in great Irish lakes.)

[5] [" We believe that the Grayling has not been found in Scotland, and certainly this hiatus in its distribution is not a little singular, the more especially, as being an Alpine fish, naturally fond of cool water, and abounding in much more northern countries on the Continent, the Highland rivers seem peculiarly adapted for it " (N. L. 1843). The southern Scottish rivers and streams have since been stocked with this graceful fish."]

Cumbrae and Brignird Point on February 7; and five obtained, in 37 fathoms, in Loch Striven on February 13 (Murray—Günther).

Family CLUPEIDÆ.

Engraulis encrasicholus (*L.*). Anchovy.

Clupea harengus *L.* Herring.

Gaelic—*Sgadan*; juv. *Scheal* (Ang. *Sile?*).

The Herring of the West Coast of Scotland has long been recognised as the finest fish of its kind, incomparably superior to those of the German Ocean, either North or South. They, however, differ materially in quality, not only according to season, but according to the nature of the ground on which they are taken. Those of Loch Hourn are notably superior amongst the outer lochs, and finer fish than those of Barray, in the season, could not be desired. It has come to be acknowledged, however, that the Herring is essentially a local fish, and a skilled dealer can tell the Herring of every principal loch on the West Coast from the general appearance. At the same time there are probably oceanic shoals spawning in deeper waters, and less subject to periodicity of movement, although at uncertain periods approaching the coast. It will ultimately be found that in place of two spawning periods, the West Coast Herring outside Mull of Kintyre spawns at various times throughout the year, and that no time is really without herring sile, except perhaps a month or two in the "dead season." Thus I have found herring sile of some size in July, August, and September, all in the same season. I have taken them in the "gutpoke"—or feeding state after spawning—condition in Loch Crinan in July. They are noted as being taken at the south end of Luing in that state in May and June. I have secured "gutpokes" in Loch Ainort (in Skye, north), in August, and west of Mull on September 21st. This gives a range of five months. In September 1886 I took in Loch-na-Keal, Mull, and off its mouth, Herring lately spawned, herring sile, and half-grown Herring some four inches long! At the beginning of the year herring sile are sold in quantity as "whitebait," and there is nothing to hinder this delicacy being provided throughout most months in the year.

My object is to prove that the West Coast Herring in youth or age is the main support of the Western fauna, not excepting the teeming population of sea-birds. The Sea-terns (*Sterna*) that build on the islets in our sea-lochs almost invariably feed their young upon herring sile, and no doubt choose their nesting-places with this object in view.

The spawned or "gutpoke" Herring, so called from its pronounced stomach filled with readily decaying material that prevents the fish curing properly, is caught with a Saithe fly, or even with plain hooks by jigging, over a great stretch of coast. It was so caught before 1820 in Craignish district, was known to be so procurable about Luing for about fifty years, and Pennant tells us "the herring-fishers never observe the remains of any kind of food in the stomachs of that fish so long as they are in good condition: so soon as they become foul or poor they will greedily rise to the fly, to be taken like the Whiting-pollock." It was often looked upon as a peculiar variety, or even species, but is well known now to be merely a herring "kelt."

The large Loch Roag in Lewis was at one time a very great herring loch, along with Lochmaddy, North Uist, the lochs of South Uist and the West of Skye, but by the date of the New Statistical Account, 1845, no fishing of any consequence remained, nor did they return until quite recent times to some of them. Ullapool was built as a herring emporium, but the fish suddenly left the vicinity, and the place became deserted. Loch Sunart was a rich herring loch, but about 1820 they deserted it, and did not return for more than twenty years. The last great herring fishing in Loch Creran occurred some twenty years ago, since which no fish have been taken of any consequence. Other fishes have equally deserted this loch since that time. It is unnecessary to multiply instances. Enough that, after successful fishing seasons, and the wasteful conduct of the fishermen, the shoals are frequently driven away for a quite indefinite time. It is possible the local shoal may be over-fished, and unable to cope with its ordinary enemies, for the absence of a herring fishery does not necessarily mean the absence entirely of Herring. I have taken young Herring in some quantity in Loch Creran, and dredged herring spawn from some fathoms, without any sign of important bodies of fish. The chances are that the local supply had, through fortunate conditions, increased sufficiently to attract attention, and that it was virtually fished out, the residue being unable to regain ascendency until similar exceptional conditions recur. If the offal were prevented being employed in polluting the waters and attracting dog-fish and coal-fish, and each loch were to spawn artificially a certain supply of Herring to maintain the stock, there is reason to believe that the result would be advantageous to the fishing community, and restrictions—impossible to be adequately carried out—be uncalled for.[1]

The comparatively shoal water to the west of Barray and the Uists seems to keep and probably breed a vast supply of the finest Herring, far superior to those of the Minch. The stormy nature of the coast, and the difficulty of handling drifts of nets, has prevented this coast being adequately fished, but boats that would fish off this coast as the Dutch busses fished—and Dutch and other boats still fish, forty or more miles off Shetland—would reap a rich harvest.

The herring fishing of the Western Highlands and Islands ought to be one of the best ordered and most productive industries of Europe, but with a population growing too ignorant and proud to eat Herring, because they are cheap, one is forced to be silent.

Clupea sprattus *L.* Sprat.

I am informed of shoals of Sprats at different periods in Loch Linnhe (*nobis*). Great shoals of Sprats on coast of Kintyre (N. S. A.). Great shoal of sprats off Oban this year (1888).

Clupea alosa *L.* Allis Shad.

Shad (?), Eilan Shona and "general" (T. A. S.). Captain Swinburne does not indicate the species.[2]

[1] Mr. John Murray supports this view. "He believes the herring of Loch Fyne never migrate beyond the entrance of the sea-loch—if they even go so far—but only from the shallows to the deeps of Inverary, where he has proved their food to consist almost entirely of crustacea dredged at 97-100 fathoms, viz., *Hippolyte securifrons*, *Pandalus annilicornis* (two species of red prawns), *Nyctyphanes* (the phosphorescent species), the rarer *Pasiphæa sivado*, and also *Ecketa* and *Calanus*" (H-B.) While insisting for many years on the local character of our finest herring shoals, I have been disappointed in obtaining Herring from stomachs of fish taken in deep water, at such times as they were absent from the nets!

[2] Very large specimens taken in Loch Fyne, 1888.

Clupea finta *L.* **Twaite Shad.**

Clupea pilchardus (*Bloch*). **Pilchard.**

I was informed of a heavy fishing of the "big-scaled herring" (Pilchard?) off Skye a generation ago, by those who had fished them then, but had not seen them since.

Family **MURÆNIDÆ.**

Anguilla vulgaris (*Flem.*). **Eel.**

Gaelic—*Easgann*, in Berneray; called *Cennache* or "the merchant."—H-B.

Plentiful and of very large size in Hebridean lochs, and throughout the region wherever there is fresh water.

Anguilla vulgaris, *Flem.* **Sharp-nosed Eel.**

Numerous in Lochavoulin close to Oban, where lately a very extensive lake-dwelling has been partly exposed. At one time, within the memory of some now living, this loch was more extensive, and was entered by the sea at high tides. Now it is little more than a moving marsh, and in the centre there is about 40 feet of soft mud, in which large numbers of eels exist, often appearing on the surface. One which Harvie-Brown examined was of the above species.

Anguilla latirostris *Risso.*

Harvie-Brown writes: "After a long hunt at the loch and burn in Berneray (Sound of Harris), we secured a specimen. It is only the Common Eel (broad-nosed?) which has become of an unusually silvery colour, from living in a loch the bottom of which is entirely of white sand.

"In a north wind, which produces a spate in the burn, these eels are carried down and stranded upon the sand below tide-mark, and are there collected and relished as a great delicacy. The aquatic life in the small warm sandy pools of the stream was marvellous, caddis worms, beetles, etc. etc."

Conger vulgaris, *Cuv.* **Conger.**

Plentiful throughout the region, often growing to great size. They manifest great fight and ferocity when "black" and living amongst the rocky ground; but if from deep water off-shore they are paler in colour and more sluggish. This fishery is growing in importance in the west. The young in little silver slips were common on the coast of Mull, and in the fresh-water streams between tide-marks in March and April 1887.[1]

Muræna helena *L.* **Murry.**

Order 5. LOPHOBRANCHII.

Family SYNGNATHIDÆ.

Siphonostoma typhle (*L.*). **Deep-nosed Pipe-fish.**

Throughout the region (*nobis*). Pipe-fish, Kilmuir, Skye; Glenshiel, Ross-shire (N. S. A.).

Syngnathus acus *L.* **Great Pipe-fish.**

Throughout the region—Lewis, west. Young in West Loch Tarbert, Knapdale, in July (*nobis*). Loch Carron (T. E. B.).

Nerophis æquoreus (*L.*). **Æquorial Pipe-fish.**

Off Moidart; off Mull.[2]

Nerophis ophidion (*L.*). **Straight-nosed Pipe-fish.**

Nerophis lumbriciformis (*L.*). **Little Pipe-fish. Worm Pipe-fish.**

Quite plentiful in Lochs Buie and Spelvie in Mull. A specimen

[1] "Conger elvers go up Severn in month of April, preceding the Shad, which is supposed to migrate into that river to feed upon them" (P.). It may be questioned how far there is any specific distinction between the marine and fresh-water eels. The fishermen capture fresh-water eels half-way over to Arran in Clyde; and the young of Conger go up fresh water freely."

[2] Taken in Clyde area.

procured for us with ova in ventral surface in July 1886, north of Mull, by Rev. J. Somerville (*nobis*). Loch Carron (T. E. B.).

Hippocampus antiquorum *Leach.* **Sea-Horse.**

Order 6. PLECTOGNATHI.

Family SCLERODERMI.

Balistes maculatus *Gm.*

Balistes capriscus *Gm.* **Filefish.**

Family GYMNODONTES.

Tetrodon lagocephalus *L.* **Crop-fish. Pennant's Globe Fish.**

"Has been obtained off north coast of Mull when 'scringing,' *i.e.* seining. Mr. J. N. Forsyth has often made them blow themselves out by stroking them, and seen them float helpless on their backs on being thrown overboard" (H-B.).

Orthagoriscus mola (*L.*). **Snort Sunfish.**

Been often seen off west of Lewis (*nobis*). "All along the coasts of Britain from Shetland to Cornwall" (N. L.). Off Barray and south of Mull (T. A. S.). Off Lewis (W. B.).

Orthagoriscus truncatus (*Retz*). **Oblong Sunfish.**

Sub-class 3. *CYCLOSTOMATA.*

Family PETROMYZONTIDÆ.

Petromyzon marinus *L.* **Lamprey.**

Loch Melfort (?) (N. S. A.). Figured by MacCulloch, found by him at Jura in 1819, adhering to the back of a gurnard—*Trigla lineata* (H-B.).

Petromyzon fluviatilis *L.* **Lampern, River-Lamprey.**

"Very large lampreys are found in a fresh-water loch in Coll, growing to 2 or even 3 feet in length (?), but difficult to obtain; are sometimes washed out in spate-time down the little ditch-like burn and on to the sands" H-B.).

Petromyzon branchialis *L.* **Pride. Sandpiper. Small Lamprey.**

Family MYXINIDÆ.

Myxine glutinosa, *L.* **Glutinous Hog. Borer Hogfish.**[1]

Sub-class 4. *LEPTOCARDI.*

Family CIRROSTOMI.

Branchiostoma lanceolata (*Pall.*). **Lancelet.**[2]

[1] [Taken at Rothesay, Clyde area. Said to be numerous, and most destructive to the line-fish, off Girvan.]

[2] [Taken plentifully in suitable ground in Clyde area.]

ABBREVIATIONS.

A. C.	*Gaelic*—Mr. Alexander Carmichael.
C. J.	Captain Campbell of Inverneil.
Dr. M'In.	Dr. M'Intosh.
F. D.	Dr. Francis Day.
H-B.	Mr. J. A. Harvie-Brown, Dunipace.
H-B. & B.	Harvie-Brown and Buckley.
J. M.	Mr. John Murray.
J. N. F.	Mr. J. N. Forsyth, Mull.
M'L. C.	Mr. M'Lean of Carsaig.
Nether Lochaber.	Rev. Alex. Stewart.
N. L.	Naturalist's Library, 1843.
N. S. A.	New Statistical Account.
O. S. A.	Old Statistical Account.
P.	Pennant's Zoology, etc.
T. A. S.	Captain Swinburne, Eilan Shona.
T. E. B.	Mr. T. E. Buckley.
W. B.	Mr. William Blackburn, Moidart.
Lewsiana,	Mr. Wm. Anderson Smith.

Where the authority is not stated, the compiler is personally responsible for the fact mentioned.

ON THE GENERAL GEOLOGICAL FEATURES OF THE OUTER HEBRIDES.

By Professor Heddle.

The term "Outer Hebrides" may be taken as including all the land which lies westward of a line drawn medially in the Sound called The Minch, from the furthest extremities of Scotland to north and south. It thus includes the water-intersected stretch of the Long Island, with Barray and Ronay, outliers to the north, St. Kilda to the west, and the Shiant Islands to the east.

If the two last islands, or groups of islands, be excepted, it may be said that the general features of the rest are very similar. This is due to the rock-mass which composes them being, for the most part, from end to end the same; and hence the whole stretch has been subjected to the same "earth-movements."

Unquestionably the most striking physical feature is an extraordinary sinuosity of coast-line, conjoined in many parts with a bareness, if not a ruggedness, of mountain crest. Both are due to the same causes,—first, an exposure at a high elevation to the agencies of decay, there operating with their greatest activity; and, secondly, a submergence of the whole to an extent which permitted the insinuation of the water of the ocean among the gorges, glens, and hollows which had been so formed.

The long duration of the exposure was due to the great age of the rock, and to its having been to only a small extent muffled up or mantled over by younger rocks. The depth and amount of the air-wasting resulted from the rock being largely formed of a substance incapable of offering a staunch resistance to those corroders which constitute the fangs of time.

An attempt has been made to correlate this rock with that at present taken to represent the oldest now visible in the earth's crust—the Laurentian of Canada. As many such speculations and correlations have had recently to be abandoned, it should suffice to correlate it with such rocks as can, in some or other way, be considered in reference to itself. When this is done, it is found to be unquestionably not only older, but very much older, than these; and the caution which former mistaken precipitancy has induced, has also led to the attaching to it of a name drawn from locality alone, so that it is now known simply as the Hebridean rock, or Hebridean gneiss.

A considerable tract of the same rock occurs stretching for some miles inland from the western shores of Sutherland and Ross; and it also forms some of the islands of the Inner Hebrides.

Aside of the sinuosity of the actual margin which separates sea and land, there is a striking rectilinearity, in a general sense, in the trend of the western shore of the mainland portion of this rock, and of the eastern shore of the insular portion. As the trend of both of these shores is nearly parallel to the great fault which intervenes in the mainland between this older rock and the great mass of newer gneiss of Scotland, it becomes a very tempting speculation which would assign the medial portion of submerged land—presently overflowed by the waters of the Minch—to other parallel faults which have "let down" that medial portion. Extended survey alone can pronounce as to this; but while there is much reason to assign this submergence to faulting—especially in presence of the fact that these shore-lines lie nearly at right angles to the general "strike" of the rock,—it unquestionably is a safer speculation to hold that the "earth-movements" which caused the rents—supposing such to exist—took place at the time when the volcanic islands which now stud the Minch were being formed.

In mineralogic constitution this rock is typical, in virtue of its very general and occasionally very large content of *hornblende*; in structural arrangements of parts, it is highly crystalline throughout. That which, however, is its most characteristic and distinctive feature is the abruptness with which the lithologic nature of its beds or layers change, and an appearance very near to actual separation which is to be seen between the beds themselves. The impression conveyed is that whatever may have been the processes which gave rise to this structure, the alternate banding now existent does not represent the layers of deposit in which the material of the rock was originally sedimented or arranged.

To such an extent does this segregatory arrangement of diverse mineral structures obtain, that there can be little question that in the more granitic districts many such bands have been erroneously set down as true granite veins; and where, in such districts, true pegmatite veins exist, it is far from easy in all cases to discriminate between the true vein and the granitic banding. Unquestionably some part of this difficulty is due to this, that where this openness of banded structure is most pronounced, true veins frequently run for long courses between the open bands,—so following the strike of the rock.

The rock of the Long Island is not so granitic as is that in the vicinity of Cape Wrath, nor does it anywhere show such hornblendic masses as that of Ben Aridh-a-Char on Loch Maree. It however exhibits a greater variety of structure and of composition than is seen upon the mainland. Speaking generally, it may be said to be more granitic in structure at the northern extremity of this group; more schistose at the southern extremity; and most truly gneissic in the central parts.

The dip of the strata also is highest in the north, being nearly vertical about the Butt of Lewis, while it is low, and in a direction the reverse of that usual to the rock at Barray Head.

In the far north outlier of Ronay the characteristic features of this gneiss as a hornblendic rock are rather indicated than well seen; indeed, for its size, this island may be said to present a greater variety of rock than does perhaps any other spot of equal extent in the whole length of the series. At its northern peninsular extremity it consists of thin-bedded and almost fissile schists; but the occurrence in these, which are almost crypto-crystallic, of huge crystalline nodules, or concretions of platey, lustrous, dark-green hornblende, prepares us for an increased amount of this ingredient in the general mass of the rock.

In the east and west hill-ridge, which constitutes the main bulk of the island, the rock is still fissile under the stroke of a hammer;

but numberless large garnets act as rivets to it where unaltered, though their occurrence seems to have an effect the opposite of that of a rivet upon the disintegrating rock.

In this hill-ridge of the island, also, we have the first appearance of the huge pegmatite veins which, in many places further south, are so marked a feature of the rock, and which even go to influence the scenic features of the country. To some extent this may be the case here also, as one of these veins forms the facing—if indeed it has not caused the line of weakness which actually formed the great sea-cliff of the north-west frontlet of the island. And this first vein which we have to notice exhibits that structure which is common to almost the whole of these veins—namely, the *graphic arrangement* of two of its constituents, the quartz and the felspar.

Another feature of the hornblendic gneiss is also well seen in Ronay—namely, the extreme suddenness with which the dip of the rock alters.

The dip of the fissile beds of the northern peninsula is about 5° to S.E.; of those of the central portion of the hill-ridge about 80° to S.E.; while at the S.W. point the dip falls again to a low angle. No actual break could be seen in the beds at the spots where the alterations of dip occur, though these are very sudden.

Sulisgeir, which is the most north-westerly outlier, shows a highly felspathic and pale-tinted variety of the rock—which has somewhat of a syenitic aspect. At the south end of this rock the bedding is distinct, with a low dip to the north; elsewhere there is crushing and local folding. The felspar here has a rose tint.

At the Butt of Lewis the rock bears much resemblance to that of Cape Wrath, with multitudinous serpent-like twistings and foldings. The beds, if they be beds, and employing the word always under protest, are nearly vertical; but they are not so independent of each other,—so separately individualised,—as upon the opposite mainland. In the outlying western Flannan Islands, however, this feature of looseness of jointing is most marked; while here also the alternation of granitic layers is a dominant feature.

For a great distance southward of this point, on account of the manner in which the rock has been ground down by ice, and after swathed in a deep mantle of peat, its features are, except upon the coast-line, concealed from view. As there disclosed, no unusual features are seen, beyond this, that at several points upon the west shore the sea has left dissevered fragments standing removed from the shore cliffs; and these more enduring portions sometimes strikingly display the extreme amount of folding and bending, if not twisting, to which, with sudden and extreme abruptness, the rock has been subjected.

Passing from Lewis to Harris, we have almost as sudden and as extreme a change in every feature, but here on a large scale.

The flat and desolate moor of eastern Lewis, with its ice-rounded hummocks of rock, and endless quagmires of lake and moss, gives place to a tumultuous assemblage of lofty hill-peaks, with a serrated outline, and a sky-line surpassing in dignity and grandeur anything which Scotland can display, not even excepting the clustered volcanoes of Rum.

The first or northern range of these hills stretches nearly in a straight line from Loch Reasort to the lower reach of Loch Seaforth. The hills roll their frontlets over to the north, very much like waves about to curl or break. Indeed some have broken ground, and show forth as bare and even overhanging rock faces.

These craggy frontlets show a fissile rock dipping to the south or south-west, at but a low angle comparatively, and so their southern faces are grass and heather-clad slopes, more fertile, if it be only in appearance, than can be seen, except in small spots, elsewhere to pertain to this rock. But further south than this, when the central tract of Harris proper is reached—where the land is again low, has again been beneath the level of the mantling ice, and the rock is again of the usual gnarled texture and hornblendic type,—a bareness and desolation surpassing even that of the ground-down flat of Lewis is seen.

There is no part of Scotland which rivals, in the hideousness of its bareness and desolation, a stretch of miles in extent which lies to the east of the middle hills of Harris. The moor of Rannoch is, in comparison therewith, a well-watered garden. This is well watered, it is true, but it is bare rock which is watered. Indeed the land surface does not dominate much over that of water. The rocks to be scaled, rivers to be crossed, and maze of lakelets to be unravelled, are so insuperable, that the one road of the island winds upon the opposite shore for about four times the distance, to evade the attempt.

The forces of degradation, disintegration, and decay, operating upon a rock which cannot yield the pabulum for plant-life, have tended only to destruction. What little of plant life there may once have been was scoured off by the ice, which seems to have scalped the land, and it has since continued bare—a Golgotha.

Southward of this waste expanse, and with extreme suddenness, there occurs, crossing nearly the whole island, a ridge of rock hardly to be looked for in a gneiss of this description. This is serpentine.

Dr. MacCulloch and Professor Nicol both vouch for the occurrence of serpentine at Loch Valamus in Lewis. We failed to find it there, but found agalmatolite,—which in appearance is similar.

Here the Serpentine extends for miles; at its eastern extremity it is associated with large quantities of asbestos, actynolite, and hornblende, and it *seems* in one sense to pass into this last; such a contention, however, in the absence of the augitic type of mineral, would be so repulsive to those who contend that serpentine is the product of change of *olivine* alone, that we hesitate to adopt it. Therefore we do no more than record that here there is no appearance of olivine at all, while of asbestos alone there is enough, in the words of the late Professor Jameson, "to load an Indiaman."

This magnesian rock, here largely charged with magnetite, can effect no change upon the sterility of the country, which continues southward until, at Rodel, there is a second geologic surprise in the occurrence of a bed of limestone, and its surrounding greenery and cultivation. Of this limestone all that need be recorded is that it carries *sahlite*, grey in colour and of a waxy lustre; also, rarely, darker coloured varieties of the same mineral. Petrologically, the limestone is similar to that of Glen Elg, and of Tarffside.

Very possibly the long stretch of Loch Langabhat, with the sequence of lakelets to the east of it, may represent a second, but here washed-out limestone bed; of direct evidence as to this, however, there is none; nor even of indirect, unless the cultivation at Borve is such.

South of this lake, a ridge of hills parallel to those spoken of in North Harris rises suddenly. These hills, granitic to the east in Roniebhal, become charged with garnet and kyanite above Obb, and still more decidedly with the last mineral at the west extremity,—Capval. The rock above Obb is true eklogite.

From end to end this ridge is traversed by huge veins of pegmatite. One in Roniebhal, carrying magnetite and huge crystals of black mica, has a tortuous course; all the others cut the rock at right angles to its strike, and continue their course through the roll of the land which forms the bluff north shore of the Sound of Harris.

In the valley or depression which lies between this southern shore-rampart and the northern eklogite ridge, a singular rotting, highly ferruginous, and fissile rock is seen, with a north-easterly dip; this seems to be the dip of the eklogite ridge also; so that, though there are numerous intermediate crumplings, there seems to be a great synclinal trough between the Totam-Clesham range and the Roniebhal and Capval ridge.

Of the granite veins—admirably seen in their outgoings on the north shore of the Sound of Harris—it has to be remarked, that along with numerous minerals, they all contain large masses of graphic granite. The vein of Ben Capval is the largest granite vein in Scotland, being nearly two miles in length, about thirty feet wide, and projecting above the surface of the neck of the peninsula like a huge rampart. It is specially rich in minerals of interest. Three of these granitic veins converge to form the rock of Stromay. This contains both graphic granite and huge crystals of nacreous felspar of a splendent lustre, and also the rare variety of albite with nacreous lustre called peristerite. These granitic veins, though most abundant along the southern shore, appear at many other points—notably at the narrows of Tarbert. A graphic vein carrying rare minerals occurs at East Loch Tarbert, and a mass of felspar of a brick-red colour protrudes from the south bank of West Loch Tarbert.

There is a fact of interest, as bearing upon the recently elucidated and demonstrated contention of the late Professor Nicol, that there is a reappearance of the Hebridean rock to the east of the great fault which occurs eastward of the main bulk of that rock, and of the Torridon sandstones. Before the elucidation as to this question was complete, it was maintained by some that the rock brought up to the east was not in all respects one and the same as the Hebridean gneiss, but was "formed out of it." Upon the north side of the road, which passes Loch Bearasta Mòr, in Luscantire Valley, there occurs in a bank a vein seemingly—for it has vein sides—of a fragmented and reagglutinated rock of a singularly jasperine appearance. It is altogether indistinguishable from the most highly fragmented and altered varieties of the so-called "Logan rock" as found in Glen Logan. This rock when polished is a highly ornamental stone. It here can hardly be assigned to any origin other than the crushing and re-cementing of the adjacent rock.

The extreme rectilinearity of the north shore of the Sound of Harris seems at first sight to indicate the existence of a great fault. As, however, this water-filled trench lies directly parallel to the line of depression at Tarbert, as well as to those at Luscantire, Loch Langabhat, and the Rodel limestone valley, there can be little question that it represents a similar valley of still greater depth. A rock which doubtless will come to be of economic interest occurs at the western foot of Aird Nishibost. This is a syenitic black-mica granite—apparently a huge vein. It has every property of a first-class building stone.

While it is a problem of very considerable difficulty to trace and follow out any regular set of beds throughout the greater part of the vast tract of The Lews, and also in the middle reaches of South Harris, there are yet occasional cross-valley trenches in both which permit of the determination that the formation as a whole has in those northern islands the same general N.W. and S.E. trend which it displays on the mainland.

In the southern islands the difficulty of recognising any determinate strike is very much greater, however. There is in them crumpling, crushing, and local faulting to a much greater extent, so that the same dip and strike is not maintained for any distance, especially in the islands of Uist. The best exposures are, as before, to be seen on the coast-lines, especially the western, when not mantled by sand drifts; the outlying islands and rocks also well display it. High dips, at least of any continuance, are also here much more rare.

In the islet of Gaskeir, off West Tarbert Loch, the dip is 80° to the east. The islet of Shillay consists of a great roll-over, terminating in a dip of 70° to east-north-east. This islet is cut in two by an east and west fault. Its northern portion shows great curved red and green beds, so boldly and clearly displayed in section by the fault that it resembles a huge diagram. On account of the exalted metamorphism resulting from the crushing, the rock is throughout excessively granular and rough, whatever direction it be fractured in. The constituent minerals—nacreous red felspar, with Haughtonite,—and quartz with dark hornblende—segregate apart. All is grit-like and sharp-angled. The southern portion of the islet is a mere heap of ruins; the shattered rock, here fissile, and of fine-grained structure, being tossed erect, and resembling a mass of scorched ruins.

Something similar on a gigantic scale, but with the fissures plugged, and the shattered masses agglutinated by epidositic matter, is seen in the Hill of Eabhal and the adjoining eastern tract of North Uist; and a similar *mélange*, but with the cement still more highly epidositic and tending to hornstone, occurs in veins at Loch Maddy and elsewhere. Petrologically, the rock of these veins resembles Logan Rock.

Passing south from Shillay on the west to Haskeir, the rock is found to be a very dark fissile hornblendic schist, with a dip of 10° to the north; while at the opposite shore at Newton the dip is 80° to the east. In this district the rock shows many varieties; pegmatite veins with graphic granite, striated albite, and huge crystals of oligoclase, being associated with it. A half-tide skerry in the Sound of Berneray is composed of the very rare rock true *actynolite slate*; while a second rock, formed of a *mélange* of foliated hornblende and a bronzy black mica, gives rise to a suspicion of the near presence of limestone.

The varying character of the rocks is here associated with a not inconsiderable number of rare minerals.

At West Monach, further to the south-west, there is again much of variety in the nature of the rocks, which are here highly crystalline, and with much isolation of their mineral constituents, and the not unfrequent occurrence of imbedded minerals. The dip here is 40° to north-east.

Passing to the eastern shores of the group, we find the rock at Loch Eport dipping near Loch Obisary to the north-north-west at an angle of 20°. At Loch Boisdale, the dip, which varies much, is to the north-east, and the rock is again traversed by granitic veins, and also by veins of quartz. Of these, one is studded with large crystals of Haughtonite, while another, which contains yellow felspar, and a kind of yellow ochre, has a decidedly auriferous appearance.

In Mingulay the dip varies from 30° to 85°, but is throughout to the north-north-east; while in Berneray, the most southerly of the group, it is 22° to the north-east.

In the more southerly islands granitic veins are much less frequent than in these northern, and the veins seldom if ever carry any white felspar—oligoclase or albite—the felspar being red, and the crystals of black mica are of much smaller size.

The Long Island is intersected from end to end by that system of dykes to which Geikie has assigned a comparatively recent date —more recent at least than the liassic volcanic islands of the Minch. These dykes, best seen upon the eastern sea-board, occur sometimes with clustered frequency, and sometimes do not appear for miles. Geikie conjectures that the whole volcanic material of this system found a passage upward through lineated rents alone— there being nowhere a circular orifice or vent of any size.

The three Maddys, which lie off the mouth of Loch Maddy, have, however, every appearance of being vents with which these dykes were directly and immediately connected. As pointed out by Dr. MacCulloch, dykes, wider in their dimensions than usual, strike inland from the shore immediately opposite the two larger of the Maddys. The rock of the Maddys, moreover, is precisely that of the greater number of the dykes.

Dr. MacCulloch, who argues for a probable connection between the two larger Maddys and the opposing dykes, hesitates, in the so doing, on account of the dyke opposite to Maddy More not being throughout dense black basalt, but having an amygdaloidal core, with merely a sheath of basalt. But this is precisely what obtains elsewhere, where there is a filamentous leader or vein to a central or dominant mass of igneous rock. That larger mass, retaining for a much more protracted period of time its heat of fluidity and viscosity, assumed during the slow cooling its crypto-crystalline character and columnar structure; while the thin and less bulky vein, chilled by close contact with its retaining walls, concreted as a scoriaceous, vesicular, and ultimately amygdaloidal paste.

This system of dykes, running very much in the line of the main strike of the gneiss, does not much disturb it or produce striking scenic effects. In the island of Mingulay, however, their influence in this respect is most marked. Dykes which strike north by east, and which have been in great part sapped out by the waves, form a number of stacks, grand both in form and in altitude; while one which strikes east and west, and which has been cut out by the buffetings of the western ocean, up to a spot where it is cross-buttressed by one of the first-mentioned series, has formed the great cliff of Mingulay, the loftiest in the Long Island.

At the neck of the peninsula of the Dun, and also in Pabbay, the effect of these dykes in causing three shrinkage-cracks parallel to themselves, and also in imparting a pseudo-basaltic structure to the gneiss in contact with them, is well seen.

When the very considerable amount of felspar which gneiss of all descriptions contains, and the large amount which is present in the huge veins of granite which permeate the rock of these islands is considered, it at first sight is very much of an enigma how or why it is that throughout their whole extent they should be so bare, so treeless, and so unproductive. The remarkable scarcity of land shells is also most striking.

Simple inspection of the condition of the rock-surface, however, suffices to show that disintegration through chemical change, that is through air-rotting, is here very slight; and that the greatest amount, at least of recent change of surface, has been by degradation through ice-scouring. The veins, again, even with their vast preponderance of felspar, have suffered still less change, for, with no exception, they protrude more or less, sometimes markedly, above the general level of the rock which encloses them. The almost total absence of pyrite in these, and also in the rock mass, may so far account for its endurance.

The contemplation of the great rolling flat of the Lewis from the elevation of any one of the Harris hills, and the dragging of the wearied limbs over the seemingly unending sands of the western "machars," alike, however, impress one with the enormous amount of destruction and of waste which has here taken place. Of waste in the most direct and immediate sense, for *whither* has been carried the enormous quantity of enriching felspar, equivalent to those wide horizon-bounded stretches of sand-grains?

A consideration of the order of sedimentation which obtains when the several resultants of the disintegration of a rock are swept off in temporary suspension in running water, affords the foremost, if not the altogether satisfactory clew.

The specific gravities of quartz and felspar are not appreciably different, but the process of the comminution of the two is altogether different. Possibly at the outset the cohesion of particles of both may have been, to a certain extent, overcome by physical impact; but ultimately, in the case of felspar, the disintegration is a chemical one, resulting in—*necessitating* the separation of the whole mass down to the primal molecule of kaolin,—the water carrying off in solution the potash salts.

In the case of quartz, on the other hand, the disintegration is a physical one alone; no chemical disintegration is here possible, for silica is, when *per se*, already a chemical residue—non-attackable. Its particles are thus reduced in size by physical impact alone; being reduced to their smallest size, either by being flung against one another, or against rock to be splintered by wave force, or by having protruding portions ground off by attrition beneath running streams.

In the chemical disintegration of felspar, the limit is the molecule of that compound which has the composition of kaolin. In the physical comminution of quartz, we see the limit in the almost uniform size of the grain of sand—uniform whatever has been the parent rock or the circumstances of its formation.

So long as the particle of quartz is of such size that in virtue of its greater gravity it sinks rapidly in water, so long will it receive the full force of a rolling billow when raised by it to be flung against its fellows on the beach—the adhesion of water to its surface having, thus long, an altogether inappreciable suspending effect. But whenever this reduction in size, through continued splintering, had arrived at the point where its surface is increased in proportion to its weight so far that the suspending adhesion of water for it becomes appreciable, then is the force of the collision diminished. Ultimately the sand-grains, caught up from the bottom in the back tow, are enveloped in the bosom of the curling breaker to such an extent that the impact, which some alone receive, does not suffice to effect further fissuring.

In moving water, therefore, sand-grains, large in comparison, settle rapidly when the motion is arrested; while suspended mud and kaolin,—almost as an *emulsion* of land and water,—are carried to lower levels to be sedimented slowly in straths, or carried as delta-formers far out to sea.

It is not the configuration of the land alone which goes to show that the straths and lower levels to which the fertilising kaolin, derived from the great waste of the long-departed highlands of the Long Island, had been carried, are now submerged; for not only are their mud-filled depressions, but also the woods and the heather which drew their sustenance therefrom, to be found now, in many localities, lying beneath the waters. An oar may be thrust deep into peat which lies beneath low water in many of the bays; little sea-cliffs of peat cincture the shores of several of the sounds; forests lie recumbent beneath the sands of the narrows and inlets around Newton and elsewhere; and sea-fishes are the pioneers of high-water inroads of the ocean into several of the lakes.

The Long Island is but the skeleton of a submerged country; it will take centuries of air-wasting before the dead bones can live again. In the present it matters little what the seed may be: if thrown upon stony ground it must straightway wither away.

In the southern portion of the Eye Peninsula of Lewis, and also along part of the opposite shore, there is a conglomerate deposit which has been conjectured to be of the same age as that on the opposite mainland, where it has received the name of the

Torridon Sandstone. It must be its geographic position more than a petrologic similitude between the rocks which has led to the correlation; for the rocks, except in tone of colouring, have not much in common. The Lewis rock appears in some respects of less age. No data have yet been formed whereby to ascertain this. The beds are cut by the same system of dykes as is the underlying gneiss, and contain some secondary veins of baryte.

Two groups of volcanic islands of Tertiary age lie, one on the Atlantic side of Lewis, another upon its opposite or eastern side.

The Atlantic group is known as—

ST. KILDA.

This cluster of rocky islets, which takes its name from the largest of the group, is composed of igneous rocks of Liassic age. As the rocks are here solely of the igneous type, they are assigned to that age through indirect correlation, more than by direct evidence. Only two varieties of rock occur; one, highly felspathic, and either in itself of or weathering to a pale colour, is altogether similar to the rock which, with no great amount of precision of nomenclature, has been termed *syenite* where it occurs in Skye, Mull, and Rum. The other, an essentially augitic rock, equally evidently is of the same type as the augitic rock which forms the serrated peaks of the Coolins in Skye, and the even rougher eminences of the central portion of Rum.

Being, in virtue of the specific identity of these two rocks, correlated with the igneous outbursts of the Inner Hebrides, all find their position in the cycle of time from the occurrence of Liassic beds intercalated with the traps of Skye and Mull.

In St. Kilda, as in Skye and Rum, the felspathic and more acidic rock underlies the augitic and basic rock. The lower specific gravity of the acidic rock determined its superior position in the volcanic caldron of molten matter, and so, of necessity, its earlier overflow of the vent or crater; and this secondarily determined its inferiority in bedding, when the heavier rock was, in sequence of events, discharged.

In physical, and it may be said in geographical features, the

St. Kilda group much resembles the partially submerged circlet of cliff-faced eminences which cincture a volcanic crater; broken away, however, at one side in great part, as is ever the case with presently existing volcanoes.

The felspathic rock lies centrally, forming the northern third part of St. Kilda proper. The intensely dark-coloured and rough augitic rock is thrown out both to north in Borreray, and to east and south in the Dune, and the southern part of St. Kilda with its dependent Soay. As the islet Levenish lies to the east, there has thus been enfolded, as it were, a space which may have been the centre of the outflow. Some doubt must, however, attach to this, as the dip of the hyperite in Soay, in its stacks, and all along Dahl Bay and the shores of Coinagher, is to E. by N.

Two thin syenitic beds which are seen intercalated with those of hyperite in the cliffs south of the Dune demonstrate that these islets were not erupted by a single continuous ejection of volcanic matter.

The pale rock, the so-called *syenite*, is somewhat featureless. It is for the most part a mass of compact felspar with very little hornblende—if it be hornblende—and still less quartz. Rarely it is porphyritic, and still more rarely drusy, with crystals of brown quartz and of pinkish felspar lining the druses. In St. Kilda quartz dominates somewhat more in this rock than it does in Skye.

The augitic rock is dark grey from compact anorthite, or almost black—the disintegration of interstitial felspar allowing the more enduring augite to protrude in hackly crystals, imparting a somewhat scoriaceous appearance, and a surface of extreme roughness. This augite has neither the occasional bronzy metalline lustre of that of the Coolins, nor the gem-like colour and transparency of that of Rum. It is of intermediate appearance, except in colour, and is serpentinised to some extent. Dykes of dense basalt cut the syenite, much in the style of those of the Long Island.

The external features of this group of islets is conferred upon them almost entirely by the nature of the felspar which the two varieties of rock contain. The pale-coloured rock which has consolidated in even-bedded slabs, which are cross-fractured with smooth-faced rents, suffers but little from the disintegration of its felspar; and so do its cliff-faces remain mural and devoid of foothold; while its hills have assumed a rounded contour.

The felspar of the augitic rock, again, on account of its large content of lime,—freely soluble in rain and other waters,—is rapidly abstracted from the surface, leaving the crystals of augite to protrude like the teeth of a rasp. The rugosities and spiked pinnacles of the latter hence afford a safe foothold for birds, which not only do not affect, but markedly shun those portions of the island which consist of syenite.

To this rapid disintegration of the felspar, and endurance of the augite, is to be assigned the aspect of extreme roughness which is the characterising features both of the cliff fronts and of the sky-line of the group as a whole,—the basic rock prevailing and cincturing the acidic. There is the appearance of a raised beach on the larger island, which stretches also some little distance upon the side of the outlier called the Dune.

THE SHIANT ISLANDS.

These islands have for long been known as containing probably the largest colony in Britain of the Puffin or Tammy Norrie. This small group is the most northerly in Britain of a line of orifices of volcanic outflow, which extends along the trench of the Minch from Antrim in Ireland, and points in the direction of Faröe.

As liassic rocks lie included in the igneous sheets, there is no question as to this eruptive cluster being contemporaneous with the not far distant Skye. On account of the small extent of the group, not many varieties of igneous rock occur. There are three islands, which loop round a deep-water bay. Of these, two—Eilean Tigh and Garbh-Eilean—are connected at low water by a shingle strand. Eilean Mhŭire, the third island, lies to the east of these. It is separated to some extent from them, but is at low water brought nearer thereto by a singular band of very dense and enduring rock. This resembles, but differs in some respects from massive basalt. When first seen in island Mhŭire, this rock thrusts itself above the general surface like a huge dyke, and at the western extremity of the island it forms the mural boundary of a picturesque "göe," with ramifying caves. It forms a far-projecting reef across one-third of the channel. On the western side of this it again appears as a great horn protruding from the main bulk of Garbh-Eilean, and thrusting itself into its mass; and it again appears at some distance to the west, in a line of lofty and steep-sided rocks.

The Liassic beds are solely connected with, and are indeed enveloped in, this band of rock. They occur in the east horn of Garbh-Eilean, and are so affected by the igneous entanglement that no fossils are to be found. Their former presence may perchance explain the occurrence here of the rare mineral *Wavellite*, a phosphate of alumina, which is found at only one other spot in Scotland.

The eastern island has a bedded disposition of its rocks, sometimes with an amygdaloidal structure. The other islands contain only one variety of igneous rock. The southern, though basaltic in structure, especially on its eastern face, does not exhibit any marked regularity in that structure; but it is from that regularity, or rather the surface-roughness consequent thereon, that the larger or northern island derives its name.

Few things are rougher than an Elizabethan collar; and a circular Elizabethan collar, held in a vertical position, well represents the rock-surface of this island. So uniformly is it girt with the plications of lofty pillars that there are but two spots at which in a circuit of perhaps two miles it can be scaled.

Though falling far short of Staffa, and especially of the north-east of Skye, in the diversified features of basaltic scenery, this island very much surpasses them, and indeed all the scenery of the west, in the altitude of its pillared cliffs; and especially in the gloomy vista produced by their regularly-extended sweep. In this last feature there is no locality in Scotland which even approaches in effect the grandly-curved colonnade, which extends for more than half a mile along the north frontlet of this island. As the eastern horn of the crescentic cliff projects boldly northwards, it cuts off the rays even of morning light, and throws both cliff-face and its pediment of waters into deep shade,—with the occasional exception of a bastion-like cluster of grouped pillars.

The western, but lower part of this colonnade, consists of a sharply-cut range of incurved pillars; this surpasses, in chasteness of flexure and in regularity, everything of the kind in Scotland, with the single exception of the clusters at the caves of Duntulm, in Skye. The height of this line of basaltic cliff is 499 feet. The pillars of Fingal's cave are but 18 feet in length.

Both the indurated clay-slate of the liassic rock, and much of the dolerite, exhibit *externally* an appearance of *sphæro-radiate* structure. Dr. MacCulloch assigned this, in the case of the former, to divergent spiculæ of Wavellite. No such structure is apparent internally, however, in either rock, when examined under the microscope. Judd finds much picrite in these islands.

Nepheline, in specimens of mineralogic recognition, were found by the writer in a cave on the south-east side of Eilean Mhŭire.

There is probably no rock structure feebler in the cohesion of its parts than that termed the basaltic. It is a system of more or less open joints, disposed in general at right angles to a more or less horizontally-bedded rock. If there be any departure from perfect verticality, gravitation must be ever tending to the opening of the joint, while each crevice receives a water-wedge, some portion of which must abide during even the warmest and driest weather. In certain cases it must be conceded that this jointed structure is one of double jointing, transverse the one to the other—the resultant of the compression of spheres of repressed crystalline arrangement. In such cases there is naturally—in all cases there comes ultimately to be—a cross-rending to the angular shafts, a never-ceasing toppling of the sectioned pillars, and, except where swept away by the scour of waters, a never-ending accumulation at the cliff-foot.

An accumulation of fragments, sharp-angled, chance-thrown, perched in all positions, propped in all directions, locked immoveably on many sides, lattice-girdered against every thrust,—cellular as charcoal, porous as a sponge, open as a sieve, intricate as a maze, more enduring than the Pyramids, as ever growing larger and never growing less—the Tammy Norries have chosen well! Little wonder that they never desert their old ancestral halls!

APPENDIX.

APPENDIX A.

SINCE concluding the text, our friend the Rev. H. A. Macpherson has sent us a copy of Fleming's *British Animals*, which contains some notes on birds. These were mostly written by the late Dr. John D. Ferguson, a connection of Mr. Mackenzie, factor at Dunvegan, and who often goes over to St. Kilda in the smack.

We give below the more important of such of these notes as relate to the O. H., many of which, however, bear internal evidence of having already been utilised by the MacGillivrays.

These notes are as follows :—

Redwing and Fieldfare common, but makes no mention of the Common Thrush, which is curious. The Dipper is common [and also the Yellow Wagtail. This latter is probably the Grey Wagtail. But if it is so, this is the only record of the species in the O. H. which we have]. The Common Sparrow was only seen at Eoligary, Barray.

The Brambling is a common visitant to S. Uist.

The Hoopoo was shot in N. Uist in 1844, and in S. Uist in 1843.

A specimen of the Snowy Owl was shot at Benbeculay, on October 21st, 1855, by Dr. Ferguson.

The Hen Harrier is common in Uist and Barray.

Five Golden Eagles were shot in S. Uist; the Sea Eagle is very common in the Hebrides.

Dr. Ferguson considers the Greenland Falcon common in the Hebrides; one was shot in Benbeculay in 1844.

A pair of Spoonbills were shot by a brother of the MacGillivrays in S. Uist.

Dr. Ferguson obtained the Gadwall, Shoveller, Scaup, and Velvet Scoter in S. Uist.

The Water-rail is common in the Hebrides; but the Water-hen is rare in S. Uist.

The Bar-tailed Godwit was shot in S. Uist in the winter of 1841-2; and a note is added that it is "frequent."

The Black-tailed Godwit was said also to have been obtained in S. Uist in the same winter, but, from its known rarity elsewhere, this can hardly be considered as sufficient evidence, without fuller particulars.

No mention is made of the Whimbrel on migration, but the Curlew as common at that time (?).

The Arctic [" and Black " ?] Terns are occasional in Skye, and frequent in S. Uist.

The capture of a Black-headed Gull in S. Uist in 1842 is mentioned as a rare occurrence.

The Glaucous and Iceland Gull and the Buffon's Skua are mentioned as having occurred in S. Uist, and Richardson's Skua is marked as common in the Hebrides.

The Storm Petrel breeds on Mingulay, and a specimen was got in the island of Berneray, Barray, in 1843-7. To this we would add, compare the MacGillivrays' and others' statements in the text.

The Great Auk is put down as having occurred in S. Uist. (We have not been able to obtain any further note regarding this stated occurrence. T. E. B. and H-B.).

APPENDIX B.

BESIDES Dr. Ferguson's notes just quoted, we have received an interesting MS. account of the birds of Benbeculay, an island of the group which has not been visited by us personally, though it was examined with considerable care by Major H. W. Feilden in 1870.

This account has been obligingly prepared by Dr. John MacRury, who also sent us the account of Pallas' Sand Grouse given in the text. We find this list includes 102 species.

Dr. MacRury—whose brother was the late Mr. Robert Gray's principal informant and correspondent in Benbeculay—speaks of the Pochard as "fairly numerous," and as a regular winter visitor to the west side of the island; but he includes the Scaup-duck as "scarce." He also adds to our positive knowledge of Velvet Scoters' localities—Drimisdale, in S. Uist—and speaks of this bird as "occasional round the coast in winter." He records some of the less common species, as Greenland Falcon, Snowy Owl, Pallas' Sand Grouse (as already stated), Gadwall, "scarce," and Shoveller "scarce," and he speaks very doubtfully of the occurrence of the Goosander. He is inclined to believe in the occurrence of two species of Tern, as seen by himself, and he speaks of having shot the Sclavonian Grebe on a fresh-water loch on one occasion only., He also includes the Short-eared Owl as breeding regularly in the island.

Dr. MacRury mentions rather an unusual exploit of a hen Grouse, which he witnessed only the day before he wrote (7th August 1888):—"In enticing a dog away from its brood, it actually sat on the water in the middle of a loch, folded its wings, and remained some 10 or 12 seconds on the surface."

APPENDIX C.

Bat. sp.?

Dr. MacRury informs us that he certainly saw a Bat (but gives no name to it), in Benbeculay, some twenty years ago, but he is not aware whether they are constantly seen there or not.

Trichechus rosmarus *L.* **Walrus.**

The fullest account of the specimen, mentioned in the text as having been obtained at Caolas Stochnis, is given in Sir William Jardine's *Naturalists' Library*, "British Quadrupeds" (vol. vii. p. 223), by which it appears Sir William had an opportunity of examining the specimen, and taking measurements and description (1838).

Balænoptera musculus (*L.*). **Common Rorqual.**

One recorded and identified by Professor J. Struthers (*Journal of Anatomy and Physiology*, vol. vii. p. 3, 1872), was found dead about 14 miles from Stornoway—an adult male; and the remains are in the Museum at Aberdeen. A full account is given in the above-named journal (*loc. cit.*).

Lepus cuniculus *L.* **Rabbit.**

Dr. MacRury writes to Harvie-Brown that "although the Rabbit has been abundant all over S. Uist for a very long time, and at low water could get across to Benbeculay dry-footed, it has never made its appearance in the latter island. A pair or two were liberated in N. Uist a few years ago, and now the whole of that island is pretty well stocked with them."

APPENDIX D.

We are still continuing to receive from a few of the best Lighthouse Stations in Scotland filled-in schedules on migrational phenomena. These are intended for our own use for faunal purposes. The principal station in the O. H. appears to us to be Monach Lighthouse, where also, very fortunately, Mr. Joseph Agnew (formerly head lighthouse-keeper at Isle of May) is now stationed, and takes as keen an interest as ever in the subject. We have schedules received from him, ranging from December 22d, 1887, to date.

On the 28th of September 1888 there was a great migration, and Mr. Agnew writes:—"Those scheduled are the only ones identified." We extract the most useful memoranda:—On Jan. 3d, 1888, six Wood Pigeons passed with S.W. fresh breeze, and in clear weather. The Tree Sparrow is recognised, 9th April 1888, as a spring migrant past Monach. A great rush of Redwings took place on 15th April, continuing for three hours after midnight, and many were seen on the island next day. The Brambling was recognised as a spring migrant passing Monach, on 29th April. Migrations of what were probably White Wagtails (*M. alba*) have been several times noted, both in 1888 and in previous years, in end of April and beginning of May. The Goldcrest—rare in the Long Island—has one single record in 1888, a female, on the 12th May.

The returns of Whinchats (*rera*) are few and far between in 1888, but they *do* pass the station. Mr. Agnew first notices them on 17th May.

The additions to the list of the birds of the O. H. are:—One Blue-throated Warbler (see separate note, p. 252), and the Redstart (*R. phœnicurus*, with which latter the rarer species was in company. This was during the first weeks in October; and the dates of Redstarts in the schedules are the 1st October, and (one prior record on) the 28th September. A Barn Owl, a single bird, perfectly recognisable from Mr. Agnew's description, was seen though not obtained.

[Besides this there is a Flycatcher mentioned on the 28th of September, but it is not clear whether it is a female Pied Flycatcher or a Spotted Flycatcher.]

Blue-throated Warbler. Ruticilla suecica.

We are now able to make this interesting addition to the migratory fauna of the Outer Hebrides, a finely plumaged male having reached us from Monach Island, which arrived there along with Redstarts on the 11th October 1888. Unfortunately, like so many birds which reach us from the remoter isles, it was in a decomposing condition, and all we could do to preserve it was to place it in spirits. Writing under date of 13th October 1888, Mr. Joseph Agnew, lighthouse-keeper at Monach, says:—"All last week it was blowing a gale from the N. and NE. with snow and hail showers. This bird was two days on the island before it died." It has since been placed in the Edinburgh Museum of Science and Art.

[Troglodytes hirtensis. St. Kilda Wren.

In a letter dated February 22d, 1888 Dr. Günther informs us that a specimen of this bird was presented to the British Museum by Sir John Campbell Orde in July 1886, from St. Kilda. We mention this, as the species, or insular variety, is so limited in its range that any particulars ought to be recorded.]

Linota cannabina *L.* Linnet.

Frequented Mingulay this past winter (1887-8), and Harvie-Brown saw a few there on May 20th.

Pyrrhula europæa *Vieill.* Bullfinch.

We find that, through inadvertence, we have omitted all our notes under this species, though the name appears in its proper place in our list. Harvie-Brown was informed by Mr. Finlayson, gamekeeper, that the Bullfinch had occasionally occurred in Rodel Glen, Harris, of late years. This is no doubt the indication of an extension, *vià* Skye.

Corvus monedula *L.* Jackdaw.

The first Jackdaw ever seen in Mingulay by Mr. Finlayson was perched on the ridge of the roof of the new schoolhouse, one day in January 1888.

Cypselus apus *L.* Swift.

We have omitted from record in the text an occurrence of the Swift in N. Uist by Sir John Campbell Orde; and an additional record of the species in St. Kilda is given by Rev. H. A. Macpherson (*Ibis*, 1887, p. 470).

Alcedo ispida *L.* Kingfisher.

We have the note that one bird of this species was found in Rodel Glen on March 1884, in an exhausted state, and after a severe gale. Our personal interview with Mr. Finlayson satisfied us of the correctness of the observation, the Dipper or Water Ouzel being perfectly well known to him also.

Nyctea scandiaca *L.* Snowy Owl.

Since the text was written and printed off, Harvie-Brown has received from Mr. Finlayson, of Mingulay, one feather of the bird mentioned in the text, and a letter from him enclosing it—"the only one remembered on the island"—which was shot there in the winter of 1886-87, or in January 1887.

Astur palumbarius *L.* Goshawk.

Buckley and Harvie-Brown have examined the specimen already mentioned in the text (*q.v.* p. 87), and it is a specimen of the European race. It was shot at Stornoway by Mr. Graham—the present gamekeeper at the Castle, and is a young male. It was sent to Mr. Macleay, Inverness, for preservation for the Castle Museum, at the end of December 1887, or beginning of January 1888.

Fulica atra *L.* Coot.

Feilden found this species abundant in South Uist and Benbeculay as shown by his "Continuation of the 1870 Journals" (see Appendix F).

Wildfowl at Rodel.

Already, (1888,) we have learned that one of the Falkland Island Geese has been shot near Stornoway, where it had been killed by one of the sailors on board Mr. Assheton-Smith's yacht in July 1887. It was identified by Dr. Gunther as one of *Anser rubidiceps*, and has been preserved in London by Mr. Gardner.[1]

[1] A pair was also shot at Bunchrew near Inverness, and are now preserved in Mr. Macleay's shop in that town. So far as we know, there are no other Falkland Island Geese of either species on any private ponds or lochs in Scotland.

Golden Eye; Gadwall; and Winter Wild Fowl.

Additional records of Golden Eye reach us from Benbeculay, Dr. MacRury having seen a pair near Milton early in September. They appeared to fly with difficulty. On the 26th September Dr. MacRury saw 5 Brent Geese, which were very tame, and evidently early migrants. Dr. MacRury never saw any of the Winter Geese arrive so early before. Wigeon were seen the same day on the sea,[1] and a few days afterwards a number of Gadwalls; and also Bernicle Geese on the 5th October.

Anas strepera. Garganey Duck.

From the MSS. and sketch-books of the late Mr. Henry Davenport Graham, which have been placed in our hands for the purposes of publication, we learn that one was shot in Barray by Mr. Colin M'Vean, C.E. (who served on the Admiralty Survey under Captain Otter, R.N., in 1863). This specimen is now in the collection of Sir John Campbell Orde, Bart., of Kilmory, "who exhibited it at a meeting of the Brit. Ornith. Union in London in the same year."

Syrrhaptes paradoxus (*Pall.*). Pallas' Sand Grouse.

Since the notes on this species were written in the body of the book, we have learned that, upon the Galston shootings in Lewis, a flock of from thirty to forty Sand Grouse has been present for the past two months; and similar birds have been seen on the Mhorsgail shootings. In a letter to Mr. Macleay, Inverness, Mr. William Mackay, Chamberlain to Lady Matheson at Stornoway, whilst announcing the above facts, also invoices a bird sent for preservation. Buckley saw this bird, which is a female, showing no signs of incubation, and the eggs in the ovary were about the size of peas.

Machetes pugnax (*L.*). Ruff.

Harvie-Brown received a Reeve from Dr. MacRury, which that gentleman had shot in Benbeculay about the 18th August 1888. "There was a pair of them," he writes, "on the machar, and they were very tame. Both were alike, so far as I could make out. I do not think I ever saw this bird here before."

[1] On the 19th September Harvie-Brown shot the "first young Wigeon of the year" on Loch Fern, Co. Donegal, and after that a good many were seen.

Sterna fluviatilis, *Naum.* **Common Tern.**

Dr. MacRury of Benbeculay shot an adult specimen of the Common Tern on the 18th August 1888, and sent the wing to Harvie-Brown. At first blush, this seemed rather a puzzle, but, after careful reference to dates of autumn migrations of Terns, and of their departures from stations in the north of Scotland, we came to the conclusion that this wing belonged to a passing migrant (see Migration Reports of Butt of Lewis; Stornoway (1883); Monach (1880); Skervuile (1881); and others). The following are a few data from the Reports:—Dhuheartach, six "Terns," seen 27.8.79 on migration. Monach, August 15th to 20th, 1880, great flocks congregating previous to departure, and all gone by 25th, except a few detained by later young ones. Also reported to have left Stornoway Light by 15th August (p. 90). On 26th August sixteen arrived at *Bahama* light-vessel. In 1881, earliest passing at Skervuile, 8th July. Latest at Little Ross, 2d September. 1882, leave Butt of Lewis about August 12th (p. 62). 1883, left Stornoway ten days later than usual. The above data could be greatly added to, showing that Terns migrate southwards at dates ranging over July, August, and September. It remains to be seen to what extent, during migrations, the Common Tern on the west coast mingles with the Arctic.

Dr. J. MacRury (*in lit.* 12.10.88), however, seems pretty confident that the Tern shot was one of a *nesting* colony of the species (Common Tern, *S. fluviatilis*), and adds what appears to be a conclusive note, that the young were just able to fly. We quote his letter:—"They were sitting on a small island, and I sent out my little dog after them, when they all got up and followed him towards me, when I shot one of the old birds." Dr. MacRury adds:—"My impression is that a large number of the Terns that breed on the Benbeculay lochs are Common Terns, but I did not shoot any more this year—as they all went away shortly after I sent you that wing." As we have not personally visited Benbeculay, we cannot, of course, differ in opinion from Dr. MacRury, but we feel very sure still of the correctness of our notes as regards Terns elsewhere in the Outer Hebrides. We still desire to obtain a few specimens of Common Tern shot in the middle of the breeding season, from any part of the Outer Hebrides, for the purpose of settling some points in its distribution at that time.

In the summer of 1888 the only Terns seen in S. Uist and elsewhere in the Long Island by Colonel Irby, and Captain Savile G. Reid, were Arctic Terns.

Fulmarus glacialis *L.* **Fulmar Petrel.**

It is of sufficient interest to record that, on Saturday, 25th November 1888, whilst crossing from Coll to Barray, and about fifteen miles west of Coll, Mr. William Donald, of the *Dunara Castle* s.s., saw a number of Fulmars flying about in their peculiar fashion. Only on one occasion before had he seen—in his long experience of the routes among the Isles—these birds, except near St. Kilda. About four years ago (say 1884) he saw a solitary bird, "just about the same place." (This is where the great fishing banks lie in about thirty fathoms."—H.-B.)

Lomvia troile (*L.*). **Common Guillemot.**

The St. Kildans hold an opinion that all the *bridled* birds are males, and Mr. F. P. Johnson, who has lately been making some interesting observations in Skye, informs us that Mr. Mackenzie, factor to Macleod of Macleod, examined three "bridled" Guillemots, all of which *proved* to be males. On the other hand, we find in our journals, written *on the spot*, the positive statement that we saw at Barray Head both varieties *in copulâ* together, the males and the females being of both varieties. This is a question, however, well deserving of further attention.

Uria grylle (*L.*). **Black Guillemot.**

It has hitherto been generally believed by ornithologists that the winter and summer plumages of the Black Guillemot are different. It may therefore be interesting to take note here that a very observant lighthouse-keeper at Noss, in Shetland, Mr. John Nichol, sends us the following note, under date of May 5th, 1888:—"In John's book (*British Birds in their Haunts*), MacGillivray says the Black Guillemot in winter is white, with a tinge of grey. I carefully noted this, and strictly watched the Guillemots here in the north, and I am able to say they do *not* change in plumage any time of the year. But young birds for the first winter are white, with a tinge of grey." In connection with this question, we find that we do not take notice of the fact of birds retaining portions of the immature plumage, and breeding in this dress, in our volume on the *Fauna of Sutherland*, etc., but we have done so in an earlier article.[1] Of three birds caught upon their eggs, two on dissection proved to be males, and the third was "curiously mottled with white all over the lower breast and belly." Such specimens we were, at that time, informed by Mr. W. Macleay, of Inverness, are not common in summer, but he had on several occasions received them for preservation. These statements, taken together, tend to show that the Black Guillemot breeds the year following its advent into the world. We mention them here as points worthy of further elucidation.

[1] *The Birds found breeding in Sutherland* (Glasgow, 1875, p. 128).

APPENDIX E.

LIST OF VERMIN KILLED UPON THE LEWIS AND HARRIS ESTATES IN ROSS-SHIRE.

Year.	Cats.	Martens.	Otters.	Rats.	Ravens.	Crows.	Peregrines.	Hawks.	Eagles.	Total each Year.
1876	20	Negative fact entered, J. A. H-B.	1	160	21	33	...	27	2	...
1877	29		17	289	51	73	...	31	5	...
1878	32		9	490	75	138	...	91	2	...
1879	28		6	325	78	79	...	47	9	...
1880	17		13	130	58	60	...	40	1	...
1881	25		4	249	55	41	...	37	...	...
1882	21		8	318	58	47	...	33	2	...
1883	27		...	318	49	53	...	16	...	...
1884	47		18	1016	129	93	...	75	7	...
1885	41		19	525	63	72	...	46	2	...
In 10 yrs.	287	...	95	3820	637	698	...	443	30	...
In 1880	20	...	12	579	67	69	10	20	2	779

The last items are from Mr. Notman's returns for 1880.

During the ten years (1870-1880) the following were killed in North Harris, vermin returns having been handed in by Mr. Murdoch MacCaulay, head forester, Abhuinnsuidh :—

Martens, 10; Otters, 6; Rats, 2000; Ravens, 100; Crows, 200; Peregrines, 3; Hawks, 15.

In the year 1881, Mr. Donald Ross, gamekeeper, killed in his first year's charge, as follows :—

Cat, 1; Otters, 2; Rats, 10; Ravens, 45; Crows, 613; Hawks, 30.

APPENDIX F.

FEILDEN'S MSS.

SINCE the date, October 29th, 1888, whilst rummaging over a store of old papers, letters, etc., with a view to clearing out useless ones, we came across the "Continuation of J. A. Harvie-Brown's and H. W. Feilden's Tour in the Long Island during May and June 1870." Feilden remained in South Uist and Benbeculay between May 30th and June 6th, after Harvie-Brown left, and the above is a very full journal of his stay. His notes refer principally to the "shell mounds" of the "machars," where he collected "considerable numbers of bone needles and pieces of animals' bones: and a large piece of a human skeleton was exposed. It appeared to be the remains of a very tall man. The most interesting relics we obtained," he continues, "were a well-carved bone spoon, and half of an elegantly-made bronze brooch. Finding these vestiges of civilisation *in situ* side by side with the charred and hatchet-splintered bones and the horn needles of the prehistoric man, for a moment somewhat shook our faith in the antiquity of the kitchen-middens; but on reflection two hypotheses were deduced for consideration:—*Firstly*, That it is just possible, though improbable, that the carved spoon and brooch might have been dropped there in comparatively recent times. This argument, however, is negatived by the fact of our disinterring these interesting remains from the actual bone bed. *Secondly*, Is it not reasonable to suppose that when the Long Island was peopled by the prehistoric men who raised these shell-mounds, a higher civilisation may have existed in parts of Europe, possibly also on the mainland of Britain, and that the spoon and brooch travelled to the Benbeculay shell-mounds by accident or barter?

"To prop up this theory, let us suppose that, in the course of centuries, the Esquimaux kitchen-middens on Melville Island or Smith's Sound were opened by some inquiring archæologist who perchance may never have heard of our Arctic Expeditions, would not the scattered relics of Franklin's and Kane's expeditions give rise to some analogous theories?"

Feilden continues—and we think his remarks too interesting to be omitted:—"We do not recommend digging for these remains in these mounds, for the spade is apt to break the bones, which, when they are first exposed, are friable and discoloured: after a strong gale from the westward the sand-drifts leave the heavier objects exposed, and these whiten and harden with exposure. We would advise persons desirous of examining these interesting remains of prehistoric times, probably the richest in Great Britain, to visit them in winter or in spring after heavy gales."

The rest of Feilden's notes principally describe a weary tramp in a south-easterly direction in Benbeculay—some ten miles in all, of bog and moor; his crossing the Sound of Wiay, and taking Arctic Terns' eggs, and his vain search for the nest or birds of Buffon's Skua. Not only do these not breed there, but he was assured by the shepherd, "who knows the Arctic Skua well," that he had never even seen one of the latter on the island, "though he had often seen them chasing the terns and gulls in the Sound of Wiay. To make sure"—continues Feilden—"we walked round the island some six miles, and not a trace of a Skua was to be seen." At another page he has the remark:—"Arctic Terns were far more abundant than the Common Terns"—thereby implying that both species were present.

FINIS.

INDEX.

s

INDEX.

VIEWS OF SCOTLAND AND THE ISLES

SHIANTELLE SERIES

1887.

No. THURSO.

1. Town Hall, Thurso.
2. Window, Old Parish Kirk.
3. Interior, do.
4. Kirk and Kirkyard, Thurso.
5. View near Holborn Head.
6. The Clett, Holborn Head.
7. Do. do., from the South.
8. East Cliff, Holborn Head.
9. Devil's Bridge, do.
10. The Clett, from the West.
11. Rift in the Rocks, Holborn Head.
12. Devil's Bridge (Interior).

STACK AND SKERRY.

13. Cormorants' Roost.
14. View of the Interior of Skerry.
15. Do. do.
16. Do. do., looking South.
17. Landing-place, Skerry Island.

NORTH RONAY ISLAND.

18. Old Chapel and Kirkyard. Breeding-place of the Fork-tailed Petrel.
19. North Face of the West Horn.
20. Do. do. East Horn.
21. Cave in Ronay, near the Fulmar-haunted Cliffs of the West Horn.
22. Kittiwake and Guillemot Colony.
23. North-East Peninsula, with Central Aiguille.
24. North-East Peninsula, from the West Horn.
25. North-West Horn.
26. West side of North-East Peninsula: Resting Colony of Puffins.
27. View from top of West Cliff.

SULISGEIR.

28. The Solans Rock.
29. Sulisgeir—top surface.
30.–33. Views of Sulisgeir from the Sea.

SHIANT ISLES.

34. East End from the Sea.
35. Rocky Beach.
36. The Basaltic North Cliffs.
37. Do. near Eagle's Eyrie.
38. View, looking East, with Puffins Colony.
39. Pinnacle of Rock in Island Mhurrie.
40. Guillemots Cliff.
41.
42. South Cliff, with Shiant main island in the distance.

LEWIS AND STORNOWAY.

43. Stornoway from the River Creed.
44. View near the Creed.
45. Glen near Stornoway.
46. Do. do.
47. View, with the Creed in distance.
48. The River Creed.
49. Gypsy Encampment (Interior of Lewis).
50.–53. Panoramic View of Interior of Lewis.

Single Views of each, Mounted, 1s. 6d.; Unmounted, 1s.; Per Set of 53, £3, 6s.; Per Dozen, Mounted, 16s.; 53 Unmounted, £1, 19s. 9d.; Per Dozen, 9s. Size of Views—Whole Plate.

Any of the above can be supplied, except such as are used in the Vertebrate Fauna of Scotland Series. Apply to Mr. NORRIE.

BOOKS

ON

SPORT & NATURAL HISTORY

PUBLISHED BY

DAVID DOUGLAS

Small Folio, price £21, with Sketches of Scenery and Animal Life by some of the best British and American Artists and Etchers.

THE

RISTIGOUCHE

AND ITS

SALMON FISHING

WITH A CHAPTER ON ANGLING LITERATURE

BY DEAN SAGE

EDINBURGH: DAVID DOUGLAS
1888

ONLY 105 COPIES PRINTED.

One Volume, Royal 8vo. 50s.

WITH 40 FULL-PAGE ILLUSTRATIONS OF SCENERY AND ANIMAL LIFE, DRAWN BY GEORGE REID, R.S.A., AND J. WYCLIFFE TAYLOR, AND ENGRAVED BY AMAND DURAND.

NATURAL HISTORY & SPORT IN MORAY

BY CHARLES ST. JOHN

AUTHOR OF "WILD SPORTS IN THE HIGHLANDS"

EDINBURGH: DAVID DOUGLAS
1882

Two Volumes, Crown 8vo, Illustrated. 21s.

A TOUR IN SUTHERLANDSHIRE

WITH EXTRACTS FROM THE FIELD BOOKS OF A SPORTSMAN AND NATURALIST

By CHARLES ST. JOHN

AUTHOR OF "NATURAL HISTORY AND SPORT IN MORAY"

SECOND EDITION

WITH AN APPENDIX ON THE FAUNA OF SUTHERLAND

By J. A. Harvie-Brown and T. E. Buckley

EDINBURGH: DAVID DOUGLAS

1884

One Volume, Demy 8vo, with Maps and Illustrations. 12s.

NOTES AND SKETCHES

FROM THE

WILD COASTS OF NIPON

WITH CHAPTER ON CRUISING AFTER PIRATES
IN CHINESE WATERS

BY CAPTAIN H. C. ST. JOHN, R.N.

EDINBURGH: DAVID DOUGLAS
1880

One Volume, Demy 8vo, with Etchings and Map. 8s. 6d.

THE

CAPERCAILLIE IN SCOTLAND

WITH SOME ACCOUNT OF THE EXTENSION OF ITS RANGE SINCE ITS RESTORATION AT TAYMOUTH IN 1837 AND 1838

BY J. A. HARVIE-BROWN, F.Z.S.

MEMBER OF THE BRITISH ORNITHOLOGISTS' UNION, ETC.

EDINBURGH: DAVID DOUGLAS. MDCCCLXXIX

Nearly Ready. In Two Volumes, Demy 8vo. To Subscribers only.
Profusely Illustrated with Etchings and Lithographs.

THE

BIRDS OF BERWICKSHIRE

WITH REMARKS ON THEIR LOCAL DISTRIBUTION MIGRATION, AND HABITS, AND ALSO ON THE FOLK-LORE, PROVERBS, POPULAR RHYMES AND SAYINGS CONNECTED WITH THEM

BY

GEORGE MUIRHEAD, F.R.S.E., F.Z.S.

MEMBER OF THE BRITISH ORNITHOLOGISTS' UNION, MEMBER OF THE BERWICKSHIRE NATURALISTS' CLUB, ETC.

EDINBURGH: DAVID DOUGLAS

1889

In the Press, One Volume, Demy 8vo, and Large Paper Edition, with additional Illustrations, Small 4to.

A HISTORY OF CURLING

SCOTLAND'S AIN GAME

AND OF FIFTY YEARS OF THE ROYAL CALEDONIAN CURLING CLUB

EDITED BY

THE REV. JOHN KERR, M.A., DIRLETON

The Spirit of Curling

EDINBURGH: DAVID DOUGLAS

1889